AF358840

Pathogenesis and Treatments of Head and Neck Cancer

Pathogenesis and Treatments of Head and Neck Cancer

Guest Editor

Marko Tarle

Basel • Beijing • Wuhan • Barcelona • Belgrade • Novi Sad • Cluj • Manchester

Guest Editor
Marko Tarle
Department of Maxillofacial
Surgery
Dubrava University Hospital
Zagreb
Croatia

Editorial Office
MDPI AG
Grosspeteranlage 5
4052 Basel, Switzerland

This is a reprint of the Special Issue, published open access by the journal *International Journal of Molecular Sciences* (ISSN 1422-0067), freely accessible at: https://www.mdpi.com/journal/ijms/special_issues/05Y6C08HE2.

For citation purposes, cite each article independently as indicated on the article page online and as indicated below:

Lastname, A.A.; Lastname, B.B. Article Title. *Journal Name* **Year**, *Volume Number*, Page Range.

ISBN 978-3-7258-4341-1 (Hbk)
ISBN 978-3-7258-4342-8 (PDF)
https://doi.org/10.3390/books978-3-7258-4342-8

Contents

About the Editor

Marko Tarle

Dr. Marko Tarle is an assistant at the Department of Maxillofacial Surgery, School of Dental Medicine, University of Zagreb, and a maxillofacial surgeon at Dubrava University Hospital in Zagreb, Croatia. He completed his medical studies with highest honors (summa cum laude) and has received multiple Dean's Awards for academic excellence and contributions to scientific and teaching reputation. Dr. Tarle obtained his specialist title in maxillofacial surgery in 2022 and defended his doctoral thesis in 2023 on the role of nuclear epidermal growth factor receptor in precancerous and malignant lesions of the oral cavity. In 2023, he completed the prestigious Postgraduate Programme in Head and Neck Surgery: Oncology organized by the European Association for Cranio-Maxillofacial Surgery, as the first Croatian specialist to do so. His scientific interests include head and neck oncology, reconstructive surgery, and translational cancer research. He is a collaborator on projects funded by the Croatian Science Foundation and the University of Zagreb and has authored or co-authored more than 40 scientific publications. He also serves as a peer reviewer for several international journals and was the Guest Editor of the Special Issue Pathogenesis and Therapy of Oral Carcinogenesis. Since 2024, he has been the secretary of the Croatian Society of Maxillofacial, Plastic and Reconstructive Head and Neck Surgery.

Preface

We are pleased to present this Special Issue reprint dedicated to the pathogenesis and treatment of head and neck squamous cell carcinoma (HNSCC), a group of malignancies that continue to pose significant clinical and biological challenges. This compilation brings together original research articles that provide valuable insights into the complex molecular landscape of HNSCC, with the goal of advancing our understanding of disease progression and improving patient outcomes.

Despite growing efforts in early diagnosis and therapeutic development, the prognosis for patients with advanced HNSCC remains poor. This reprint addresses a wide range of topics, including metabolic reprogramming, immune evasion, inflammation-driven tumor progression, and resistance to therapy. Particular attention is given to the roles of enzymes such as IDO1 and IL4I1 in immunosuppression, the prognostic value of genes like SLC2A3 and SDHA, and the relevance of inflammatory and metabolic biomarkers in predicting treatment response and recurrence.

Several articles examine novel therapeutic strategies, such as the targeting of ASPH and modulation of tumor metabolism, as well as the interplay between the tumor microenvironment, immune cell function, and disease progression. Additional studies investigate the diagnostic and prognostic significance of serum markers and explore how gene expression signatures can stratify patients and guide more personalized treatment approaches.

We believe this reprint will be of particular interest to researchers, clinicians, and healthcare professionals working in oncology, molecular biology, and translational medicine. It serves as a testament to the collective effort of the scientific community to address the pressing needs in head and neck cancer research. We thank all authors for their high-quality contributions and the reviewers for their critical insights in shaping this Special Issue.

We hope this collection inspires further innovation and collaboration in the field of HNSCC.

Marko Tarle
Guest Editor

Article

Potential Role of the Intratumoral Microbiota in Prognosis of Head and Neck Cancer

Masakazu Hamada [1,*], Hiroaki Inaba [2], Kyoko Nishiyama [1], Sho Yoshida [2], Yoshiaki Yura [1], Michiyo Matsumoto-Nakano [2] and Narikazu Uzawa [1]

[1] Department of Oral & Maxillofacial Oncology and Surgery, Osaka University Graduate School of Dentistry, Suita 565-0871, Japan; kyon.0419@gmail.com (K.N.); narayura630@gmail.com (Y.Y.); uzawa.narikazu.dent@osaka-u.ac.jp (N.U.)

[2] Department of Pediatric Dentistry, Okayama University Graduate School of Medicine, Dentistry and Pharmaceutical Sciences, Okayama 700-8558, Japan; hinaba@okayama-u.ac.jp (H.I.); syoshida@okayama-u.ac.jp (S.Y.); mnakano@cc.okayama-u.ac.jp (M.M.-N.)

* Correspondence: hamada.masakazu.dent@osaka-u.ac.jp; Tel.: +81-6-6879-2941

Abstract: The tumor microbiome, a relatively new research field, affects tumor progression through several mechanisms. The Cancer Microbiome Atlas (TCMA) database was recently published. In the present study, we used TCMA and The Cancer Genome Atlas and examined microbiome profiling in head and neck squamous cell carcinoma (HNSCC), the role of the intratumoral microbiota in the prognosis of HNSCC patients, and differentially expressed genes in tumor cells in relation to specific bacterial infections. We investigated 18 microbes at the genus level that differed between solid normal tissue ($n = 22$) and primary tumors ($n = 154$). The tissue microbiome profiles of *Actinomyces, Fusobacterium,* and *Rothia* at the genus level differed between the solid normal tissue and primary tumors of HNSCC patients. When the prognosis of groups with rates over and under the median for each microbe at the genus level was examined, rates for *Leptotrichia* which were over the median correlated with significantly higher overall survival rates. We then extracted 35 differentially expressed genes between the over- and under-the-median-for-*Leptotrichia* groups based on the criteria of >1.5 fold and $p < 0.05$ in the Mann–Whitney U-test. A pathway analysis showed that these *Leptotrichia*-related genes were associated with the pathways of Alzheimer disease, neurodegeneration-multiple diseases, prion disease, MAPK signaling, and PI3K-Akt signaling, while protein–protein interaction analysis revealed that these genes formed a dense network. In conclusion, probiotics and specific antimicrobial therapy targeting *Leptotrichia* may have an impact on the prognosis of HNSCC.

Keywords: intratumor microbiome; oral bacteria; *Leptotrichia*; head and neck squamous cell carcinoma; RNA sequencing; TCGA; TCMA

1. Introduction

The Cancer Genome Atlas (TCGA) is a large comprehensive cancer genome project initiated in 2006 that aims to catalog and discover major cancer-causing genome alterations through multi-dimensional analyses for the creation of new cancer treatments, diagnostics, and prevention methods for more than 20 cancer types [1–3]. In 2015, TCGA profiled cases of head and neck squamous cell carcinoma (HNSCC) to comprehensively characterize genomic alterations [4]. HNSCC includes tumors that arise in the lip, oral cavity, pharynx, larynx, and paranasal sinuses. Treatment depends on the location of the tumors that develop, but the standard treatment for HNSCC is a combination of surgery, radiation therapy, and chemotherapy; the 5-year survival rate is only 40–50% despite advances in treatment [5]. Various studies have since been conducted, including the identification of novel prognostic biomarkers for HNSCC using TCGA [6–9].

Alcohol and tobacco abuse are the most common etiologies of oral, pharyngeal, and laryngeal cancers. In addition, evidence has accumulated supporting the involvement of the microbiome in the developmental process of HNSCC and its susceptibility to chemoradiotherapy. For example, human papillomavirus (HPV)-positive malignancies have been shown to account for approximately 20% of HNSCCs and 55% of those originating in the oropharynx [10]. Patients with HPV-positive-related oropharyngeal squamous cell carcinoma have a better prognosis than HPV-negative patients when treated with chemoradiotherapy [11,12]. Furthermore, the relationship between HNSCC and oral bacteria, such as the periodontopathogen *Porphyromonas gingivalis*, and the caries pathogen *Streptococcus mutans* was recently investigated. Several oral bacteria have been shown to promote tumor progression [13–15]. The relationship between the oral microbiome and HNSCC may also have important implications for the prevention and early detection of HNSCC [16].

Microbes in cancer are rapidly being developed as a potentially powerful new toolkit for improving patient care [17,18]. The intratumoral microbiota affects tumor progression through several mechanisms, including DNA damage, the activation of oncogenic pathways, the induction of immunosuppression, and the metabolism of drugs [18]. The Cancer Microbiome Atlas (TCMA) was recently published and includes the curated tissue-endemic microbial profiles of 3689 unique samples from 1772 patients from five TCGA projects and 21 anatomical sites [16]. This may enable a multi-omics analysis with systematic microbe–host matching and has been actively used in studies on various cancers, such as gastric cancer, colon cancer, and HNSCC [19–21].

However, the relationship between the abundance of different bacteria in the intratumor microbiome and the prognosis of HNSCC patients in TCMA database remains unclear. Therefore, the present study investigated the intratumoral microbiota of HNSCC and its prognostic impact on patient survival and examined differentially expressed genes in tumor cells associated with specific bacterial infections.

2. Results

2.1. Microbiome Profiling of 18 Selected Microbes at the Genus Level

We selected 154 patients who were present in both TCGA and TCMA datasets (Figure 1A), and the following microbes were identified to at least the genus level in TCMA database: *Actinomyces, Aggregatibacter, Alloprevotella, Campylobacter, Capnocytophaga, Fusobacterium, Granulicatella, Haemophilus, Lactobacillus, Leptotrichia, Mycoplasma, Neisseria, Porphyromonas, Prevotella, Rothia, Streptococcus, Treponema,* and *Veillonella* (Figure 1B). In solid normal tissue, the population with the highest percentage among the 18 selected microbes was *Prevotella*, followed by *Streptococcus* and *Fusobacterium*. In primary tumors, the population with the highest percentage was *Prevotella*, followed by *Fusobacterium* and *Streptococcus*. We then investigated differences in the 18 microbes at the genus level between the solid normal tissue (n = 22) and primary tumors (n = 154) of HNSCC patients. The results obtained showed that the tissue microbiome profiles of *Actinomyces, Fusobacterium,* and *Rothia* differed between solid normal tissue and primary tumors (Figure 1C).

2.2. Prognostic Significance of 18 Selected Microbes at the Genus Level in HNSCC Patients in TCGA Database

We examined the relationship between the 18 selected microbes at the genus level and the prognosis of TCGA-HNSCC patients. Patients were divided into two groups based on the rate of occurrence of a microbe being over and under the median. Differences detected in survival times using the Kaplan–Meier method were analyzed with the generalized Wilcoxon test and log-rank test (Figure 2). The generalized Wilcoxon test and log-rank test showed that among the 18 microbes examined, a rate over the median for *Leptotrichia* (Figure 2J) correlated with significantly higher overall survival rates. A rate over the median for *Campylobacter* (Figure 2D) or *Capnocytophaga* (Figure 2E) correlated with significantly higher overall survival rates only in the generalized Wilcoxon test or log-rank test. In contrast, the generalized Wilcoxon test showed that a rate over the median for *Lactobacillus*

(Figure 2I) correlated with lower overall survival rates. On the other hand, rates over the median for *Actinomyces* (Figure 2A), *Aggregatibacter* (Figure 2B), *Alloprevotella* (Figure 2C), *Fusobacterium* (Figure 2F), *Granulicatella* (Figure 2G), *Haemophilus* (Figure 2H), *Mycoplasma* (Figure 2K), *Neisseria* (Figure 2L), *Porphyromonas* (Figure 2M), *Prevotella* (Figure 2N), *Rothia* (Figure 2O), *Streptococcus* (Figure 2P), *Treponema* (Figure 2Q), and *Veillonella* (Figure 2R) were not associated with survival rates.

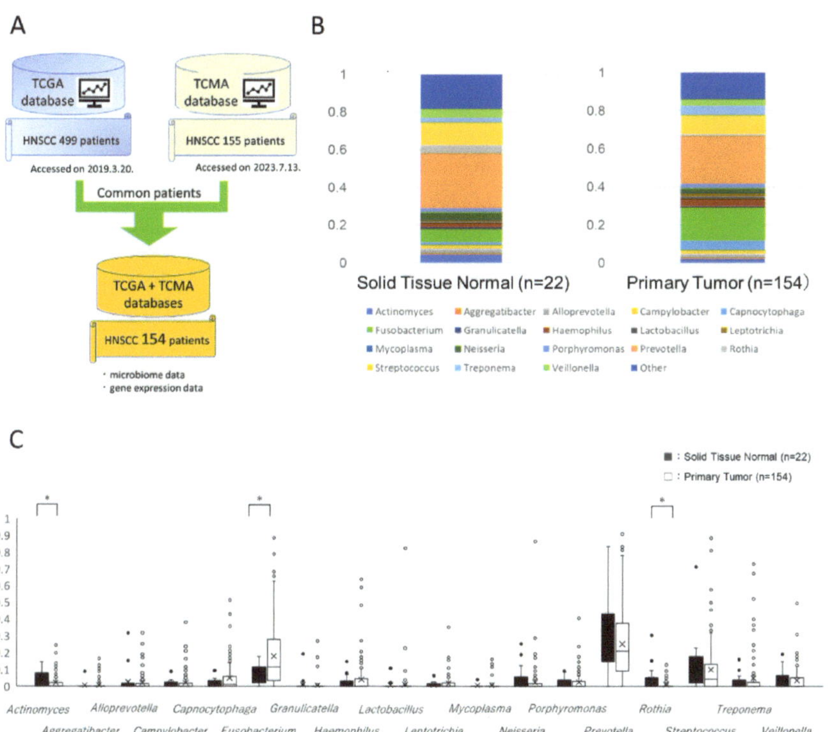

Figure 1. Microbiome profile analysis of 18 selected microbes at the genus level in solid normal tissue and primary tumors of HNSCC patients. (**A**) The extraction schedule of common patients from TCGA and TCMA databases. (**B**) A breakdown of the 18 selected microbes at the genus level, including *Actinomyces, Aggregatibacter, Alloprevotella, Campylobacter, Capnocytophaga, Fusobacterium, Granulicatella, Haemophilus, Lactobacillus, Leptotrichia, Mycoplasma, Neisseria, Porphyromonas, Prevotella, Rothia, Streptococcus, Treponema,* and *Veillonella*. (**C**) Comparison of the 18 selected microbes at the genus level in the solid normal tissue and primary tumors of HNSCC patients. * $p < 0.05$.

2.3. Relationships between Classical Prognostic Factors and Survival Rates Associated with *Leptotrichia* in TCGA-HNSCC Patients

Since a rate over the median for *Leptotrichia* (Figure 2J) correlated with significantly higher overall survival rates, the effects of drinking (Figure 3A,B), smoking (Figure 3C,D), HPV status (Figure 3E,F), sex (Figure 3G,H), the presence of lymph node metastasis (Figure 3I,J), and tumor size (Figure 3K,L) on survival rates associated with *Leptotrichia* were examined. Survival curves were not affected by the population of *Leptotrichia* in the absence of drinking (Figure 3B), without lymph node metastasis (Figure 3J), in females (Figure 3H), and in T1–T2 tumors (Figure 3K) but were markedly affected by a rate over the median for *Leptotrichia* with other factors.

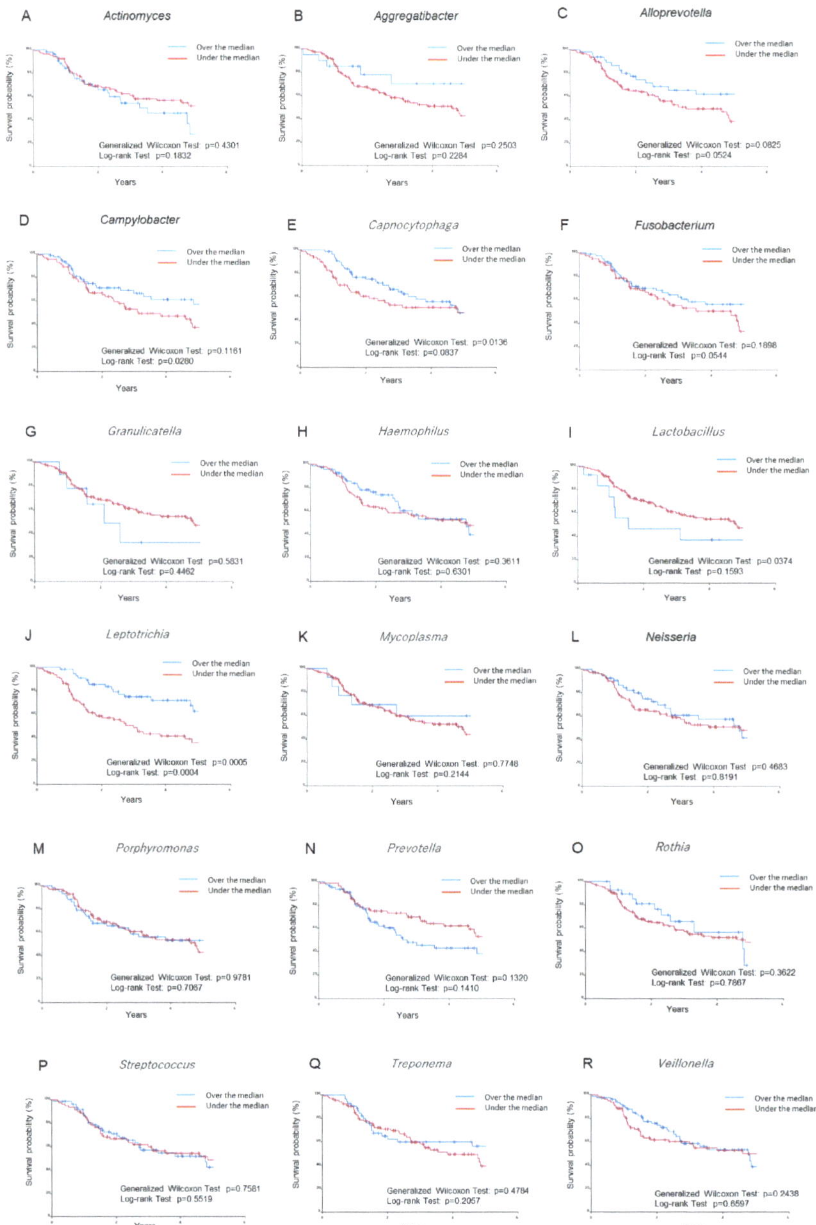

Figure 2. Prognostic significance of 18 selected microbes at the genus level in HNSCC patients in TCGA database. The relationships between the overall survival of TCGA-HNSCC patients and the 18 selected microbes at the genus level were assessed using the Kaplan–Meier method. Differences detected in survival times were then analyzed with the generalized Wilcoxon test and log-rank test. (**A**): *Actinomyces*, (**B**): *Aggregatibacter*, (**C**): *Alloprevotella*, (**D**): *Campylobacter*, (**E**): *Capnocytophaga*, (**F**): *Fusobacterium*, (**G**): *Granulicatella*, (**H**): *Haemophilus*, (**I**): *Lactobacillus*, (**J**): *Leptotrichia*, (**K**): *Mycoplasma*, (**L**): *Neisseria*, (**M**): *Porphyromonas*, (**N**): *Prevotella*, (**O**): *Rothia*, (**P**): *Streptococcus*, (**Q**): *Treponema*, (**R**): *Veillonella*. Differences were considered significant at $p < 0.05$.

2.4. Cox Regression Analysis of Relationships of 18 Selected Microbes at the Genus Level and Classical Prognostic Factors with Survival in TCGA-HNSCC Patients

The 18 selected microbes and their correlations were analyzed in more detail. Univariate and multivariate analyses (Cox proportional hazard model) were performed using the 18 selected microbes and classical risk factors, such as sex, HPV, smoking, age, and TNM stage, as independent variables. In the univariate analysis, *Leptotrichia*_High (vs. Low) (HR = 0.380, 95% CI = 0.215–0.669, p = 0.001) correlated with the prognosis of TCGA-HNSCC patients (Table 1). In addition, the multivariate analysis showed that *Leptotrichia*_High (vs. Low) (HR = 0.273, 95% CI = 0.116–0.645, p = 0.003) correlated with an improved prognosis in TCGA-HNSCC patients (Table 1).

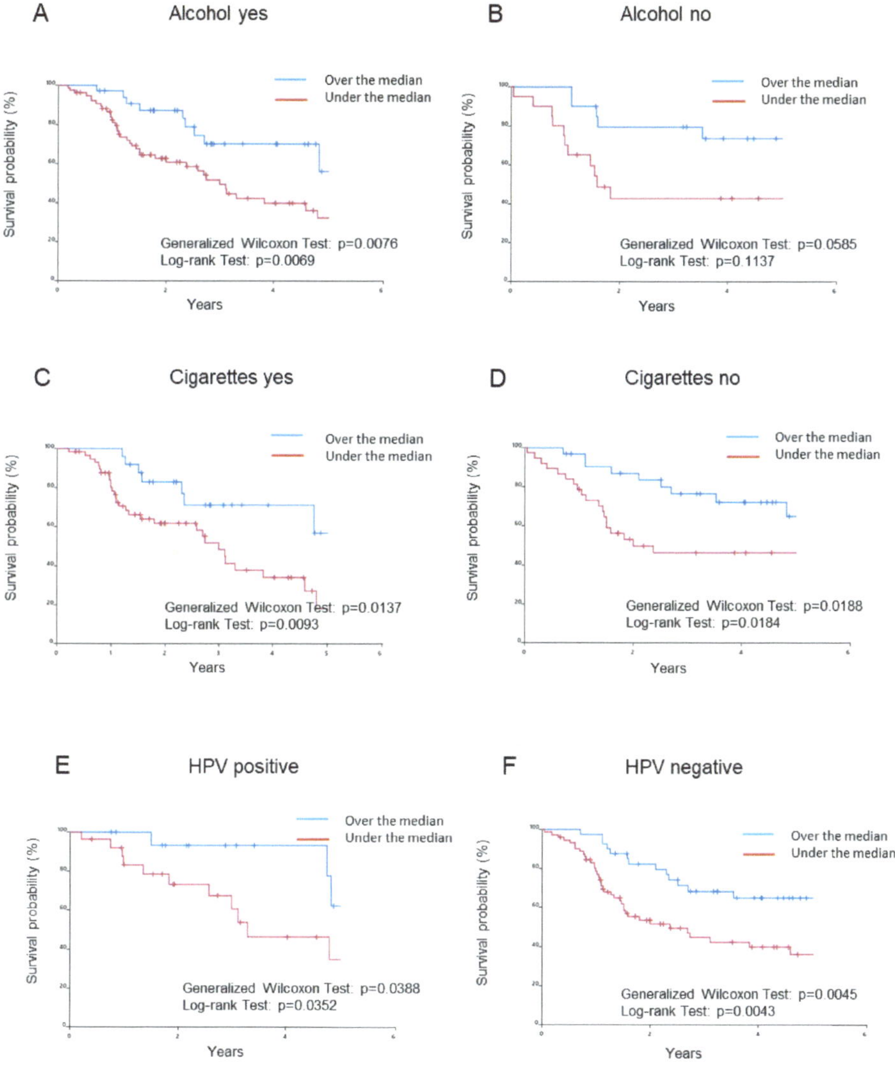

Figure 3. *Cont.*

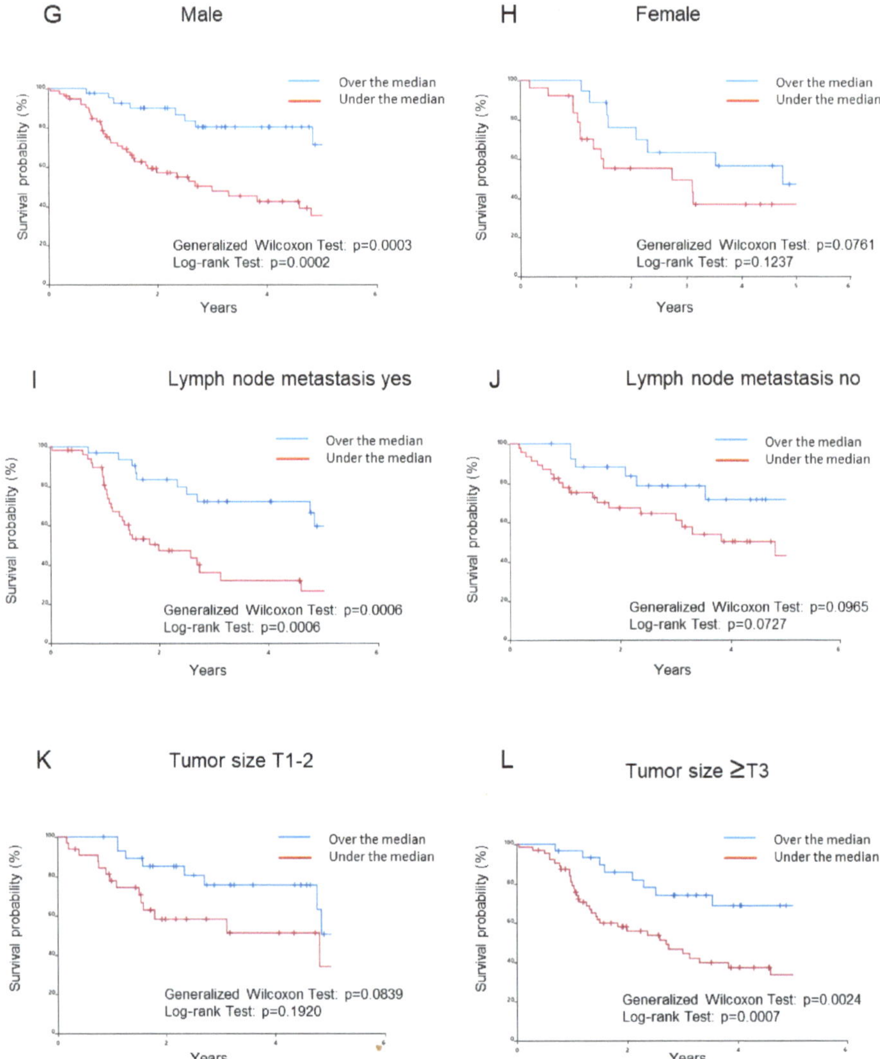

Figure 3. Relationships between classical prognostic factors and survival rates associated with *Leptotrichia* in TCGA-HNSCC patients. Survival curves were recalculated based on the population of *Leptotrichia* in consideration of classical prognostic factors, such as drinking, smoking, HPV status, sex, the presence of lymph node metastasis, and tumor sizes. (**A**) A history of drinking. (**B**) No history of drinking. (**C**) A history of smoking. (**D**) No history of smoking. (**E**) A history of HPV infection. (**F**) No history of HPV infection. (**G**) Males. (**H**) Females. (**I**) Lymph node metastasis. (**J**) No lymph node metastasis. (**K**) Tumor sizes T1–T2. (**L**) Tumor sizes ≥ T3. Differences were considered significant at $p < 0.05$.

Table 1. Univariate and multivariate analyses of the relationship of the 18 microbes at the genus level and classical prognostic factors with survival in TCGA-HNSCC patients.

	Univariate				Multivariate				
	HR	**95% CI**			*p*-**Value**	**HR**	**95% CI**			*p*-**Value**
*Actinomyces*_High (vs. Low)	1.350	0.813	-	2.240	0.246	4.090	1.553	-	10.774	**0.004**
*Aggregatibacter*_High (vs. Low)	0.508	0.204	-	1.266	0.146	0.546	0.202	-	1.478	0.234
*Alloprevotella*_High (vs. Low)	0.578	0.324	-	1.030	0.063	0.620	0.283	-	1.359	0.233
*Campylobacter*_High (vs. Low)	0.644	0.389	-	1.067	0.088	0.560	0.291	-	1.079	0.083
*Capnocytophaga*_High (vs. Low)	0.714	0.438	-	1.162	0.175	0.766	0.392	-	1.496	0.435
*Fusobacterium*_High (vs. Low)	0.722	0.442	-	1.178	0.192	1.446	0.701	-	2.981	0.318
*Granulicatella*_High (vs. Low)	1.503	0.602	-	3.750	0.383	5.873	1.495	-	23.079	**0.011**
*Haemophilus*_High (vs. Low)	0.890	0.536	-	1.476	0.651	0.809	0.373	-	1.753	0.590
*Lactobacillus*_High (vs. Low)	1.747	0.796	-	3.832	0.164	1.463	0.407	-	5.259	0.560
*Leptotrichia*_High (vs. Low)	0.380	0.215	-	0.669	**0.001**	0.273	0.116	-	0.645	**0.003**
*Mycoplasma*_High (vs. Low)	0.758	0.303	-	1.895	0.554	1.305	0.437	-	3.891	0.633
*Neisseria*_High (vs. Low)	0.880	0.523	-	1.480	0.630	1.429	0.677	-	3.016	0.349
*Porphyromonas*_High (vs. Low)	0.978	0.591	-	1.618	0.931	1.435	0.765	-	2.692	0.260
*Prevotella*_High (vs. Low)	1.589	0.971	-	2.600	0.065	2.378	1.123	-	5.034	**0.024**
*Rothia*_High (vs. Low)	0.876	0.457	-	1.677	0.689	1.114	0.375	-	3.313	0.846
*Streptococcus*_High (vs. Low)	1.037	0.637	-	1.689	0.883	0.723	0.322	-	1.621	0.430
*Treponema*_High (vs. Low)	0.805	0.475	-	1.366	0.422	0.792	0.337	-	1.863	0.593
*Veillonella*_High (vs. Low)	0.915	0.562	-	1.492	0.722	0.453	0.183	-	1.121	0.087
Age (per 1 year)	1.000	0.979	-	1.021	0.993	1.022	0.991	-	1.055	0.168
Sex_male (vs. female)	0.825	0.491	-	1.388	0.469	0.936	0.464	-	1.889	0.854
HPV status_Positive (vs. Negative)	0.693	0.384	-	1.253	0.225	0.499	0.206	-	1.208	0.123
Alcohol_history_Yes (vs. No)	1.302	0.736	-	2.304	0.365	1.054	0.460	-	2.410	0.902
Cigarettes per day_>0 (vs. 0)	1.308	0.797	-	2.146	0.288	1.069	0.547	-	2.091	0.845
M stage_m1 (vs. m0)	8.793	1.166	-	66.316	**0.035**	45.367	1.018	-	2022.711	**0.049**
N stage (Continuous variable per 1)	1.093	0.966	-	1.236	0.158					
N stage (Category)										
n0	1.000		ref			1.000		ref		
n1	1.121	0.563	-	2.230	0.746	1.855	0.769	-	4.472	0.169
n2	2.924	1.119	-	7.640	**0.029**	3.238	0.684	-	15.332	0.139
n2a	2.774	0.835	-	9.215	0.096	3.365	0.780	-	14.517	0.104
n2b	0.767	0.332	-	1.768	0.533	1.328	0.468	-	3.767	0.594
n2c	2.360	1.097	-	5.077	**0.028**	2.215	0.822	-	5.967	0.116
n3	2.110	0.499	-	8.926	0.310	4.001	0.400	-	39.979	0.238
T stage (Continuous variable per 1)	1.160	0.895	-	1.503	0.262					
T stage (Category)										
t1	1.000		ref			1.000		ref		
t2	0.852	0.252	-	2.882	0.797	1.542	0.264	-	9.025	0.631
t3	0.912	0.264	-	3.155	0.884	1.988	0.319	-	12.393	0.462
t4a	1.198	0.363	-	3.952	0.767	2.101	0.345	-	12.794	0.421
t4b	1.752	0.181	-	16.908	0.628	1.760	0.034	-	91.851	0.779

HR: hazard ratio; 95% CI: 95% confidence interval; ref: reference value. The hazard ratio refers to a high/low survival. A multivariate analysis was performed with forced insertion of all variables. Bold type indicates $p < 0.05$.

2.5. Extraction of Differentially Expressed Genes between over- and under-the-Median-for-Leptotrichia Groups

We focused on *Leptotrichia* at the genus level, which significantly improved the prognosis of TCGA-HNSCC patients, and investigated whether intratumoral *Leptotrichia* affected the expression of genes in HNSCC cells. The population of *Leptotrichia* at the genus level was divided into two groups: a rate over and under the median. Based on the criteria of >1.5 fold and $p < 0.05$ in the Mann–Whitney U-test, we extracted 35 differentially expressed genes between the over- and under-the-median groups, which included both up- and down-regulated genes in tumor cells (Figure 4A). We produced a heat map to show the up-

or down-regulated expression profiles of the 35 differentially expressed genes between the over- and under-the-median groups (Figure 4B).

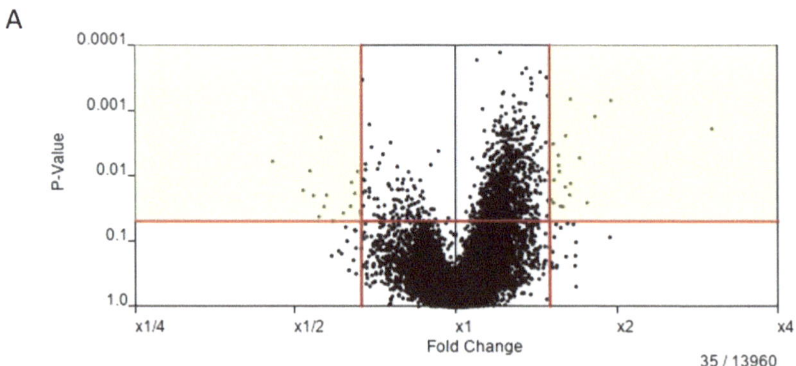

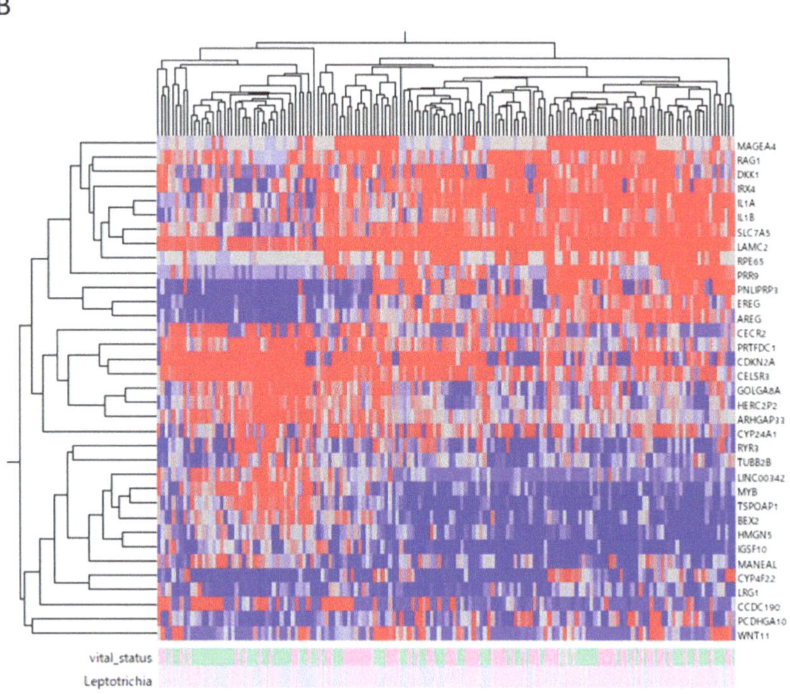

Figure 4. Extraction of genes related to *Leptotrichia*. (**A**) Based on the criteria of >1.5 fold and $p < 0.05$ in the Mann–Whitney U-test, 35 differentially expressed genes between the over- and under-the-median-for-*Leptotrichia* groups were extracted. (**B**) Heat map and hierarchical clustering of the 35 extracted genes. Colors from blue to red indicate low to high expression levels. The vital status of TCGA-HNSCC patients and the rate of *Leptotrichia* are color-coded as follows: alive ( ), dead ( ), over the median ( ), and under the median ( ), respectively.

2.6. Functional and Protein–Protein Interaction (PPI) Analyses of Leptotrichia-Related Genes

The 35 differentially expressed genes between the over- and under-the-median-for-*Leptotrichia* groups extracted based on the criteria of >1.5 fold and $p < 0.05$ in the Mann–Whitney U-test were subjected to functional and PPI analyses. Gene Ontology (GO) terms and Kyoto Encyclopedia of Genes and Genomes (KEGG) pathway analyses were performed to investigate biological properties and potential signaling pathways. In the GO enrichment analysis, enriched terms were as follows: the positive regulation of gene expression, the positive regulation of mitotic nuclear division, the positive regulation of cell division, the positive regulation of cytokine production, fever generation, the positive regulation of heart induction via the negative regulation of the canonical Wnt signaling pathway, the positive regulation of cell proliferation, the positive regulation of immature T cell proliferation in the thymus, the positive regulation of interleukin-6 production, the ERBB2-EGFR signaling pathway, the regulation of nitric oxide synthase activity, the positive regulation of prostaglandin secretion, the positive regulation of epidermal growth factor-activated receptor activity, the positive regulation of transcription, DNA-templating, the cytokine-mediated signaling pathway, the positive regulation of keratinocyte proliferation, the positive regulation of angiogenesis, ectopic germ cell programmed cell death, the positive regulation of glial cell proliferation, the negative regulation of cell proliferation, the cellular response to lipopolysaccharide, the negative regulation of apoptotic process, the positive regulation of vascular endothelial growth factor production, and cell–cell signaling (Figure 5A). The KEGG analysis revealed that the 35 extracted genes were significantly enriched in the pathways of Alzheimer disease, the pathways of neurodegeneration-multiple diseases, prion disease, the MAPK signaling pathway, and the PI3K-Akt signaling pathway (Figure 5B). The PPI network analysis showed that AREG, DKK1, EREG, IL1A, IL1B, LAMC2, RAG1, SLC7A5, CDKN2A, CECR2, CYP24A1, HMGN5, MYB, and WNT11 formed a dense network among these genes (Figure 5C). Among these genes, CDKN2A, CECR2, CYP24A1, HMGN5, MYB, and WNT11 were up-regulated with a fold change (over/under the median) > 1.5, while AREG, DKK1, EREG, IL1A, IL1B, LAMC2, RAG1, and SLC7A5 were down-regulated with a fold change < 0.66.

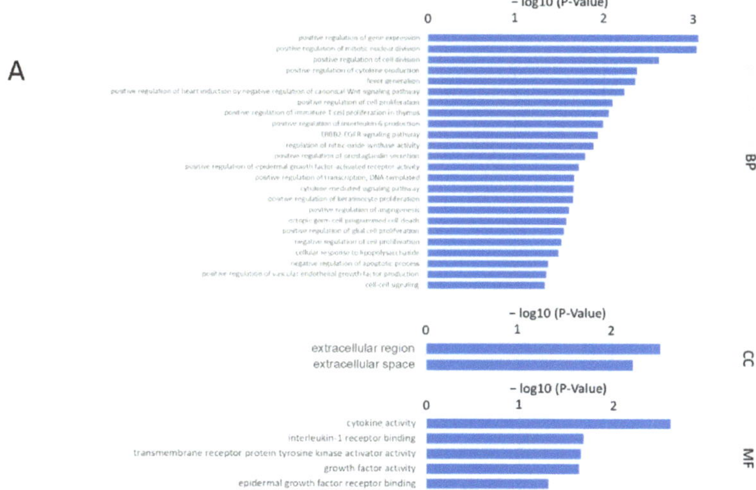

Figure 5. *Cont.*

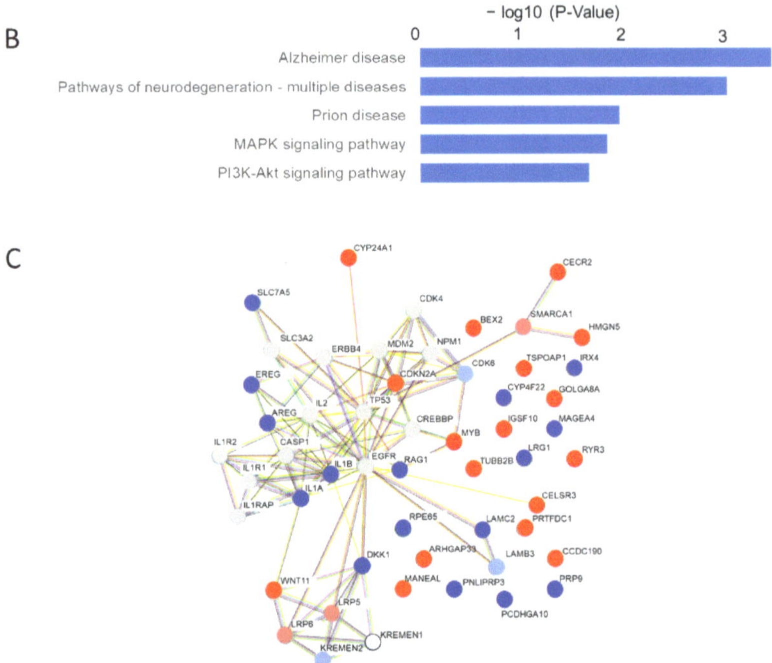

Figure 5. Functional and protein–protein interaction analyses of *Leptotrichia*-related genes. (**A**) A list of GO terms identified through the GO enrichment analysis of 35 differentially expressed genes between the over and under the median for *Leptotrichia* groups extracted based on the criteria of >1.5 fold and $p < 0.05$ in the Mann–Whitney U-test. BP, biological process; CC, cellular composition; MF, molecular function. (**B**) List of molecular pathways identified via the KEGG pathway enrichment analysis of the 35 extracted genes. (**C**) Proteins encoded by the 35 extracted genes were subjected to a PPI network analysis. Genes with fold changes (over/under the median) > 1.5, 1.5 > fold change > 1.2, 1.2 > fold change > 0.83, 0.83 > fold change > 0.66, and 0.66 > fold change are color-coded as red (■), pale red (■), gray (■), pale blue (■), and blue (■), respectively. White color means none of the above.

3. Discussion

The oral microbiome is a complex ecological environment comprising 750 microorganisms, including bacteria, archaea, protozoa, fungi, and viruses [22,23]. TCMA, which has led to the discovery of prognostic species and blood signatures of mucosal barrier injury and enables multi-omics analyses with systematic microbe–host matching, was recently published [24]. TCMA is also used in HNSCC to search for novel microbial markers and causative bacteria of the inflammatory tumor microenvironment [25,26]. In the present study, 221 microbes were detected at the genus level and were narrowed down to 18 microbes based on their percentages. *Fusobacterium* was found to be among the most abundant species in both normal and tumor tissues, while *Porphyromonas* was among the least abundant species in both normal and tumor tissues (Figure 1). *Fusobacterium* and *Porphyromonas* species, which are associated with the prognosis of cancer, did not affect patient prognosis in an analysis of two classified groups, a rate over and under the median, while the genus *Leptotrichia* was shown to improve patient prognosis (Figure 2). The reason why *Fusobacterium* and *Porphyromonas* species were not clearly associated with the prognosis of HNSCC patients may be related to the amount of bacteria in the selected tissues.

Leptotrichia species are biochemically anaerobic Gram-negative rods that belong to the normal flora of humans and are generally present in the oral cavity, intestines, and human female genitalia [27]. *Leptotrichia* were found to be significantly more abundant in allergic rhinitis (AR) and allergic rhinitis with asthma (ARAS) [28]; the composition ratios of *Leptotrichia* species were higher in AR and ARAS (5.9 and 5.2%, respectively) than in healthy controls (3.5%). Another study reported a relationship between the use of dentures and *Candida albicans* [29]. The composition ratios of *Leptotrichia* were 3–4% in dentures and dental plaque, and, thus, the genus *Leptotrichia*, which negatively correlated with *C. albicans*, may be useful in antifungal therapy to control the growth of *C. albicans* [29]. *C. albicans*, the most common oral commensal, is also associated with cancer and has been suggested to exert tumorigenic effects and affect PD-L1 expression [30,31]. In the present study, patients in the over-the-median-for-*Leptotrichia* group had a better prognosis; therefore, *C. albicans* may have been less abundant under that condition.

Leptotrichia and *Fusobacterium* have been implicated in the development of colon cancer [25,32]. Furthermore, intratumoral *Leptotrichia* has been identified as a novel microbial marker of a favorable clinical outcome in HNSCC patients [25]. *Leptotrichia* species correlated with higher overall survival rates and were significantly more abundant in early-stage patients than in advanced-stage patients, suggesting the protective effects of *Leptotrichia* species in the HNSCC tumor microenvironment [25]. We herein investigated the effects of classical risk factors, such as drinking, smoking, HPV status, sex, lymph node metastasis, and tumor size, in two groups: a rate over and under the median for *Leptotrichia* species. Survival curves were not affected by the population of *Leptotrichia* species in the absence of drinking, without lymph node metastasis, in females, and in T1–T2 tumors, but were markedly changed when the rate of *Leptotrichia* species was over the median with other classical prognostic factors examined in the present study. The result on tumor sizes is not consistent with previous findings [22]. The present results suggest that the genus *Leptotrichia* was less protective in the HNSCC tumor microenvironment in early-stage patients.

Microbial and molecular differences in HNSCC sites were recently studied [33]. Through KEGG pathway analysis, it was found that, in oral cancers, positively correlated genes were prion diseases, Alzheimer disease, Parkinson disease, *Salmonella* infection, and pathogenic *Escherichia coli* infection. In non-oral cancers, positively correlated genes were herpes simplex virus 1 infection and spliceosome, while negatively correlated genes were the PI3K-Akt signaling pathway, focal adhesion, the regulation of actin cytoskeleton, ECM–receptor interaction, and dilated cardiomyopathy. In the present study, genes differentially expressed between the over- and under-the-median-for-*Leptotrichia* groups were extracted to assess the effects of intratumoral *Leptotrichia* on the behavior of HNSCC (Figure 4). KEGG analysis showed that *Leptotrichia*-related genes were significantly altered in Alzheimer disease, pathways of neurodegeneration - multiple diseases, prion disease, the MAPK signaling pathway, and the PI3K-Akt signaling pathway, consistent with previous reports not limited to the genus *Leptotrichia*. Pathways of neurodegeneration-multiple diseases and MAPK signaling, which were identified for the first time in the present study, may be hallmarks of the effects of *Leptotrichia* on HNSCC. Furthermore, the *Leptotrichia*-related genes selected in the present study, which were significantly up- or down-regulated and formed a dense PPI network, were associated with a favorable prognosis in HNSCC patients. However, bioinformatics-based research on the microbiota and HNSCC is in the early stages, and further research will help to elucidate the prognosis and progression of HNSCC.

4. Materials and Methods

4.1. Data Collection from TCGA and TCMA Databases

We obtained the RNA-seq count data (HTSeq version) of TCGA-HNSC (499 primary-tumor and 45 solid-tissue normal samples) from the GDC Data Portal [34] (https://portal.gdc.cancer.gov/ accessed on 20 March 2019) with the Subio Platform (https://www.subioplatform.com accessed on 17 October 2023). We also obtained the intratumor microbiome compositions of 177 TCMA-HNSC samples (155 primary tumor and 22 solid tissue normal samples) at the genus level from TCMA [35] (https://tcma.pratt.duke.edu/ accessed on 13 July 2023). We selected 154 patients in both TCGA and TCMA (Figure 1A).

4.2. Filtering of TCMA Genus Microbes

We used Subio Platform [36] software v1.24.5859 (Subio Inc. Aichi, Japan) to filter microbes. A total of 221 microbes were defined at TCMA genus level, but most were undetected; therefore, we excluded those with a rate < 0.1 in 175 out of 177 samples. A total of 18 microbes remained, and we summed the rates of the 203 other microbes and labeled them as "Other".

4.3. Kaplan–Meier Survival Analysis

Regarding each of the filter-passed 18 microbes, TCGA-HNSC primary tumor samples were divided into two groups: a rate over and under the median. Subio Platform software was used to generate Kaplan–Meier survival curves for comparisons of the outcomes of the over and under the median groups for each microbe.

4.4. Extraction of Differentially Expressed Genes between over and under the Median for Leptotrichia Groups

We normalized RNA-Seq count data at the 90th percentile, replaced all non-zero counts less than 50 with 50, and replaced 0 with 32 as a low signal cut-off. We turned normalized counts to log2 ratios against the average of solid normal tissue samples. We excluded genes if their counts were too low (counts < 50 in all samples) or too stable (log2 ratios between -1 and 1 in all samples.)

We extracted 35 differentially expressed genes between over and under the median for *Leptotrichia* groups based on the criteria of >1.5 fold and $p < 0.05$ in the Mann–Whitney U-test (Figure 4A). We applied hierarchical clustering to 35 genes (Figure 4B).

4.5. Functional Pathway and PPI Analyses

The Database for Annotation, Visualization, and Integrated Discovery (DAVID) server was used to examine the molecular pathways of the selected genes for GO terms and KEGG pathways. GO enrichment was performed over three primary levels: cellular components (CC), biological processes (BP), and molecular functions (MF). Based on the STRING online database (https://string-db.org/ accessed on 5 September 2023), these genes were used to establish a PPI network. We then visualized the most significant modules in the PPI network.

4.6. Statistical Analyses

Statistical analyses were performed using Student's *t*-test with Microsoft Excel (Microsoft, Redmond, WA, USA). Results were expressed as the mean ± SD. Differences were considered significant at $p < 0.05$. In the survival analysis shown in Table 1, the hazard ratio (HR) relative to the indicated reference (ref) value, its 95% confidence interval (CI), and the *p*-value (those < 0.05 are indicated in bold) for the Cox hazard model are shown. The HR and its 95% CI were calculated using a Cox regression analysis after the proper evaluation of assumptions of Cox regression models using the survival package.

5. Conclusions

We used TCMA and The Cancer Genome Atlas and examined microbiome profiling in HNSCC. The tissue microbiome profiles of *Actinomyces*, *Fusobacterium*, and *Rothia* at the genus level differed between the solid normal tissue and primary tumors of HNSCC patients. However, when the prognosis of groups with rates over and under the median for each microbe at the genus level was examined using a Cox regression analysis and the Kaplan–Meier method, a rate over the median for *Leptotrichia* was found to be correlated with significantly higher overall survival rates in TCGA-HNSCC patients. *Leptotrichia* species-targeted probiotics and specific antibacterial therapy may have an impact on the prognosis of HNSCC.

Author Contributions: Conceptualization, M.H. and H.I.; methodology, M.H., H.I., K.N. and S.Y.; formal analysis, M.H., H.I., Y.Y., M.M.-N. and N.U.; writing—original draft preparation, M.H., H.I. and Y.Y.; writing—review and editing, M.H., H.I. and Y.Y.; supervision, Y.Y., M.M.-N. and N.U. All authors reviewed the manuscript. All authors have read and agreed to the published version of the manuscript.

Funding: This work was supported in part by a Grant-in-Aid for Scientific Research from the Japan Society for the Promotion of Science (No. 21K10111, No. 22K17202, and No. 23K09137).

Institutional Review Board Statement: Not applicable.

Informed Consent Statement: Not applicable.

Data Availability Statement: Data are available from the corresponding author upon reasonable request.

Acknowledgments: We thank Subio Inc. (Aichi, Japan) for technical support on the Subio Platform and SATISTA Inc. (Kyoto, Japan.) for supporting statistical analyses.

Conflicts of Interest: The authors declare no conflict of interest.

References

1. The Cancer Genome Atlas Research Network. Comprehensive genomic characterization defines human glioblastoma genes and core pathways. *Nature* **2008**, *455*, 1061–1068. [CrossRef] [PubMed]
2. Garraway, L.A.; Lander, E.S. Lessons from the cancer genome. *Cell* **2013**, *153*, 17–37. [CrossRef] [PubMed]
3. Nakagawa, H.; Fujita, M. Whole genome sequencing analysis for cancer genomics and precision medicine. *Cancer Sci.* **2018**, *109*, 513–522. [CrossRef]
4. The Cancer Genome Atlas Network. Comprehensive genomic characterization of head and neck squamous cell carcinomas. *Nature* **2015**, *517*, 576–582. [CrossRef] [PubMed]
5. Jemal, A.; Bray, F.; Center, M.M.; Ferlay, J.; Ward, E.; Forman, D. Global cancer statistics. *CA Cancer J. Clin.* **2011**, *61*, 69–90. [CrossRef]
6. Wong, N.; Khwaja, S.S.; Baker, C.M.; Gay, H.A.; Thorstad, W.L.; Daly, M.D.; Lewis, J.S., Jr.; Wang, X. Prognostic microRNA signatures derived from The Cancer Genome Atlas for head and neck squamous cell carcinomas. *Cancer Med.* **2016**, *5*, 1619–1628. [CrossRef] [PubMed]
7. Zhao, Y.; Ruan, X. Identification of PGRMC1 as a Candidate Oncogene for Head and Neck Cancers and Its Involvement in Metabolic Activities. *Front. Bioeng. Biotechnol.* **2019**, *7*, 438. [CrossRef]
8. Wang, X.; Yin, Z.; Zhao, Y.; He, M.; Dong, C.; Zhong, M. Identifying potential prognostic biomarkers in head and neck cancer based on the analysis of microRNA expression profiles in TCGA database. *Mol. Med. Rep.* **2020**, *21*, 1647–1657. [CrossRef]
9. Hamada, M.; Inaba, H.; Nishiyama, K.; Yoshida, S.; Yura, Y.; Matsumoto-Nakano, M.; Uzawa, N. Prognostic association of starvation-induced gene expression in head and neck cancer. *Sci. Rep.* **2021**, *11*, 19130. [CrossRef]
10. Gillison, M.L.; Koch, W.M.; Capone, R.B.; Spafford, M.; Westra, W.H.; Wu, L.; Zahurak, M.L.; Daniel, R.W.; Viglione, M.; Symer, D.E.; et al. Evidence for a causal association between human papillomavirus and a subset of head and neck cancers. *J. Natl. Cancer Inst.* **2000**, *92*, 709–720. [CrossRef]
11. Bird, T.; De Felice, F.; Michaelidou, A.; Thavaraj, S.; Jeannon, J.P.; Lyons, A.; Oakley, R.; Simo, R.; Lei, M.; Guerrero Urbano, T. Outcomes of intensity-modulated radiotherapy as primary treatment for oropharyngeal squamous cell carcinoma—A European singleinstitution analysis. *Clin. Otolaryngol.* **2017**, *42*, 115–122. [CrossRef] [PubMed]
12. De Felice, F.; Tombolini, V.; Valentini, V.; de Vincentiis, M.; Mezi, S.; Brugnoletti, O.; Polimeni, A. Advances in the Management of HPV-Related Oropharyngeal Cancer. *J. Oncol.* **2019**, *2019*, 9173729. [CrossRef] [PubMed]
13. Inaba, H.; Sugita, H.; Kuboniwa, M.; Iwai, S.; Hamada, M.; Noda, T.; Morisaki, I.; Lamont, R.J.; Amano, A. *Porphyromonas gingivalis* promotes invasion of oral squamous cell carcinoma through induction of proMMP9 and its activation. *Cell. Microbiol.* **2014**, *16*, 131–145. [CrossRef] [PubMed]
14. Zhou, Y.; Sztukowska, M.; Wang, Q.; Inaba, H.; Potempa, J.; Scott, D.A.; Wang, H.; Lamont, R.J. Noncanonical activation of β-catenin by *Porphyromonas gingivalis*. *Infect. Immun.* **2015**, *83*, 3195–3203. [CrossRef]
15. Tsai, M.S.; Chen, Y.Y.; Chen, W.C.; Chen, M.F. *Streptococcus mutans* promotes tumor progression in oral squamous cell carcinoma. *J. Cancer* **2022**, *13*, 3358–3367. [CrossRef]
16. Orlandi, E.; Iacovelli, N.A.; Tombolini, V.; Rancati, T.; Polimeni, A.; De Cecco, L.; Valdagni, R.; De Felice, F. Potential role of microbiome in oncogenesis, outcome prediction and therapeutic targeting for head and neck cancer. *Oral Oncol.* **2019**, *99*, 104453. [CrossRef]

17. Sepich-Poore, G.D.; Zitvogel, L.; Straussman, R.; Hasty, J.; Wargo, J.A.; Knight, R. The microbiome and human cancer. *Science* **2021**, *371*, eabc4552. [CrossRef]
18. Xie, Y.; Xie, F.; Zhou, X.; Zhang, L.; Yang, B.; Huang, J.; Wang, F.; Yan, H.; Zeng, L.; Zhang, L.; et al. Microbiota in Tumors: From Understanding to Application. *Adv. Sci.* **2022**, *9*, e2200470. [CrossRef]
19. Li, R.; Gao, X.; Sun, H.; Sun, L.; Hu, X. Expression characteristics of long non-coding RNA in colon adenocarcinoma and its potential value for judging the survival and prognosis of patients: Bioinformatics analysis based on The Cancer Genome Atlas database. *J. Gastrointest. Oncol.* **2022**, *13*, 1178. [CrossRef]
20. Lehr, K.; Nikitina, D.; Vilchez-Vargas, R.; Steponaitiene, R.; Thon, C.; Skieceviciene, J.; Schanze, D.; Zenker, M.; Malfertheiner, P.; Kupcinskas, J.; et al. Microbial composition of tumorous and adjacent gastric tissue is associated with prognosis of gastric cancer. *Sci. Rep.* **2023**, *13*, 4640. [CrossRef]
21. Wang, Y.; Wang, Y.; Wang, J. A comprehensive analysis of intratumor microbiome in head and neck squamous cell carcinoma. *Eur. Arch. Otorhinolaryngol.* **2022**, *279*, 4127–4136. [CrossRef]
22. Metsäniitty, M.; Hasnat, S.; Salo, T.; Salem, A. Oral Microbiota-A New Frontier in the Pathogenesis and Management of Head and Neck Cancers. *Cancers* **2021**, *14*, 46. [CrossRef]
23. Mosaddad, S.A.; Tahmasebi, E.; Yazdanian, A.; Rezvani, M.B.; Seifalian, A.; Yazdanian, M.; Tebyanian, H. Oral microbial biofilms: An update. *Eur. J. Clin. Microbiol. Infect. Dis.* **2019**, *38*, 2005–2019. [CrossRef]
24. Dohlman, A.B.; Arguijo Mendoza, D.; Ding, S.; Gao, M.; Dressman, H.; Iliev, I.D.; Lipkin, S.M.; Shen, X. The cancer microbiome atlas: A pan-cancer comparative analysis to distinguish tissue-resident microbiota from contaminants. *Cell Host Microbe* **2021**, *29*, 281–298.e285. [CrossRef]
25. Yu, S.; Chen, J.; Yan, F.; Chen, X.; Zhao, Y.; Zhang, P. Intratumoral Leptotrichia is a novel microbial marker for favorable clinical outcomes in head and neck cancer patients. *MedComm* **2023**, *4*, e344. [CrossRef]
26. Qiao, H.; Li, H.; Wen, X.; Tan, X.; Yang, C.; Liu, N. Multi-Omics Integration Reveals the Crucial Role of Fusobacterium in the Inflammatory Immune Microenvironment in Head and Neck Squamous Cell Carcinoma. *Microbiol. Spectr.* **2022**, *10*, e0106822. [CrossRef]
27. Eribe, E.R.; Olsen, I. Leptotrichia species in human infections. *Anaerobe* **2008**, *14*, 131–137. [CrossRef]
28. Pérez-Losada, M.; Castro-Nallar, E.; Laerte Boechat, J.; Delgado, L.; Azenha Rama, T.; Berrios-Farías, V.; Oliveira, M. The oral bacteriomes of patients with allergic rhinitis and asthma differ from that of healthy controls. *Front. Microbiol.* **2023**, *14*, 1197135. [CrossRef]
29. Fujinami, W.; Nishikawa, K.; Ozawa, S.; Hasegawa, Y.; Takebe, J. Correlation between the relative abundance of oral bacteria and Candida albicans in denture and dental plaques. *J. Oral Biosci.* **2021**, *63*, 175–183. [CrossRef]
30. Theofilou, V.I.; Alfaifi, A.; Montelongo-Jauregui, D.; Pettas, E.; Georgaki, M.; Nikitakis, N.G.; Jabra-Rizk, M.A.; Sultan, A.S. The oral mycobiome: Oral epithelial dysplasia and oral squamous cell carcinoma. *J. Oral Pathol. Med.* **2022**, *51*, 413–420. [CrossRef]
31. Wang, X.; Zhao, W.; Zhang, W.; Wu, S.; Yan, Z. Candida albicans induces upregulation of programmed death ligand 1 in oral squamous cell carcinoma. *J. Oral Pathol. Med.* **2022**, *51*, 444–453. [CrossRef]
32. Nakatsu, G.; Li, X.; Zhou, H.; Sheng, J.; Wong, S.H.; Wu, W.K.; Ng, S.C.; Tsoi, H.; Dong, Y.; Zhang, N.; et al. Gut mucosal microbiome across stages of colorectal carcinogenesis. *Nat. Commun.* **2015**, *6*, 8727. [CrossRef]
33. Kim, Y.K.; Kwon, E.J.; Yu, Y.; Kim, J.; Woo, S.Y.; Choi, H.S.; Kwon, M.; Jung, K.; Kim, H.S.; Park, H.R.; et al. Microbial and molecular differences according to the location of head and neck cancers. *Cancer Cell Int.* **2022**, *22*, 135. [CrossRef]
34. Available online: https://portal.gdc.cancer.gov (accessed on 20 March 2019).
35. Available online: https://tcma.pratt.duke.edu/ (accessed on 13 July 2023).
36. Available online: https://www.subioplatform.com (accessed on 20 March 2019).

International Journal of
Molecular Sciences

Article

Necrotic Cells from Head and Neck Carcinomas Release Biomolecules That Are Activating Toll-like Receptor 3

Tea Vasiljevic [1], Marko Tarle [2,3], Koraljka Hat [2], Ivica Luksic [2], Martina Mikulandra [4], Pierre Busson [5] and Tanja Matijevic Glavan [1,*]

[1] Laboratory for Personalized Medicine, Division of Molecular Medicine, Rudjer Boskovic Institute, Bijenicka 54, 10000 Zagreb, Croatia
[2] Department of Maxillofacial Surgery, Dubrava University Hospital, School of Medicine, University of Zagreb, Gojko Šušak Avenue 6, 10000 Zagreb, Croatia; tarlemarko1@gmail.com (M.T.)
[3] School of Dental Medicine, University of Zagreb, Gunduliceva 5, 10000 Zagreb, Croatia
[4] Division of Oncology and Radiotherapy, University Hospital for Tumors, Sestre Milosrdnice University Hospital Center, Vinogradska Cesta 29, 10000 Zagreb, Croatia
[5] CNRS-UMR 9018-METSY, Gustave Roussy Institute, Université Paris-Saclay, 39 rue Camille Desmoulins, 94805 Villejuif CEDEX, France
* Correspondence: tmatijev@irb.hr

Abstract: Tumor necrosis is a recurrent characteristic of head and neck squamous cell carcinomas (HNSCCs). There is a need for more investigations on the influence of biomolecules released by these necrotic foci in the HNSCC tumor microenvironment. It is suspected that a fraction of the biomolecules released by necrotic cells are damage-associated molecular patterns (DAMPs), which are known to be natural endogenous ligands of Toll-like receptors (TLRs), including, among others, proteins and nucleic acids. However, there has been no direct demonstration that biomolecules released by HNSCC necrotic cells can activate TLRs. Our aim was to investigate whether some of these molecules could behave as agonists of the TLR3, either in vitro or in vivo. We chose a functional approach based on reporter cell exhibiting artificial TLR3 expression and downstream release of secreted alkaline phosphatase. The production of biomolecules activating TLR3 was first investigated in vitro using three HNSCC cell lines subjected to various pronecrotic stimuli (external irradiation, serum starvation, hypoxia and oxidative stress). TLR3 agonists were also investigated in necrotic tumor fluids from five oral cancer patients and three mouse tumor grafts. The release of biomolecules activating TLR3 was demonstrated for all three HNSCC cell lines. External irradiation was the most consistently efficient stimulus, and corresponding TLR3 agonists were conveyed in extracellular vesicles. TLR3-stimulating activity was detected in the fluids from all five patients and three mouse tumor grafts. In most cases, this activity was greatly reduced by RNAse pretreatment or TLR3 blocking antibodies. Our data indicate that TLR3 agonists are consistently present in necrotic fluids from HNSCC cells and mainly made of dsRNA fragments. These endogenous agonists may induce TLR3, which might lead to a protumorigenic effect. Regarding methodological aspects, our study demonstrates that direct investigations—including functional testing—can be performed on necrotic fluids from patient tumors.

Keywords: toll-like receptor 3; endogenous ligands; head and neck cancer; exosomes; HEKBlue cells

1. Introduction

Head and neck squamous cell carcinomas (HNSCCs) account for one-sixth of all human malignancies worldwide. They represent a major public health problem. Alcohol and tobacco abuse are among the main etiological factors [1]. However, a fraction of HNSCCs is related to human papillomavirus infections, especially among oropharyngeal and tonsil carcinomas [2]. Anoxia and necrosis are more common and severe in HNSCCs than in many other categories of human malignancies [3,4].

Cell necrosis is defined as an unprogrammed form of cell death that occurs in response to overwhelming chemical or physical distress. The occurrence of necrotic cells is one of the most consistent characteristics of all human solid malignancies [5]. To a large extent, it is linked to the quantitative and qualitative mismatch between tumor angiogenesis and the proliferation of malignant cells. Insufficient vascularization results in acute shortages of oxygen and nutrients, inducing the formation of tumor necrotic foci of various sizes. Necrotic cells undergo cell lysis and release various types of biomolecules, either in native or altered configurations. A number of these diffusible necrotic biomolecules induce changes in the phenotype and behavior of distinct living cells with the status of target cells (for a review, see [6]). Many of these changes result from the binding of necrotic biomolecules to Toll-like receptors (TLRs) expressed by the target cells.

Initially described as actors associated with embryonic development in invertebrates, TLRs were later recognized as important players in the innate immune response in mammals. They can be found in immune cells, most often in dendritic cells and macrophages [7]. Nowadays, they are known to be expressed in a wide range of tissues and to play a role in the homeostasis of these tissues [8]. They are also often expressed by malignant cells in various types of tumors [9]. TLRs belong to a larger protein family called pattern recognition receptors (PRRs). Each of them recognizes a specific category of molecular motifs often shared by biomolecules derived from several groups of pathogens. Those motifs belong to bacteria, viruses, parasites or fungi and are called pathogen-associated molecular patterns (PAMPs). Besides PAMPs, which are externally derived molecules, TLRs also recognize damage-associated molecular patterns (DAMPs) that have an endogenous origin, for example, tissue damage and/or cell death. Both PAMPs and DAMPs can directly activate TLR signaling pathways [10,11]. Stimulation of TLR receptors by their ligands activates different signaling pathways and transcription factors: Nuclear factor kappa-light-chain-enhancer of activated B cells (NF-κB), Interferon regulatory factor 3 (IRF3), Interferon regulatory factor 7 (IRF7), Activator protein 1 (AP-1) and cAMP-response element binding protein (CREB) [12]. Today, 10 TLRs have been recognized in humans: TLR1, 2, 4, 5, 6, 10 and 11 reside, to a large extent, on the plasma membrane, whereas TLR3, 7, 8 and 9 are mostly found in endosomes and the endoplasmic reticulum [7]. Among DAMPs activating TLRs, we can mention heat shock proteins (HSPs) and High mobility group box 1 (HMGB1) for TLR2; HSPs, fibrinogen, heparan sulfate, fibronectin, hyaluronic acid and HMGB1 for TLR4; self ssRNA for TLR7; and self-DNA for TLR9 [13].

Like other TLRs, TLR3 can be activated by PAMPs and DAMPs. Pieces of double-strand RNA (dsRNA) of minimal size (40–50 bp) are the most well-known ligands of TLR3. Under physiological conditions, these dsRNAs are rare inside the cells and circulate through intracellular routes distinct from those of TLR3. In contrast, they are abundant and can intersect the routes of TLR3 during viral infections. TLR3 can also be activated by synthetic analogs of dsRNAs such as Polyinosinic:polycytidylic acid (poly(I:C)) or Polyadenylic–polyuridylic acid (poly(A:U)). Kariko et al. and Cavassani et al. previously reported indirect evidence that TLR3 ligands can originate from necrotic cells in vivo [14,15].

In this context, the first aim of our study was to explore whether in vitro exposition of HNSCC cells to various types of stress conditions could induce the release of TLR3-activating DAMPs. The next aim was to determine whether these TLR3-related DAMPs were contained in exosomes. Finally, we wanted to provide evidence that the same type of DAMPs was present in necrotic fluids obtained from mouse tumor models and oral cancer clinical specimens.

2. Results

2.1. Under In Vitro Stress Conditions, HNSCC Cells Release Endogenous Ligands That Activate TLR3 Reporter Cells

To determine whether endogenous ligands released from head and neck carcinoma cells exposed to different stressors can activate TLR3, we used the HEKBlue-TLR3 reporter cell line (Invivogen). It was produced by stable co-transfection of the human *TLR3* gene into HEK293 cells, along with an inducible *SEAP* (secreted embryonic alkaline phosphatase) reporter gene. The *SEAP* gene was placed under the control of the IFN-β minimal promoter fused to five NF-κB- and AP-1-binding sites. Stimulation with artificial or natural TLR3 ligands activates NF-κB and AP-1 signaling pathways, which upregulate the production of SEAP. Conditioned culture media were collected from three HNSCC cell lines (Detroit 562, FaDu and SQ20B) following challenges by irradiation, oxidative stress (H_2O_2), serum deprivation or hypoxia. The capacity of these conditioned media to activate TLR3 was tested using HEKBlue-TLR3 cells as targets. The enzymatic activity of SEAP released by target HEKBlue-TLR3 cells was used as an index of TLR3 activation. Because NF-κB and AP-1 pathways can be activated in target cells through receptors distinct from TLR3, HEKBlue-null cells were used as negative controls. To further clarify our study design and results, a flow chart of this study is presented in Figure 1.

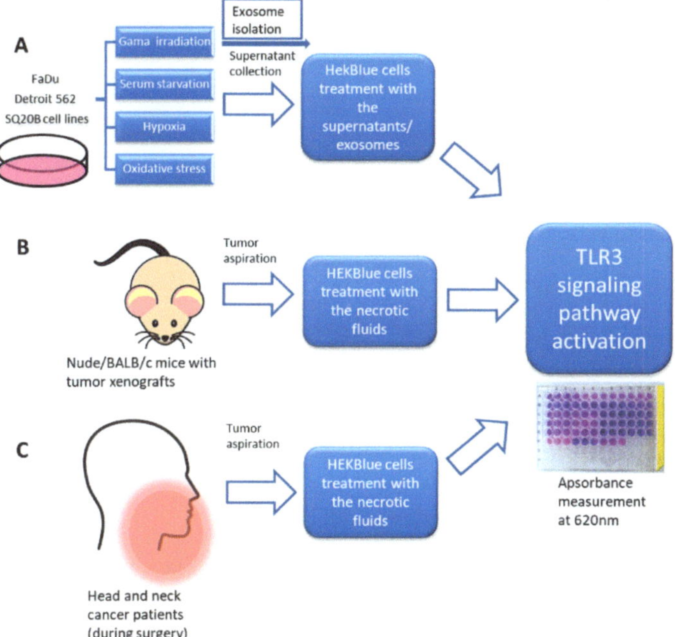

Figure 1. Flow chart showing the experimental design of the study. Conditioned media were collected from cells stressed with irradiation, serum starvation, hypoxia and oxidative stress. In addition, extracellular vesicles were isolated from conditioned media of irradiated cells. Both crude conditioned media and isolated extracellular vesicles were used to stimulate HEKBlue cells in order to measure the activation of the TLR3 signaling pathway (**A**). Additionally, necrotic fluids were collected from tumor grafts carried by nude or Balb/C mice (**B**) and from tumor tissues of patients bearing head and neck carcinomas (**C**). These fluids were also tested on HEKBlue cells in order to measure the activation of the TLR3 signaling pathway.

As shown in Figure 2, for each cell line, SEAP production was induced in the presence of conditioned medium from unchallenged cells, suggesting a constitutive release of TLR3 ligands by these cells. However, for each cell line, there was an increase in SEAP production by target cells for at least one type of challenge imposed on HNSCC cells: irradiation for all three cell lines, oxidative stress and hypoxia for Detroit 562, serum deprivation for FaDu and hypoxia for SQ20B. No substantial increases in SEAP production were observed when HEKBlue-null cells were exposed to the same supernatants, favoring an increase in the release of endogenous TLR3 ligands by malignant HNSCC cells under various types of stressful conditions, especially external irradiation. For Detroit 562 and SQ20B but not FaDu, a significant decrease in SEAP production was observed when the conditioned media from irradiated cells were treated with RNAse before application on target cells. This strengthens the idea that irradiated HNSCC cells release endogenous TLR3 ligands and suggests that these ligands are—at least in part—made of extracellular RNA fragments. The same reduction in the induction of SEAP production was recorded for conditioned medium from FaDu subjected to serum deprivation (Figure 2).

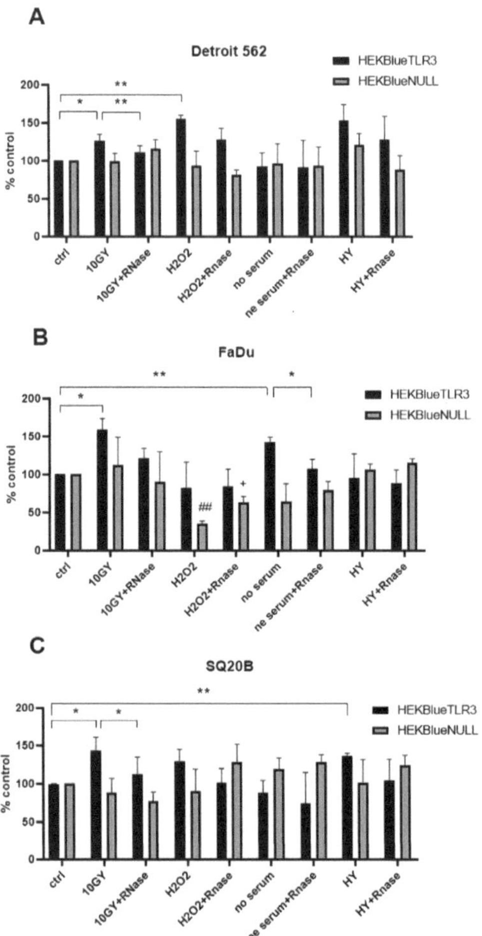

Figure 2. TLR3 activation in HEKBlue-TLR3 and HEKBlue-null reporter cells after treatment with supernatants from Detroit 562 (**A**), FaDu (**B**) and SQ20B cells (**C**) stressed with irradiation, H_2O_2, serum deprivation and hypoxia. Ctrl: cells were treated only with fresh culture medium; HY: hypoxia treatment. * $p < 0.05$, ** $p < 0.01$, ## $p < 0.01$ (HEKBlue-TLR3 in relation to HEKBlue-null control), + $p < 0.05$. (HEKBlue-null in relation to HEKBlue-null control). Statistical significance was determined by *t*-test.

2.2. Endogenous Ligands That Activate TLR3 Reporter Cells Are Contained in Extracellular Vesicles

To determine whether endogenous ligands that activate TLR3 reporter cells are contained in extracellular vesicles (EVs), head and neck carcinoma cells were subjected to irradiation, and their extracellular vesicles (mostly exosomes) were isolated from their conditioned media after 48 h of incubation. Irradiation was chosen for this investigation for two reasons: (1) it was the only stimulus inducing an excess of TLR3 activation in all three cell lines and (2) we recently reported that poly(I:C) and cisplatin stimulation leads to radiosensitization of Detroit 562 cells [16]. Exosome size was first checked by Nanoparticle tracking analysis (NTA) (Figure 3). The size of exosomes was 150–200 nm for control and irradiated Detroit 562 cells, 150–300 nm for FaDu and 50–200 nm for SQ20B. Some larger particles—400 to 800 nm in diameter—were observed outside these ranges, especially in SQ20B-conditioned medium, but they were probably exosomes aggregates. Exosome size and morphology was next evaluated by Transmission electron microscopy (TEM) (Figure 4A). Exosome material derived from all three cell types, either control or irradiated, shared similar morphological characteristics: a regular circular shape and a size distribution ranging from 50 to 220 nm, consistent with most literature data. However, there were substantial size variations depending on the cell line, with a range of 80–130 nm for Detroit 562, 160–220 nm for FaDu and only 50–70 nm for SQ20B. To assess the quality of exosome enrichment, we compared protein extracts from whole cells and exosome preparations using Western blots, investigating the amounts of two proteins: one known to be abundant in exosomes (tetraspanin CD63) and the other, calnexin, which is an endoplasmic reticulum marker virtually absent from the exosomes. As expected and as shown in Figure 4B,C, CD63 was detected in exosome preparations but not in whole-cell extracts, whereas the result was the opposite for calnexin.

Using these extracellular vesicles preparations, reporter cells were exposed either to crude-conditioned media or exosomes that were isolated from both unchallenged and irradiated head and neck carcinoma cells. For Detroit 562 and SQ20B cells, using material from unchallenged cells, exposure to exosomes was more efficient than exposure to crude media in the induction of SEAP production, whereas the opposite was observed for FaDu cells. However, when using material from irradiated cells, much greater SEAP production was observed using exosomes in comparison with both crude-conditioned medium and exosomes from unchallenged cells. This result was statistically significant for all three cell lines. Greater responses were recorded for the SQ20B cell line (Figure 4D,E).

2.3. Endogenous Ligands Contained in Necrotic Material from Fresh Oral Squamous Cell Carcinoma Activate TLR3 Reporter Cells

Inappropriate angiogenesis, hypoxia and cell necrosis are nearly constant features of solid malignancies in the context of patient tumors, as well as experimental tumors from animal models. Like other investigators in the past, we empirically observed that necrotic material behaves differently from live cells when fragments of clinical tumor specimens or experimental tumors are minced in ordinary culture medium immediately or shortly after ex vivo collection. Cells from necrotic foci lose their cohesion and form a suspension of isolated cells or small aggregates much more easily than live cells. Simultaneously, necrotic fluid resulting from earlier or concomitant cell lysis is released in the medium. No enzymatic treatment is necessary to release this necrotic material, which is generally considered more of an undesirable contaminant. In contrast, for us, rapid tumor mincing was a simple way to prepare fluid samples highly enriched in necrotic material. Such samples were obtained from surgical pieces of oral squamous cell carcinoma for five patients. The amounts of RNA and proteins contained in these samples are also shown in Figure 5. Patients' clinicopathological data and histopathological features are shown in Table 1. It is important to emphasize that tumor mincing is used not to induce necrosis but to separate the pre-existing necrotic material from the healthy cells.

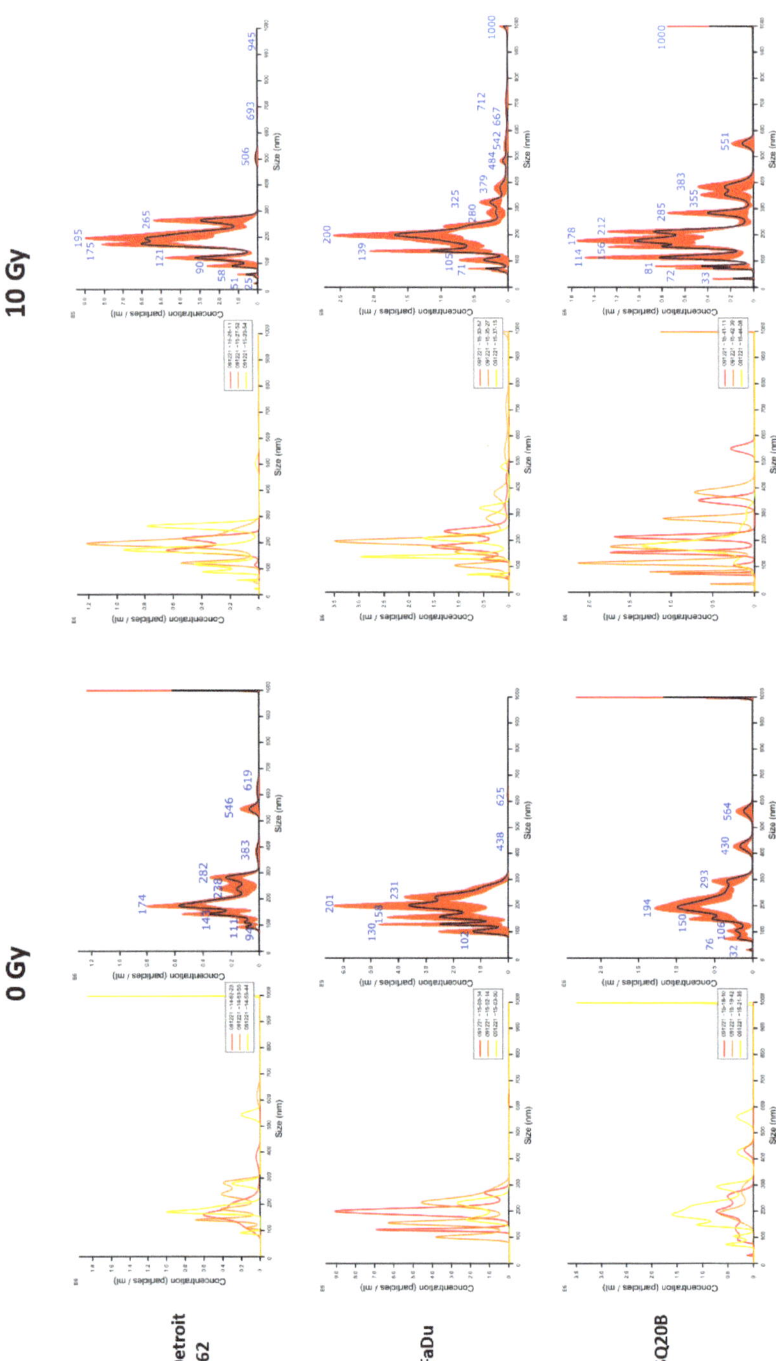

Figure 3. Exosome characterization. Nanoparticle tracking analysis (NTA) of exosomes derived from Detroit 562, FaDu and SQ20B cell supernatants from control non-irradiated cells and cells after irradiation with 10 Gy. The left side shows three runs, with Finite track length adjustment (FTLA) concentration indicated by yellow, orange and red lines, and the graph on the right shows the averaged FTLA concentration for each sample.

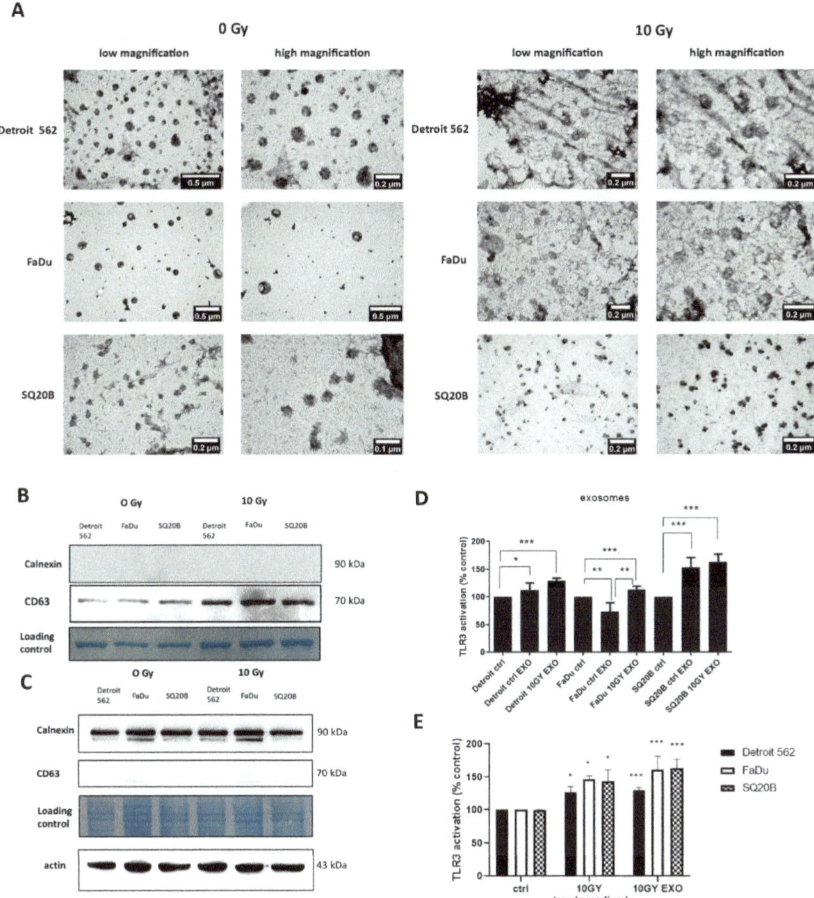

Figure 4. Exosome characterization and the treatment of HEKBlue-TLR3 reporter cells with exosomes. Transmission electron microscopy (TEM) images of exosomes derived from Detroit 562, FaDu and SQ20B cell supernatants from control non-irradiated cells and cells after irradiation with 10 Gy (**A**). Western blot of proteins isolated from exosomes (**B**) and cell lysate. The loading control is the membrane stained with amido black (**C**). TLR3 activation after the treatment of HEKBlue-TLR3 cells with exosomes (800 ng) isolated from control non-irradiated cells and cells after irradiation with 10 Gy (**D**). Comparison between TLR3 activation with crude medium and exosomes isolated after irradiation (**E**). * $p < 0.05$, ** $p < 0.01$, *** $p < 0.001$.

As shown in Figure 5, all of them induced activation of HEKBlue-TLR3 cells in a statistically significant manner. A certain increase in SEAP production was observed when HEKBlue-null cells were exposed to the same fluids, but it was relatively limited (in the range of 60 to 180% and not exceeding 200%). In contrast, TLR3 activation was at a very high level in comparison with control fluids when the HEKBlue-TLR3 cells were the targets: 728%, 763% and 796% for patients 1, 4 and 5, respectively. In addition, in each case, previous treatments with RNAse or the addition of a TLR3-blocking antibody resulted in a substantial decrease in the activation of HEKBlue-TLR3 cells. All these observations were favor a specific role of the TLR3 pathway.

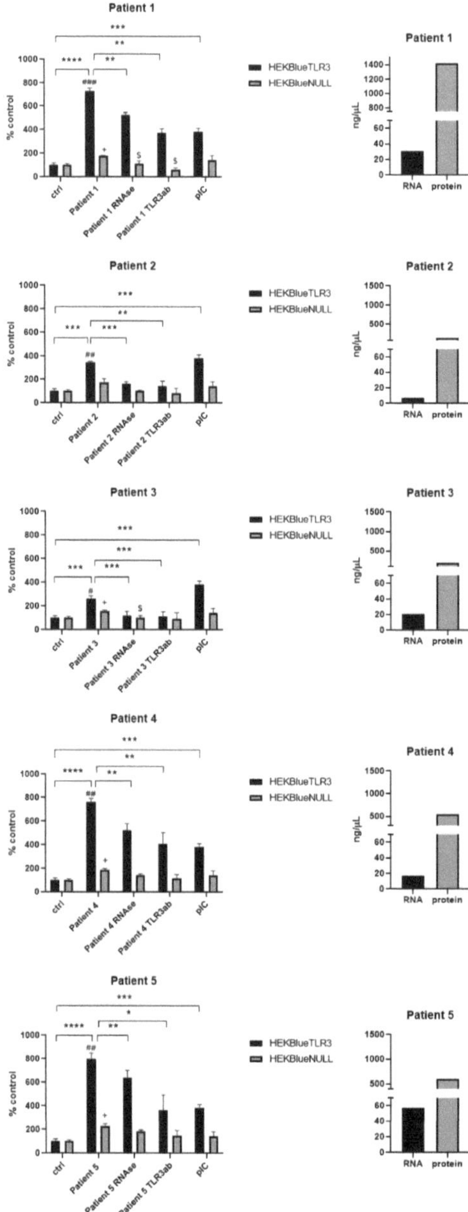

Figure 5. TLR3 activation after the treatment of HEKBlue-TLR3 and HEKBlue-null cells with necrotic fluids derived from oral squamous cell carcinoma patients: patient 1, patient 2, patient 3, patient 4 and patient 5. Ctrl—cells were treated only with fresh culture medium; RNAse—sample pretreated with RNAse; TLR3ab—sample treated with monoclonal TLR3 antibody; pIC—cells treated with 100 ng/mL poly(I:C). The amount of RNA and protein in each sample is also presented. * $p < 0.05$, ** $p < 0.01$, *** $p < 0.001$, **** $p < 0.001$, # $p < 0.05$, ## $p < 0.01$, ### $p < 0.001$ (HEKBlue-TLR3 in relation to HEKBlue-null control). + $p < 0.05$ (HEKBlue-null patient in relation to HEKBlue-null control), \$ $p < 0.05$ (HEKBlue-null treatment in relation to HEKBlue-null patient).

Table 1. Clinicopathological data about patients, tumor location, type and risk factors.

Patient No.	Sex	PHD	Tumor Localization	Tumor Differentiation	Ptnm	Age Group	Smoking	Alcohol	Hepatitis	HPV	EBV	FA/DC	Diabetes
1	M	OSCC	retromolar trigone	well (I)	T4An0M0	70–75	Yes	Yes	No	No	No	No	No
2	M	OSCC	mandibular alveolar ridge	well (I)	T4An0M0	70–75	Yes	Yes	No	No	No	No	No
3	F	OSCC	tongue	well (I)	T1N2Bm0	40–45	No	No	No	No	No	No	No
4	M	OSCC	floor of mouth	well (I)	T4An3Bm0	60–65	Yes	Yes	No	No	No	No	No
5	F	OSCC	floor of mouth	well (I)	T4An3Bm0	55–60	Yes	No	No	No	No	No	No

PCC–planocellular carcinoma; FA—Fanconi anemia; DC—dyskeratosis congenita; pTNM—postsurgical histopathological classification; OSCC—oral squamous cell carcinoma.

2.4. Endogenous Ligands Contained in Necrotic Material Derived from Experimental Mouse Tumors Activate TLR3 Reporter Cells

Samples of necrotic tumor material were also prepared from mouse tumors either from nude mice xenografts (C17 and C18) or from a syngeneic graft on immunocompetent mice (Renca cells on C57BL/6). Renca cells were used because we were willing to perform investigations on necrotic fluids not only from human tumors xenografted on nude mice but also from murine syngeneic tumor models. However, for technical reasons, it was not possible to use a syngeneic model of HNSCC. Therefore, we resorted to the Renca model, which is available in many laboratories and research institutions. The amounts of RNA and proteins in these samples are also presented in Figure 6.

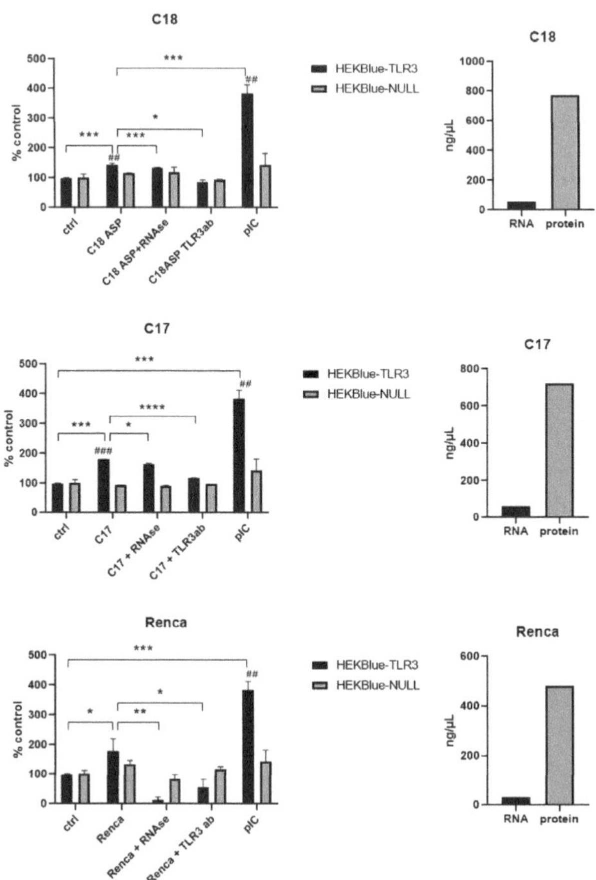

Figure 6. TLR3 activation after the treatment of HEKBlue-TLR3 and HEKBlue-null cells with necrotic fluids derived from different mice models: C18, C17 and Renca. Ctrl—cells treated only with medium; RNAse—sample pretreated with RNAse; TLR3ab—sample treated with monoclonal TLR3 antibody; pIC—cells treated with 100 ng/mL poly(I:C). The amount of RNA and protein in each sample is also presented. * $p < 0.05$, ** $p < 0.01$, *** $p < 0.001$, **** $p < 0.001$, ## $p < 0.01$, ### $p < 0.001$ (HEKBlue-TLR3 in relation to HEKBlue-null control).

All these samples induced the activation of HEKBlue TLR3 cells in a statistically significant manner when compared to the crude culture media used as controls. C17 and Renca samples achieved the highest levels of HEKBlue-TLR3 activation (180% and 179%, respectively). C18 activation was 143%. For the C17 and C18 samples, the induction of

SEAP production was very low when the HEKBlue-null cells were used as target cells (with a statistically significant difference). In addition, HEKBlue-TLR3 activation was substantially reduced when these fluids were pretreated with RNAse or when they were combined with a TLR3-blocking antibody. For the Renca samples, there was a mild increase in SEAP when HEKBlue-null cells were used as targets. However, there was a dramatic reduction in SEAP production from HEKBlue-TLR3 under RNAse pretreatment or in combination with the TLR3-blocking antibody, confirming a specific contribution of TLR3 in the SEAP response.

3. Discussion

It has been reported in previous publications that diffusible biomolecules released from necrotic cells can activate TLR3 in neighboring live cells [14,15]. However, these observations were obtained using experimental systems with limited relevance to cancer physiopathology (mechanical necrosis of HEK293 cells subjected to several freeze–thaw cycles and mouse cecal ligation and puncture, respectively). To the best of our knowledge, we are the first to provide a direct demonstration that necrotic biomolecules released by HNSCC cells can behave as agonists of TLR3 expressed by target living cells. In our experimental setting, in vitro, the release of necrotic biomolecules by malignant HNSCC cells was stimulated by stressing conditions mimicking those imposed on malignant cells undergoing pathological growth in situ (shortage of oxygen or growth factors and oxidative stress) or subjected to treatment procedures (external irradiation). External irradiation was the stimulus that was found to result in the most consistent production of TLR3 agonists, as demonstrated for Detroit 562, FaDu and SQ20B cell lines. We also showed that at least a fraction of the biomolecules released by necrotic HNSCC cells and behaving like TLR3 agonists can be conveyed by extracellular vesicles (EVs). This is consistent with an increasing number of publications suggesting that DAMPs can be found in EVs [17–22]. We demonstrated the association of TLR3 agonists with EVs by using external irradiation as a pronecrotic stimulus. In future studies, it will be interesting to investigate whether the same observation can be made using other necrotic stimuli, like hypoxia or oxidative stress. Finally, we showed that biomolecules triggering TLR3 activation are present in necrotic fluids derived from mouse tumor grafts and fresh tumor biopsies collected from patients. Regarding mouse tumor grafts, two were human PDX (patient-derived xenografts) from nasopharyngeal carcinomas (C17 and C18), while the last one, Renca, was not related to HNSCC but is a syngeneic murine tumor of renal origin propagated on immunocompetent mice (C57BL/6). As for C17 and C18, the necrotic biomolecules from Renca were effective on TLR3 reporter cells. Regardless of the category of necrotic biofluids, the evidence of TLR3 involvement was based on the comparison of the reporter response obtained with HEK-TLR3 and HEK-null cells. In addition, for several biofluids from patient or mouse tumor grafts, there was a substantial reduction in the reporter response in the presence of a TLR3-blocking antibody.

More studies will be required for precise identification of the necrotic biomolecules behaving like TLR3 agonists. However, we already have substantial evidence that RNA molecules are the main players. There was a consistent and dramatic reduction in the effects of the conditioned media from HNSCC cells subjected to pronecrotic stimuli when these media were pretreated with RNAse prior to application on reporter cells. This is consistent with our knowledge of double-stranded RNA being the main specific PAMP ligand for TLR3 [23,24]. More specifically, this is consistent with a previous study on TLR3 ligands resulting from mechanical necrosis of HEK 293 cells. These ligands were identified as double-stranded RNAs [15]. However, we cannot formally exclude the possibility that other necrotic biomolecules are involved in TLR3 activation, especially when the effects on reporter cells are only slightly reduced by RNAse treatment, for example, for the necrotic fluid from the C17 xenografted tumor. Necrotic proteins that can bind TLR3, like HMGB1, might be involved in its activation [25].

Finally, there is one important question to determine: what are the consequences of the release of TLR3 agonists by necrotic cells for tumor growth? If we first consider the classical role of TLR3 agonist in connection with cells of innate immunity, we can expect some enhancement of the antitumoral immune response resulting from enhanced activity of innate immune cells like natural killer (NK) cells and dendritic cells [26–29]. However, this is contradictory to the fact that tumors with large amounts of necrosis are generally the most aggressive. This might be explained to some extent by the high rate of necrosis, reflecting a very rapid proliferation, with angiogenesis lagging well behind, but it might also be explained by the deleterious effects of some categories of biomolecules released by necrotic cells [6]. In this regard, it is interesting to note that TLR3 is frequently expressed by tumor cells in human malignancies. This has been reported for prostate, hepatocellular, cervical (HPV-positive) and breast carcinomas, as well as multiple myeloma and melanoma [30–35]. TLR3 is also frequently expressed by HNSCC cells, fresh tumor specimens and most HNSCC cell lines propagated in vitro. In contrast, it is generally absent in healthy tissues adjacent to tumors [36]. The influence of TLR3 expression on the tumor outcome is variable. It is associated with a positive prognosis in hepatocellular carcinomas, although this is an exception [37]. In contrast, high TLR3 expression is associated with poor prognosis in prostate, breast, lung, ovarian, gastric, esophageal and oral carcinomas [37]. This might be surprising in view of some reports about apoptosis induced in vitro by TLR3 stimulation in various cell types [38,39]. However, the proapoptotic effect of synthetic TLR3 agonists is often restricted to a limited number of cell lines and often requires very high concentrations in the range of 50 μg/mL. In contrast, we and others have shown that under various experimental conditions, TLR3 activation can also promote cancer cell metabolic reprogramming, proliferation, migration and invasiveness [40–46].

Overall, our data converge with published data to support the idea that necrotic biomolecules released in the tumor microenvironment by malignant cells—especially dsRNAs—contribute to tumor growth through autocrine TLR3 stimulation. This hypothesis deserves further investigations using experimental tumor models and clinical material, especially in connection with HNSCCs. To this end, it will be useful to work on syngeneic HNSCC models instead of the Renca model, which represents a renal carcinoma. Right now, it is interesting to observe that among the five patients from whom we collected samples of tumor necrotic fluid, the three samples with the highest level of stimulation on TLR3 reporter cells were from patients with very aggressive diseases presenting extranodal extension and bone invasion.

Beyond the possible involvement of TLR3 in the progression of HNSCCs, one merit of our study is that it shows that functional studies are possible on tumor necrotic fluids collected intraoperatively. This procedure, which can be performed with ordinary needles and syringes, is easy, rapid and not harmful for the patients. In the future, combinations of biochemical and functional investigations on tumor necrotic fluids might provide useful information for improved personalized treatments.

4. Materials and Methods

4.1. Cells and Reagents

Human head and neck cancer cell lines (SQ20B, FaDu and Detroit 562) were maintained in Dulbecco's modified Eagle medium (DMEM) (Life technologies, Gaithersburg, MD, USA) supplemented with 2 mM L-glutamine and 10% fetal calf serum in a humidified chamber at 37 °C in 5% CO_2. Detroit 562 (batch No. 70004014) and FaDu (batch No. 63372030) cell lines were obtained from the ATCC (LGC Standards). The SQ20B line was provided by Prof. Eric Deutsch (Gustave Roussy, Villejuif, France). HEKBlue™ hTLR3 and HEKBlue™ Null1 cells were purchased from InvivoGen (San Diego, CA, USA; batch No. S01-4101). The cells were cultured at 37 °C in 5% CO_2 in 10 cm plastic dishes using DMEM containing glutamine, heat-inactivated fetal bovine serum (FBS), penicillin/streptomycin, and Normocin (InvivoGen). Selection of the plasmids in HEKBlue hTLR3 cells required the use of blasticidin (InvivoGen), and HEKBlue Null1 cells required the use of zeocin (InvivoGen). Cell counts and viabilities were determined by a hemocytometer and Trypan Blue exclusion. Poly(I:C) and poly(A:U) were obtained from InvivoGen (San Diego, CA, USA).

4.2. Cell Lines and PDX

Renca is a murine renal carcinoma cell line syngeneic with Balb-C mice (https://www.cellosaurus.org/CVCL_2174, PMID 31220119, PMID 3486710, accessed on 1 April 2023). In vitro, it was propagated in plastic flasks using RPMI 1640 culture medium with 10% FBS and 5 µg/mL gentamicin. Tumor growth in Balb-c mice was obtained by dorsal subcutaneous injections of 2×10^6 cells in 200 µL PBS. Then, 10 to 15 days later, mice were sacrificed prior to tumor collection in order to keep the tumor volume below a maximum of 1.7 cm^3. C17 and C18 tumors are patient-derived xenografts (PDX) initially grown from fragments of human metastatic nasopharyngeal carcinomas (PMID: 2971626; PMID:24618637). C17 was grown from the biopsy of a cutaneous metastasis taken from a multi-treated 38-year-old male patient. C18 was derived from the biopsy of an early lymph node metastasis with primary resistance to chemotherapy taken from a 48-year-old male patient. The C17 and C18 PDX were propagated on nude mice by iterative dorsal subcutaneous inoculation of tumor fragments: inoculation of 150 mg of tumor fragments and collection of tumors after 3 to 5 weeks when they were below the ethical threshold of 1.7 cm^3.

4.3. Animal Care

Tumor expansion was conducted in female mice aged 6 to 8 weeks: immunocompetent Balb/C or nude mice (mostly of Swiss genetic background) for Renca cells and C17/C18 PDX, respectively. The animals were supplied by Janvier Labs (Le Genet-St Isle, France) and hosted in batches of five in regulatory plastic cages ($27 \times 22 \times 15$ cm) under controlled conditions (day/night cycle of 12 h, 22 ± 1 °C and 55% humidity) with water and food ad libitum. All breeding operations and experiments were conducted in accordance with a protocol validated by the Gustave Roussy ethics committee and deposited in the APAFiS platform of the French Ministry of Agriculture (Apafis #12147-201711081244492 and #31464-2021050819051109 for Renca cells on Balb/C and C17/C18 PDX on nude mice, respectively).

4.4. Preparation of Necrotic Material Fluids from Ex Vivo Tumors

Necrotic material, i.e., dead cells and biomolecules released by dead cells, was prepared from fragments derived from mouse experimental tumors and from clinical tumor specimens immediately or shortly after ex vivo collection. Enrichment in necrotic material was based on the mincing of tumor fragments with scissors and forceps in ordinary culture medium. In our experience, this is sufficient to trigger the release of abundant necrotic material from the large majority of solid tumor types, regardless of their anatomic origin and histology.

Experimental tumors were removed under sterile conditions from the backs of recipient mice immediately after their sacrifice by cervical dislocation to avoid prolonged tumor anoxia. They were immediately weighed and minced into fragments with diameters of about 3 mm in RPMI-1640 medium in Petri dishes (5 mL of medium for 500 mg of tumor). The residual tumor fragments were then decanted for 2 min in 15 mL Falcon tubes, and the supernatant was subjected to $500 \times g$ centrifugation for 7 min at 20 °C to remove dead cells and debris. The resulting clarified supernatant was then aliquoted and kept frozen at -80 °C until its use for in vitro experiments.

The same procedure for enrichment in necrotic material was applied to tumor fragments derived from surgical samples from five patients bearing oral squamous cell carcinoma and admitted to the Clinical Hospital Dubrava, Zagreb, Croatia. The collection of tumor fragments and the following procedure were performed according to ethical protocols approved by the Institutional Ethics Committee. Informed patient consent was obtained prior to surgery. The study was performed in accordance with the ethical standards laid down in the 1964 Declaration of Helsinki and its later amendments or comparable ethical standards.

4.5. Stimulation and HEKBlue Detection Assay

Cells were grown to 50–70% confluence and plated at a cell density of 3×10^4 viable cells in 150 µL per well in a 96-well flat-bottomed tissue culture plate prior to treatment in HEKBlue detection medium (Invivogen). Then, 50 µL of supernatants from different head and neck cancer cells after different stresses was added to wells. Head and neck cancer cells were stressed as follows: serum deprivation and irradiation for 48 h and hypoxia and oxidative stress for 24 h. For serum deprivation, cells were washed twice with medium without serum and incubated with the same medium. For gamma irradiation experiments, cells were exposed to ^{60}Co γ-irradiation through a panoramic source (2 Gy/min, Rudjer Boskovic Institute, Radiation Chemistry and Dosimetry Laboratory) to a total of 10 Gy. Hypoxia was performed in a hypoxia chamber according to the manufacturer's instructions (Stemcell Technologies, Vancouver, BC, Canada). Oxidative stress was induced by the addition of 500 µM H_2O_2. Additionally, supernatants were incubated with 20 µg of RNAse (Sigma-Aldrich, Schnelldorf, Germany) for 2 h at 37 °C as described previously [14]. Relative NF-κB activity was determined by measuring the SEAP activity that accumulated in the culture media following overnight incubations with the supernatants. SEAP activity was measured at 620 nm using an Infinite® 200 PRO microplate reader spectrophotometer (Tecan, Männedorf, Switzerland). HEKBlue-Null1 cells served as negative controls for TLR3 activation, as they should be non-responsive to TLR3 ligands. For the treatment with exosomes, 10 µg of exosomes was incubated with 150 µL of HEKBlue-TLR3 and HEKBlue-null cells (3×10^4 cells) in HEKBlue detection medium and measured as cell lines. For the treatment with mice and patient aspirates, 50 µL of aspirate was incubated with 150 µL of HEKBlue-TLR3 and HEKBlue-null cells (3×10^4 cells) in HEKBlue detection medium and measured as cell lines. RNAse treatment was performed with 40 µg of RNAse (Sigma) for 2 h at 37 °C for patients' and mice supernatants or with 20 µg of RNAse (Sigma) for 2 h at 37 °C for cell supernatants. TLR3 antibody (Santa Cruz) was added at a concentration of 20 µg/mL. In all experiments, Poly(I:C) was used as a positive control at a concentration of 100 ng/mL.

4.6. Exosome Isolation

Cells were seeded at a density of 0.5×10^6 cells. The following day, cells were washed with PBS, and the medium was replaced with a medium containing exosome-depleted FBS (Gibco). Cells were exposed to 60Co γ-irradiation from a panoramic source (2 Gy/min) for a total dose of 10 Gy. Then, 48 h later, irradiated exosomes were isolated using a PureExo® Exosome Isolation kit (101 Bio, Mountain View, CA, USA) according to the manufacturer's instructions. Proteins were quantified using a DC protein assay kit (Biorad, Hercules, CA, USA).

4.7. Transmission Electron Microscopy (TEM)

Purified exosome preparations (5 µL) were placed on a Formvar®/carbon copper grid and air-dried for 20 min. Afterwards, they were contrasted with 2% uranyl acetate for 7 min, then washed several times. The morphology of isolated exosomes was visualized with TEM (FEI MORGAGNI 268D).

4.8. Nanoparticle Tracking Analysis

Quantification of total particles was performed using a NanoSight LM10 instrument equipped with an sCMOS camera and a red laser (Malvern Panalytical Ltd., Malvern, UK). Nanoparticle tracking analysis (NTA) was performed on samples that were diluted at a ratio of 1:25 in qH$_2$O to obtain 10–100 particles in the field of view.

4.9. Western Blot Analysis

Proteins were isolated with RIPA buffer and transferred onto a 0.2 μm nitrocellulose membrane as described previously [40]. The membranes were blocked with 5% nonfat dry milk and stained with primary antibodies: anti-CD63 (Abcam, Cambridge, UK), calnexin (Cell signaling technology, Beverly, MA, USA) and β-actin (Cell signaling technologies). Membranes were also stained with amido black as a loading control. Afterwards, the membranes were stained with peroxidase-conjugated secondary antibody (Cell signaling technology) and visualized with the following chemiluminescent systems: Western Lightning® Plus ECL (Perkin Elmer, Waltham, MA, USA) for cellular proteins and Super-Signal™ West Atto (Thermo Scientific, Waltham, MA, USA) for exosomal proteins. Images were obtained using an Alliance Q9 mini instrument (UVitec, Cambridge, UK).

4.10. Statistics

Statistical significance was assessed with two-tailed Student's *t*-test, and the results are presented as the mean $\pm$ SD.

5. Conclusions

Overall, our data indicate that TLR3 agonists are consistently present in necrotic fluids from HNSCC cells and tumors and are most likely mainly made of dsRNA fragments. This is an important finding because TLR3 is not always antitumorigenic. We previously reported that it can also have a protumorigenic effect in HNSCC [40,43,45]. Therefore, it is important to keep in mind that no external ligands are necessary for TLR3 stimulation and its potential contribution to cancer progression. Endogenous TLR3 ligands can be provided by the tumor itself. This is consistent with previous observations showing that necrotic tumors are more aggressive. It is also consistent with our findings that the three samples of the necrotic fluid with the highest levels of TLR3 stimulation were from patients with very aggressive cancers. In addition, we report that at least a fraction of the tumor necrotic endogenous TLR3 ligands are contained in extracellular vesicles. This means that they can be spread in various organs through the body and may possibly be involved in processes facilitating metastatic spread. Regarding methodological aspects, our study demonstrates that direct investigations, including functional testing, can be performed on necrotic fluids from patient HNSCC lesions.

Author Contributions: Conceptualization, T.M.G. and P.B.; methodology, T.V., M.T., K.H., M.M. and I.L.; formal analysis, T.M.G. and T.V.; writing—original draft preparation, T.M.G., P.B. and T.V.; writing—review and editing, T.M.G. and P.B.; visualization, T.M.G.; supervision, T.M.G. and P.B.; funding acquisition, T.M.G. All authors have read and agreed to the published version of the manuscript.

Funding: This work was supported by Croatian Science Foundation (project number: IP-2020-02-4225 Toll-like receptor 3 in the development and treatment of human head and neck cancer: the role of endogenous ligands) and the Young Researchers' career development project—training of doctoral students (Croatian Science foundation).

Institutional Review Board Statement: This study was conducted in accordance with the Declaration of Helsinki and approved by the Ethics Committee of Clinical Hospital Dubrava, Zagreb, Croatia (2020/2602-02, 26 February 2020) for studies involving humans. The animal study protocol was approved by the Ethics Committee of Gustave Roussy Institute and deposited in the APAFiS platform of the French Ministry of Agriculture (Apafis #12147-201711081244492 and # 31464-2021050819051109 for Renca cells on Balb/C and the C17/C18 PDX on nude mice, respectively).

Informed Consent Statement: Informed consent was obtained from all subjects involved in the study.

Data Availability Statement: The data presented in this study are available upon request from the corresponding author.

Conflicts of Interest: The authors declare no conflict of interest.

References

1. Luce, D.; Guenel, P.; Leclerc, A.; Brugere, J.; Point, D.; Rodriguez, J. Alcohol and tobacco consumption in cancer of the mouth, pharynx, and larynx: A study of 316 female patients. *Laryngoscope* **1988**, *98*, 313–316. [CrossRef] [PubMed]
2. Sabatini, M.E.; Chiocca, S. Human papillomavirus as a driver of head and neck cancers. *Br. J. Cancer* **2020**, *122*, 306–314. [CrossRef] [PubMed]
3. Bredell, M.G.; Ernst, J.; El-Kochairi, I.; Dahlem, Y.; Ikenberg, K.; Schumann, D.M. Current relevance of hypoxia in head and neck cancer. *Oncotarget* **2016**, *7*, 50781–50804. [CrossRef]
4. Kuhnt, T.; Mueller, A.C.; Pelz, T.; Haensgen, G.; Bloching, M.; Koesling, S.; Schubert, J.; Dunst, J. Impact of tumor control and presence of visible necrosis in head and neck cancer patients treated with radiotherapy or radiochemotherapy. *J. Cancer Res. Clin. Oncol.* **2005**, *131*, 758–764. [CrossRef] [PubMed]
5. Pollheimer, M.J.; Kornprat, P.; Lindtner, R.A.; Harbaum, L.; Schlemmer, A.; Rehak, P.; Langner, C. Tumor necrosis is a new promising prognostic factor in colorectal cancer. *Hum. Pathol.* **2010**, *41*, 1749–1757. [CrossRef]
6. Zapletal, E.; Vasiljevic, T.; Busson, P.; Matijevic Glavan, T. Dialog beyond the Grave: Necrosis in the Tumor Microenvironment and Its Contribution to Tumor Growth. *Int. J. Mol. Sci.* **2023**, *24*, 5278. [CrossRef]
7. Kawasaki, T.; Kawai, T. Toll-like receptor signaling pathways. *Front. Immunol.* **2014**, *5*, 461. [CrossRef]
8. Burgueno, J.F.; Abreu, M.T. Epithelial Toll-like receptors and their role in gut homeostasis and disease. *Nat. Rev. Gastroenterol. Hepatol.* **2020**, *17*, 263–278. [CrossRef]
9. Mokhtari, Y.; Pourbagheri-Sigaroodi, A.; Zafari, P.; Bagheri, N.; Ghaffari, S.H.; Bashash, D. Toll-like receptors (TLRs): An old family of immune receptors with a new face in cancer pathogenesis. *J. Cell Mol. Med.* **2021**, *25*, 639–651. [CrossRef]
10. Piccinini, A.M.; Midwood, K.S. DAMPening inflammation by modulating TLR signalling. *Mediat. Inflamm.* **2010**, *2010*, 672395. [CrossRef]
11. Vijay, K. Toll-like receptors in immunity and inflammatory diseases: Past, present, and future. *Int. Immunopharmacol.* **2018**, *59*, 391–412. [CrossRef] [PubMed]
12. Wen, A.Y.; Sakamoto, K.M.; Miller, L.S. The role of the transcription factor CREB in immune function. *J. Immunol.* **2010**, *185*, 6413–6419. [CrossRef] [PubMed]
13. Sato, Y.; Goto, Y.; Narita, N.; Hoon, D.S. Cancer Cells Expressing Toll-like Receptors and the Tumor Microenvironment. *Cancer Microenviron.* **2009**, *2* (Suppl. 1), 205–214. [CrossRef]
14. Cavassani, K.A.; Ishii, M.; Wen, H.; Schaller, M.A.; Lincoln, P.M.; Lukacs, N.W.; Hogaboam, C.M.; Kunkel, S.L. TLR3 is an endogenous sensor of tissue necrosis during acute inflammatory events. *J. Exp. Med.* **2008**, *205*, 2609–2621. [CrossRef]
15. Kariko, K.; Ni, H.; Capodici, J.; Lamphier, M.; Weissman, D. mRNA is an endogenous ligand for Toll-like receptor 3. *J. Biol. Chem.* **2004**, *279*, 12542–12550. [CrossRef] [PubMed]
16. Mikulandra, M.; Kobescak, A.; Verillaud, B.; Busson, P.; Matijevic Glavan, T. Radio-sensitization of head and neck cancer cells by a combination of poly(I:C) and cisplatin through downregulation of survivin and c-IAP2. *Cell. Oncol.* **2019**, *42*, 29–40. [CrossRef]
17. Chanteloup, G.; Cordonnier, M.; Isambert, N.; Bertaut, A.; Hervieu, A.; Hennequin, A.; Luu, M.; Zanetta, S.; Coudert, B.; Bengrine, L.; et al. Monitoring HSP70 exosomes in cancer patients' follow up: A clinical prospective pilot study. *J. Extracell. Vesicles* **2020**, *9*, 1766192. [CrossRef]
18. Hurwitz, S.N.; Rider, M.A.; Bundy, J.L.; Liu, X.; Singh, R.K.; Meckes, D.G., Jr. Proteomic profiling of NCI-60 extracellular vesicles uncovers common protein cargo and cancer type-specific biomarkers. *Oncotarget* **2016**, *7*, 86999–87015. [CrossRef]
19. Li, W.; Deng, M.; Loughran, P.A.; Yang, M.; Lin, M.; Yang, C.; Gao, W.; Jin, S.; Li, S.; Cai, J.; et al. LPS Induces Active HMGB1 Release from Hepatocytes Into Exosomes Through the Coordinated Activities of TLR4 and Caspase-11/GSDMD Signaling. *Front. Immunol.* **2020**, *11*, 229. [CrossRef]
20. Prieto, D.; Sotelo, N.; Seija, N.; Sernbo, S.; Abreu, C.; Duran, R.; Gil, M.; Sicco, E.; Irigoin, V.; Oliver, C.; et al. S100-A9 protein in exosomes from chronic lymphocytic leukemia cells promotes NF-kappaB activity during disease progression. *Blood* **2017**, *130*, 777–788. [CrossRef]
21. Silvers, C.R.; Miyamoto, H.; Messing, E.M.; Netto, G.J.; Lee, Y.F. Characterization of urinary extracellular vesicle proteins in muscle-invasive bladder cancer. *Oncotarget* **2017**, *8*, 91199–91208. [CrossRef] [PubMed]
22. Tauro, B.J.; Greening, D.W.; Mathias, R.A.; Mathivanan, S.; Ji, H.; Simpson, R.J. Two distinct populations of exosomes are released from LIM1863 colon carcinoma cell-derived organoids. *Mol. Cell Proteom.* **2013**, *12*, 587–598. [CrossRef]
23. Alexopoulou, L.; Holt, A.C.; Medzhitov, R.; Flavell, R.A. Recognition of double-stranded RNA and activation of NF-kappaB by Toll-like receptor 3. *Nature* **2001**, *413*, 732–738. [CrossRef]
24. Bell, J.K.; Askins, J.; Hall, P.R.; Davies, D.R.; Segal, D.M. The dsRNA binding site of human Toll-like receptor 3. *Proc. Natl. Acad. Sci. USA* **2006**, *103*, 8792–8797. [CrossRef]
25. Yanai, H.; Ban, T.; Wang, Z.; Choi, M.K.; Kawamura, T.; Negishi, H.; Nakasato, M.; Lu, Y.; Hangai, S.; Koshiba, R.; et al. HMGB proteins function as universal sentinels for nucleic-acid-mediated innate immune responses. *Nature* **2009**, *462*, 99–103. [CrossRef] [PubMed]
26. Sharma, S.; Zhu, L.; Davoodi, M.; Harris-White, M.; Lee, J.M.; St John, M.; Salgia, R.; Dubinett, S. TLR3 agonists and proinflammatory antitumor activities. *Expert. Opin. Ther. Targets* **2013**, *17*, 481–483. [CrossRef]
27. Glavan, T.M.; Pavelic, J. The exploitation of Toll-like receptor 3 signaling in cancer therapy. *Curr. Pharm. Des.* **2014**, *20*, 6555–6564. [CrossRef] [PubMed]

28. Matsumoto, M.; Takeda, Y.; Tatematsu, M.; Seya, T. Toll-Like Receptor 3 Signal in Dendritic Cells Benefits Cancer Immunotherapy. *Front. Immunol.* **2017**, *8*, 1897. [CrossRef]
29. Sivori, S.; Falco, M.; Della Chiesa, M.; Carlomagno, S.; Vitale, M.; Moretta, L.; Moretta, A. CpG and double-stranded RNA trigger human NK cells by Toll-like receptors: Induction of cytokine release and cytotoxicity against tumors and dendritic cells. *Proc. Natl. Acad. Sci. USA* **2004**, *101*, 10116–10121. [CrossRef]
30. Chiron, D.; Pellat-Deceunynck, C.; Amiot, M.; Bataille, R.; Jego, G. TLR3 ligand induces NF-{kappa}B activation and various fates of multiple myeloma cells depending on IFN-{alpha} production. *J. Immunol.* **2009**, *182*, 4471–4478. [CrossRef]
31. DeCarlo, C.A.; Rosa, B.; Jackson, R.; Niccoli, S.; Escott, N.G.; Zehbe, I. Toll-like receptor transcriptome in the HPV-positive cervical cancer microenvironment. *Clin. Dev. Immunol.* **2012**, *2012*, 785825. [CrossRef] [PubMed]
32. Gonzalez-Reyes, S.; Fernandez, J.M.; Gonzalez, L.O.; Aguirre, A.; Suarez, A.; Gonzalez, J.M.; Escaff, S.; Vizoso, F.J. Study of TLR3, TLR4, and TLR9 in prostate carcinomas and their association with biochemical recurrence. *Cancer Immunol. Immunother.* **2011**, *60*, 217–226. [CrossRef]
33. Gonzalez-Reyes, S.; Marin, L.; Gonzalez, L.; Gonzalez, L.O.; del Casar, J.M.; Lamelas, M.L.; Gonzalez-Quintana, J.M.; Vizoso, F.J. Study of TLR3, TLR4 and TLR9 in breast carcinomas and their association with metastasis. *BMC Cancer* **2010**, *10*, 665. [CrossRef]
34. Salaun, B.; Lebecque, S.; Matikainen, S.; Rimoldi, D.; Romero, P. Toll-like receptor 3 expressed by melanoma cells as a target for therapy? *Clin. Cancer Res.* **2007**, *13*, 4565–4574. [CrossRef] [PubMed]
35. Yuan, M.M.; Xu, Y.Y.; Chen, L.; Li, X.Y.; Qin, J.; Shen, Y. TLR3 expression correlates with apoptosis, proliferation and angiogenesis in hepatocellular carcinoma and predicts prognosis. *BMC Cancer* **2015**, *15*, 245. [CrossRef]
36. Pries, R.; Hogrefe, L.; Xie, L.; Frenzel, H.; Brocks, C.; Ditz, C.; Wollenberg, B. Induction of c-Myc-dependent cell proliferation through toll-like receptor 3 in head and neck cancer. *Int. J. Mol. Med.* **2008**, *21*, 209–215. [CrossRef] [PubMed]
37. Muresan, X.M.; Bouchal, J.; Culig, Z.; Soucek, K. Toll-Like Receptor 3 in Solid Cancer and Therapy Resistance. *Cancers* **2020**, *12*, 3227. [CrossRef]
38. Bernardo, A.R.; Cosgaya, J.M.; Aranda, A.; Jimenez, L.A.M. Synergy between RA and TLR3 promotes type I IFN-dependent apoptosis through upregulation of TRAIL pathway in breast cancer cells. *Cell Death Dis.* **2013**, *4*, e479. [CrossRef]
39. Salaun, B.; Coste, I.; Rissoan, M.C.; Lebecque, S.J.; Renno, T. TLR3 can directly trigger apoptosis in human cancer cells. *J. Immunol.* **2006**, *176*, 4894–4901. [CrossRef]
40. Matijevic Glavan, T.; Cipak Gasparovic, A.; Verillaud, B.; Busson, P.; Pavelic, J. Toll-like receptor 3 stimulation triggers metabolic reprogramming in pharyngeal cancer cell line through Myc, MAPK, and HIF. *Mol. Carcinog.* **2017**, *56*, 1214–1226. [CrossRef]
41. Bugge, M.; Bergstrom, B.; Eide, O.K.; Solli, H.; Kjonstad, I.F.; Stenvik, J.; Espevik, T.; Nilsen, N.J. Surface Toll-like receptor 3 expression in metastatic intestinal epithelial cells induces inflammatory cytokine production and promotes invasiveness. *J. Biol. Chem.* **2017**, *292*, 15408–15425. [CrossRef] [PubMed]
42. Jia, D.; Yang, W.; Li, L.; Liu, H.; Tan, Y.; Ooi, S.; Chi, L.; Filion, L.G.; Figeys, D.; Wang, L. beta-Catenin and NF-kappaB co-activation triggered by TLR3 stimulation facilitates stem cell-like phenotypes in breast cancer. *Cell Death Differ.* **2015**, *22*, 298–310. [CrossRef] [PubMed]
43. Matijevic, T.; Pavelic, J. The dual role of TLR3 in metastatic cell line. *Clin. Exp. Metastasis* **2011**, *28*, 701–712. [CrossRef]
44. Paone, A.; Galli, R.; Gabellini, C.; Lukashev, D.; Starace, D.; Gorlach, A.; De Cesaris, P.; Ziparo, E.; Del Bufalo, D.; Sitkovsky, M.V.; et al. Toll-like receptor 3 regulates angiogenesis and apoptosis in prostate cancer cell lines through hypoxia-inducible factor 1 alpha. *Neoplasia* **2010**, *12*, 539–549. [CrossRef] [PubMed]
45. Veyrat, M.; Durand, S.; Classe, M.; Glavan, T.M.; Oker, N.; Kapetanakis, N.I.; Jiang, X.; Gelin, A.; Herman, P.; Casiraghi, O.; et al. Stimulation of the toll-like receptor 3 promotes metabolic reprogramming in head and neck carcinoma cells. *Oncotarget* **2016**, *7*, 82580–82593. [CrossRef] [PubMed]
46. Chuang, H.C.; Chou, M.H.; Chien, C.Y.; Chuang, J.H.; Liu, Y.L. Triggering TLR3 pathway promotes tumor growth and cisplatin resistance in head and neck cancer cells. *Oral. Oncol.* **2018**, *86*, 141–149. [CrossRef]

International Journal of
Molecular Sciences

Article

Effective Radiosensitization of HNSCC Cell Lines by DNA-PKcs Inhibitor AZD7648 and PARP Inhibitors Talazoparib and Niraparib

Jacob Mentzel [1,2], Laura S. Hildebrand [1,2], Lukas Kuhlmann [1,2], Rainer Fietkau [1,2] and Luitpold V. Distel [1,2,*]

1   Department of Radiation Oncology, Universitätsklinikum Erlangen, Friedrich-Alexander-Universität Erlangen-Nürnberg (FAU), 91054 Erlangen, Germany; jacobmentzel@gmx.de (J.M.); laura.hildebrand@uk-erlangen.de (L.S.H.); lukas.kuhlmann@uk-erlangen.de (L.K.); rainer.fietkau@uk-erlangen.de (R.F.)
2   Comprehensive Cancer Center Erlangen-Europäische Metropolregion Nürnberg (CCC ER-EMN), 91054 Erlangen, Germany
*   Correspondence: luitpold.distel@uk-erlangen.de; Tel.: +49-9131-853-2312; Fax: +49-9131-853-9335

Abstract: (1) Head and neck squamous cell carcinoma (HNSCC) is common, while treatment is difficult, and mortality is high. Kinase inhibitors are promising to enhance the effects of radiotherapy. We compared the effects of the PARP inhibitors talazoparib and niraparib and that of the DNA-PKcs inhibitor AZD7648, combined with ionizing radiation. (2) Seven HNSCC cell lines, including Cal33, CLS-354, Detroit 562, HSC4, RPMI2650 (HPV-negative), UD-SCC-2 and UM-SCC-47 (HPV-positive), and two healthy fibroblast cell lines, SBLF8 and SBLF9, were studied. Flow cytometry was used to analyze apoptosis and necrosis induction (AnnexinV/7AAD) and cell cycle distribution (Hoechst). Cell inactivation was studied by the colony-forming assay. (3) AZD7648 had the strongest effects, radiosensitizing all HNSCC cell lines, almost always in a supra-additive manner. Talazoparib and niraparib were effective in both HPV-positive cell lines but only consistently in one and two HPV-negative cell lines, respectively. Healthy fibroblasts were not affected by any combined treatment in apoptosis and necrosis induction or G2/M-phase arrest. AZD7648 alone was not toxic to healthy fibroblasts, while the combination with ionizing radiation reduced clonogenicity. (4) In conclusion, talazoparib, niraparib and, most potently, AZD7648 could improve radiation therapy in HNSCC. Healthy fibroblasts tolerated AZD7648 alone extremely well, but irradiation-induced effects might occur. Our results justify in vivo studies.

Keywords: HNSCC; DNA damage response inhibitor; PARP inhibitor; DNA-PK inhibitor; kinase inhibitor; radiosensitivity; ionizing radiation; cell lines

1. Introduction

Head and neck squamous cell carcinoma (HNSCC) is the eighth most common cancer worldwide, with over 878,000 new cases and over 444,000 deaths in 2020 [1]. Its 5-year survival rate is approximately 50%, ranging from 80% for stage I to as low as 20% for stages III/IV [2]. The association of late diagnosis with low survival is especially problematic, since no screening strategy has proved to be effective to this day [3]. The major risk factors are consumption of tobacco and alcohol, environmental pollutants, the chewing of areca nut products and infection with the Epstein-Barr virus (EBV) and the human papilloma virus (HPV) [3–5]. While the use of tobacco and alcohol is mainly associated with cancers of the oral cavity and the larynx and the use of areca nut products is also linked to oral cavity cancer, infections with HPV and EBV are associated with oropharyngeal and nasopharyngeal cancer, respectively [3]. Tobacco and alcohol consumption not only increase the HNSCC risk on their own but also, multiplicatively, in heavy consumers of both, by 38-fold [6]. The chewing of areca nut products makes for high HNSCC incidence in some

Asian countries such as India, China and Papua New Guinea [1,7]. Viral infections with EBV and, most prominently, with HPV play an important role in HNSCC development, being responsible for 70% of new cases in Europe and North America [4,8,9]. HNSCC can be divided into HPV-associated (HPV-positive) and non-HPV-associated (HPV-negative) disease. As the incidence of HPV-positive HNSCC has risen [4], there is growing hope that, as in cervical cancer [10], primary prevention with HPV vaccination programs might drastically decrease HPV-positive HNSCC incidence in the future. This could potentially shift the focus to HPV-negative HNSCC, which is generally more resistant to the current treatment modalities, including being more radioresistant and thus associated with worse prognosis, even in lower TNM (tumor, node, metastasis) stages [9,11,12]. Nevertheless, radiation therapy plays an important role in the treatment of HNSCC [3,13].

Therefore, finding ways to make HNSCC more radiosensitive is of great interest. Irradiation induces DNA damage, including single-strand breaks (SSBs) and double-strand breaks (DSBs) [14]. DSBs are considered the most lethal DNA lesion, leading to cell cycle arrest or cell death if left unrepaired [14–16]. The DNA damage response (DDR) system is responsible for repairing DNA damage. A method to enhance radiosensitivity is to inhibit DDR and simultaneously induce DNA damage, for example, through irradiation [17,18]. The DNA-damage which then cannot be repaired accumulates, ultimately leading to cell death [19,20]. Most SSBs are repaired, the reparation process being carried out by the SSB repair system [14,21], while repairing DSBs is more complex. The two major systems of DDR to repair DSBs are the highly precise homologous recombination (HR) and non-homologous end-joining (NHEJ), which is more error-prone [17,22,23], since it ligates broken DNA ends together without a template [15–17,22,24,25]. Tumor cells are often HR-deficient and heavily rely on NHEJ for DSB repair, since impaired DDR and especially DSB repair results in genomic instability, which is a hallmark of cancer [19,21,26–29]. Therefore, combining irradiation with the inhibition of either the SSB repair system or NHEJ is a promising method for treating cancer, including HNSCC. In contrast to cancer cells, healthy cells are normally HR-sufficient and can tolerate a certain amount of irradiation-induced DNA damage [29]. In theory, this could limit unwanted side effects.

The protein poly(ADP-ribose) polymerase (PARP) is a sensor for DNA damage. Despite PARP1 being first isolated in 1971, PARP's involvement in DSB repair, especially alternative-NHEJ, and other cellular processes beyond DNA damage repair, is still the subject of active debate [21,30,31]. Meanwhile, its role in SSB repair is much better understood. PARP1 detects SSBs, then binds to the DNA, resulting in a conformational change that activates PARP1. Activated PARP1 cleaves off ADP-ribose from NAD+ multiple times, creating poly(ADP-ribose) chains (PAR chains) and attaching them to nuclear proteins. The chains are negatively charged, thus electrostatically attracting XRCC1 and repulsing PARylated histones and PARP1 [21,31,32]. The SSB repair is then continued without PARP1. PARP1 might also be involved in the repair of damaged bases, in a process called base excision repair (BER), although existing data are contradictory [32–37]. In the process of base excision, an SSB is created [14]. The repair pathway of these self-created SSBs is closely related to regular SSB repair [32].

PARP inhibition impairs SSB repair, although not all mechanisms involved are fully understood [38]. Unrepaired SSBs can turn into DSBs in the S-phase, which in turn occurs more often in rapidly proliferating cells [14,17]. Apart from simply blocking PARP´s ability to repair SSBs, PARP inhibition also prevents PARP from dissociating from the DNA in a process called PARP-trapping, which leads to the collapse of the replication fork and subsequently results in replication-dependent DSBs [31,38,39]. In HR-deficient cells, especially in the BRCA-deficient breast cancer cells, there is a dramatically increased sensitivity to PARP inhibitors [40]. In these cells, the error-prone NHEJ is responsible for repairing DSBs, resulting in increased rates of unrepaired DSBs and cell death [31,41]. Multiple PARP inhibitors (PARPis; singular: PARPi) have already been market-approved in different gynecological tumor entities [42–47], and preclinical evidence indicates the radiosensitizing effects of PARPis in HNSCC as well [48–52].

The DNA-dependent protein kinase (DNA-PK) catalytic subunit (DNA-PKcs) is essential for NHEJ [16,21,53]. DNA-PK binds to previously detected DNA DSBs and is activated by autophosphorylation. This results in conformational and positional changes of DNA-PK, enabling end-processing and ligation enzymes to repair the DSB [14,21]. NHEJ occurs in all cell cycle phases [14,54]. DNA-PKcs, which is a member of the phosphatidylinositol 3-kinase-related kinase (PIKK) family, also has some lesser known functions critical to cellular survival and proliferation, such as regulating transcription, progressing through cell cycle or maintaining telomers [55–57].

The PIKK family also includes Ataxia Teleangiectasia Mutated (ATM) and Ataxia Teleangiectasia Related (ATR), which are similar essential proteins involved in the DNA damage response. A further protein is mTOR, as a regulator of cell cycle control and proliferation. Because of their central role in the DNA damage response, they are interesting targets for kinase inhibitors [58].

Both kinase inhibitor types share the characteristic that they have no effect on cells without DNA damage. Since cancer cells tend to have higher levels of DNA damage due to oxidative stress, these cells are more likely to be targeted by the inhibitors [59]. However, with IR, DNA damage can be specifically induced in cancers at much higher levels, making the inhibitors more effective.

In the context of radiosensitization, the inhibition of NHEJ through DNA-PKcs inhibition is especially relevant, since HR-deficient cells are dependent on NHEJ to repair DSBs and, in turn, to survive [19,21]. Combining a DNA-PKcs inhibitor (DNA-PKcsi) with ionizing radiation (IR) is therefore promising in enhancing radiosensitivity in tumors [15,60]. Some DNA-PKcsis have previously shown radiosensitizing capabilities in HNSCC cells both in vitro and in vivo [61–65], and several clinical trials are underway, involving DNA-PKcsis plus IR in solid tumors, in some cases HNSCC (NCT03907969, NCT02516813, NCT04533750), and also evaluating safety and efficacy in solid tumors (NCT01353625) [66].

In this study, our goal was to evaluate whether the PARPis talazoparib and niraparib, as well as the highly selective DNA-PKcsi AZD7648, radiosensitize HNSCC cell lines in vitro and, therefore, may have the potential to improve radiation therapy in HNSCC in the future. Seven HNSCC cell lines were investigated, two of which were HPV-positive, while five were HPV-negative, which are generally more radioresistant [9,11,12]. We also included two healthy fibroblast cell lines to estimate the severity of the side effects that can be expected. This is useful, since radiation therapy can never completely spare healthy tissue in a clinical setting, and by investigating treatment effects on healthy fibroblasts representative of healthy tissue exposed to the treatment, the risk of potential side effects can be estimated. We assessed apoptotic and necrotic cell death induction and cell cycle arrest in the G2/M-phase using flow cytometry. Additionally, we performed colony formation assays as the gold-standard for evaluating in vitro radiosensitivity [67,68]. The endpoints that were particularly of interest were cell death, the G2/M-phase arrest and cell inactivation, since, combined, they provide a comprehensive image of the effects and potential of a treatment. Interesting from a clinical perspective was primarily whether combining a kinase inhibitor (KI) with IR is more effective than IR alone. Supra-additivity was also calculated.

2. Results

Three KIs that interfere with DNA repair were evaluated for efficacy in combination with IR. Two of the compounds were the PARPis talazoparib (Figure 1A) and niraparib (Figure 1B). The third was the DNA-PKcsi AZD7648 (Figure 1C). Nine cell lines, including two fibroblast cell lines as normal tissue cell lines and seven tumor cell lines from HNSCC, were used (Figure 1D).

Cell death and G2/M-phase arrest were analyzed by flow cytometry using Annexin V to estimate apoptotic death, 7AAD to estimate necrotic death (Figure 1E) and Hoechst to estimate cell cycle distribution (Figure 1F). Colony formation was performed in Petri dishes and stained with methylene blue (Figure 1G). To characterize potential synergistic effects of the combined treatments, we calculated supra-additive effects, which indicate that the effect of a combined treatment exceeds the additive effects of the inhibitor treatment and the IR treatment. We used the two-tailed Mann–Whitney U test, as described in more detail in the Materials and Methods Section, to ensure that the supra-additive effects were statistically significant.

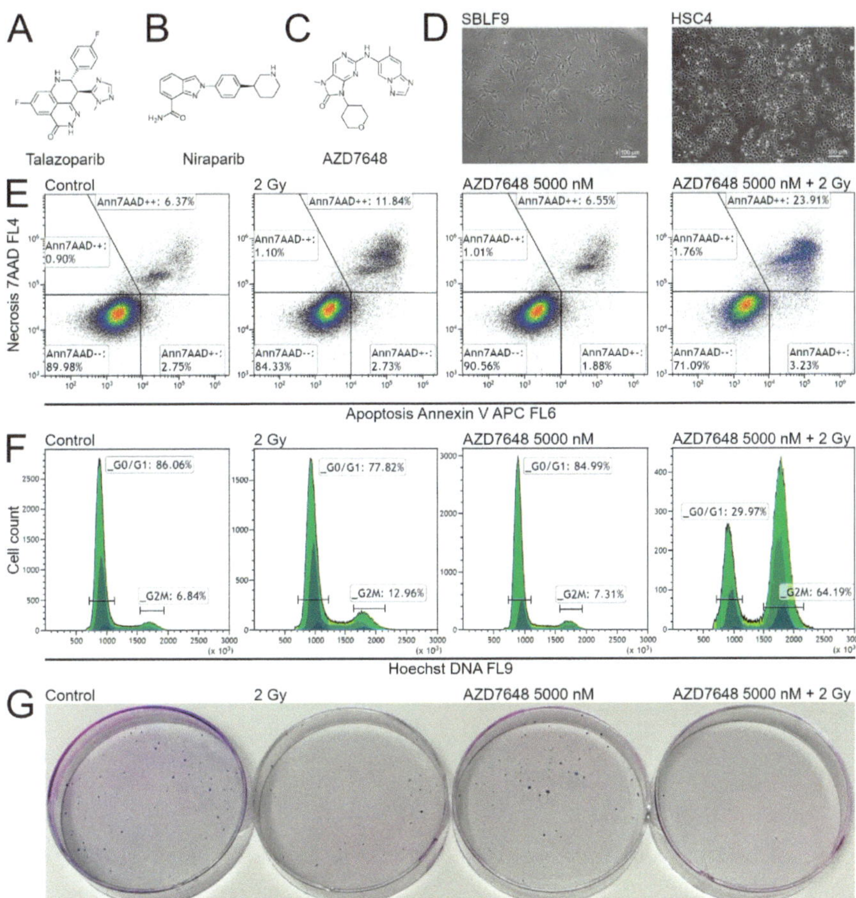

Figure 1. Kinase Inhibitors, cell lines and methods. (**A**) Talazoparib (50 nM), (**B**) niraparib (2500 nM) and (**C**) AZD7648 (5000 nM) were evaluated, each both with and without IR. (**D**) Microscopic images of SBLF9 and HSC4 are depicted representatively for the two healthy fibroblast cell lines and the seven HNSCC cell lines examined. The white bar equals 100 µm. (**E**) Representative scatter plot of the flow cytometric measurement of apoptosis and necrosis induction by Annexin V and 7AAD in UM-SCC-47. (**F**) Representative histograms of the Hoechst 33342-stained and flow cytometrically analyzed cell cycle distribution in UM-SCC-47 cell line. (**G**) Representative Petri dishes for colony formation assay containing Cal33 colonies. For all methods, the displayed samples are control, 2 Gy, AZD7648 5000 nM and AZD7648 5000 nM + 2 Gy. For all three methods, half of the samples were irradiated with 2 Gy IR, 3 h after the inhibitors were added to all samples. After a further 48 h, the inhibitors were removed, and either flow cytometric measurements were performed or standard medium was added to the colony formation assay.

2.1. Induction of Apoptotic- and Necrotic Death Varies Widely between Different Cell Lines

In the following analysis, apoptosis and necrosis rates are summarized as cell death. If not specified differently, the treatments are compared to the mono-treatment with IR. In the two healthy fibroblast cell lines, SBLF8 and SBLF9, none of the three KIs, nor any treatment with KI plus IR caused a significant difference in cell death compared to the IR mono-treatment (Figure 2A,B). Niraparib and AZD7648 alone slightly increased cell death in SBLF9 compared to the control. The HNSCC cell lines responded heterogeneously to the combined treatments of KI plus IR. In the two HPV-positive HNSCC cell lines, UD-SCC-2 and UM-SCC-47, as well as in the HPV-negative Detroit 562, the three combined treatments, talazoparib plus IR, niraparib plus IR and AZD7648 plus IR, all resulted in a significant increase in cell death compared to the IR mono-treatment (Figure 2E,H,I). Notably, the niraparib mono-treatment also increased the cell death rate in UD-SCC-2. In RPMI2650, one of this study's five HPV-negative HNSCC cell lines, there was no significant difference between any treatment and the IR mono-treatment (Figure 2G). The remaining three HPV-negative HNSCC cell lines, Cal33, CLS-354 and HSC4, responded inconsistently to the three combined treatments (Figure 2C,D,F). Talazoparib plus IR resulted in a clear increase in cell death compared to the IR mono-treatment in Cal33 and CLS-354, but not in HSC4. AZD7648 plus IR increased cell death in all three cell lines, and niraparib plus IR did so only in HSC4, but not in Cal33 and CLS-354. It is worth mentioning that significant supra-additive effects were observed for talazoparib plus IR in CLS-354, for niraparib plus IR in HSC4, as well as for AZD7648 plus IR in four out of seven HNSCC cell lines, namely Cal33, CLS-354, Detroit 562 and UM-SCC-47. Further data on supra-additive effects on the induction of apoptosis and necrosis are provided in Supplementary Table S1.

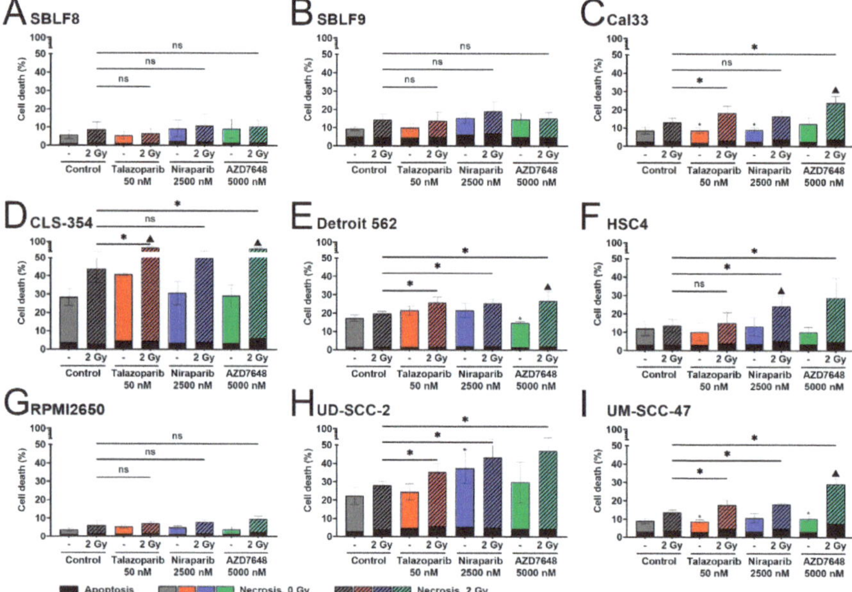

Figure 2. Flow cytometry analysis of apoptosis and necrosis. Healthy fibroblast cells (**A**) SBLF8, (**B**) SBLF9, HPV-negative HNSCC cells, (**C**) Cal33, (**D**) CLS-354, (**E**) Detroit 562, (**F**) HSC4, (**G**) RPMI2650 and HPV-positive HNSCC cells, (**H**) UD-SCC-2, (**I**) UM-SCC-47 were treated either with talazoparib (50 nM), niraparib (2500 nM) or AZD7648 (5000 nM) or combined with 2 Gy IR. After 48 h, apoptosis apoptosis(Annexin V) and necrosis (7AAD) were measured using flow cytometry. Apoptosis is presented in black, necrosis is color-coded for different inhibitors, and irradiated samples are hatched.

Each value represents mean $\pm$ SD ($n \geq 4$). Combined treatments and KI alone were compared to IR alone, and significant changes from IR alone are indicated for combined treatments by asterisks above the corresponding line and for KI alone by small asterisks above the respective bar. The abbreviation "ns" above the corresponding line represents changes that are not significant. A triangle represents supra-additivity. Significance was determined using a two-tailed Mann–Whitney U test with * $p \leq 0.05$.

In addition to cell death, cell cycle distribution and the proportion of cells in the G2/M-phase are important, as the cells there are particularly sensitive to radiation [69,70]. Since a pronounced G2/M block has a favorable effect on daily fractionated irradiation, the proportion of cells in G2/M was examined next.

2.2. Combined Treatment Causes a Pronounced G2/M Arrest in the Tumor Cell Lines

In the two healthy fibroblast cell lines SBLF8 and SBLF9, all three combined treatments of KI plus IR did not result in a distinct increase in G2/M arrest compared to IR mono-treatment (Figure 3A,B). However, mono-treatments with talazoparib, and especially niraparib, both increased G2/M arrest in SBLF8 compared to the control and the IR mono-treatment. Six out of the seven HNSCC cell lines responded with a clear increase in G2/M arrest to all three combined treatments, although the increase was more distinct in some cell lines than others (Figure 3C,E–I). The only exception was CLS-354, in which only AZD7648 plus IR was effective, with a clear G2/M arrest (Figure 3D). The two PARPis, talazoparib plus IR and niraparib plus IR, had no effect on G2/M arrest in CLS-354.

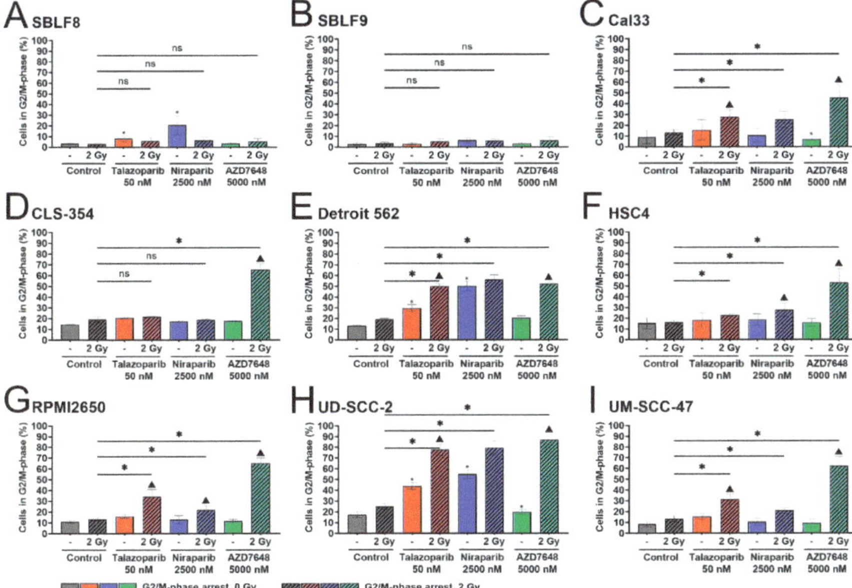

Figure 3. Flow cytometry analysis of cell cycle distribution. Fraction of G2/M-phase arrest. Healthy fibroblast cells (**A**) SBLF8, (**B**) SBLF9, HPV-negative HNSCC cells, (**C**) Cal33, (**D**) CLS-354, (**E**) Detroit 562, (**F**) HSC4, (**G**) RPMI2650 and HPV-positive HNSCC cells, (**H**) UD-SCC-2, (**I**) UM-SCC-47 were treated either with talazoparib (50 nM), niraparib (2500 nM) or AZD7648 (5000 nM) or combined with 2 Gy IR. Incubation lasted for 48 h, and afterward, cells were stained with Hoechst, and cell cycle was analyzed using flow cytometry. Different inhibitors are displayed in color, and irradiated samples are hatched. Each value represents mean $\pm$ SD ($n \geq 4$). Combined treatments and KI alone were compared to IR alone, and significant changes from IR alone are indicated for combined treatments

by asterisks above the corresponding line and for KI alone by small asterisks above the respective bar. The abbreviation "ns" above the corresponding line represents changes that are not significant. A triangle represents supra-additivity. Significance was determined using a two-tailed Mann–Whitney U test with * $p \leq 0.05$.

This can be further described by studying whether these observed effects are statistically significantly supra-additive. That was the case in five of six cell lines for talazoparib plus IR, sparing HSC4. For niraparib plus IR, this effect occurred only in the two cell lines HSC4 and RPMI2650. AZD7648 plus IR showed supra-additive effects in all the examined HNSCC cell lines. Further data on supra-additive effects on the induction of G2/M-phase arrest is provided in Supplementary Table S2.

Additionally, mono-treatment with talazoparib and niraparib increased G2/M arrest compared to mono-treatment with IR in SBLF8, Detroit 562 and UD-SCC-2. Although this effect was not as strong as that of the combined treatments in HNSCC cell lines, it was still relevant. In SBLF8, G2/M arrest was higher with talazoparib, and especially with niraparib mono-treatment, than with the respective combined treatments, which did not show a significant increase compared to the IR mono-treatment. Next, the colony-forming assay was performed, as this best reflects the effect of the treatment.

2.3. In Colony Formation Assay AZD7648 Plus IR Is the Most Effective

To assess the long-term ability of cells to proliferate after treatment, the colony formation assay is a crucial tool. Mono-treatment with both PARPis was toxic in most cell lines. This included normal tissue cell lines. There was no toxicity in the RPMI2650 and UM-SCC-47 cell lines, and talazoparib did not reduce the survival fraction in Cal33 and HSC4 (Figure 4A–I). At the same time, there was only a small effect of AZD7648 mono-treatment, including both healthy fibroblasts cell lines, with an SF of 0.83 in SBLF8 and an SF of 0.85 in SBLF9. The lowest survival rate in tumor cell lines was in the UD-SCC-2, at SF 0.79 (median). The only exception was CLS-354, which was already very responsive. IR alone was effective in all cell lines and was most effective in both the fibroblasts cell lines, with an SF of 0.34 in SBLF8 and an SF of 0.30 in SBLF9. Among the cancer cell lines, the SF was lowest in the cell line CLS-354 with 0.50. All three combined treatments of KI plus IR reduced the survival fraction compared to the IR mono-treatment in almost all cell lines. The only exceptions are niraparib plus IR in SBLF9 and talazoparib plus IR in Cal33. This also means that the combined treatment with AZD7648 plus IR resulted in a significant decrease in the survival fraction in all cell lines. Talazoparib had supra-additive and, therefore, radiosensitizing effects in SBLF8 and SBLF9 and the cancer cell lines, while sparing Cal33, CLS-354 and Detroit 562. Niraparib radiosensitized the two HPV-positive cell lines, UD-SCC-2 and UM-SCC-47. Further data on supra-additive effects on cell inactivation in the colony formation assay is provided in Supplementary Table S3.

Out of the three KIs investigated, AZD7648 plus IR had the most potent radiosensitizing effect in each cell line, with normalized survival fractions ranging from 0.05 in RPMI2650 to 0.002 in UD-SCC-2. This effect was supra-additive in all cell lines. It is important to note that this includes the healthy fibroblast cell lines SBLF8 and SBLF9.

A visual overview of the efficacy of each combined treatment is displayed as a heat map, covering all methods and cell lines (Figure 5). The higher color intensity represents a stronger effect of the combined treatment compared to IR alone. The combined treatments affected the two fibroblast cell lines significantly less compared to the HNSCC cell lines. The HPV-positive cell lines, but especially UD-SCC-2, seem to be very reactive to all the combined treatments, and AZD7648 stands out as more consistently effective compared to talazoparib and niraparib.

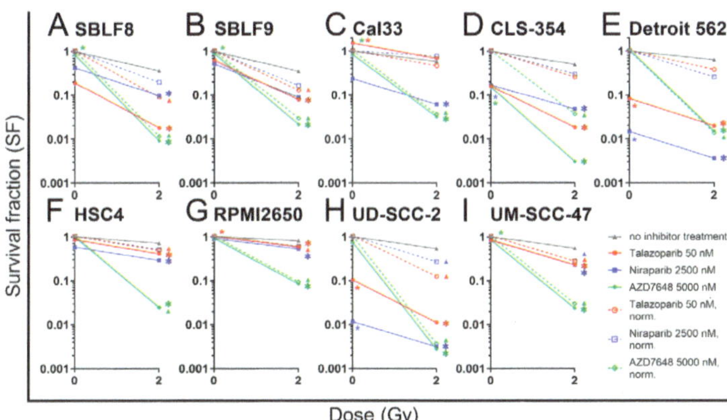

Figure 4. Colony formation assay. Survival fraction (SF) is displayed. Healthy fibroblast cells (**A**) SBLF8, (**B**) SBLF9, HPV-negative HNSCC cells (**C**) Cal33, (**D**) CLS-354, (**E**) Detroit 562, (**F**) HSC4, (**G**) RPMI2650 and HPV-positive HNSCC cells (**H**) UD-SCC-2, (**I**) UM-SCC-47 were treated either with talazoparib (50 nM), niraparib (2500 nM) or AZD7648 (5000 nM) or combined with 2 Gy IR. After treatment, solitary cells were allowed to form colonies for 10–14 days. Colonies containing at least 50 cells were counted. Inhibitors are displayed in color according to the legend. Solid lines display the data of control and KI with and without IR. Dashed lines represent a normalization, where the effect of KI alone is eliminated, and the difference between the normalized KI + IR and IR alone represents supra-additivity. Each value represents mean ± SD ($n \geq 4$, two technical replicates each). Compared to IR alone, the significance of combined treatments is represented by * on the right and supra-additivity by a triangle on the right. On the left, * represents a significant difference of KI alone compared to IR alone. When * is above the line, the effect of KI alone is weaker than that of IR alone and when * is below the line, the effect of KI alone is stronger. Significance was determined using a two-tailed Mann–Whitney U test with * $p \leq 0.05$.

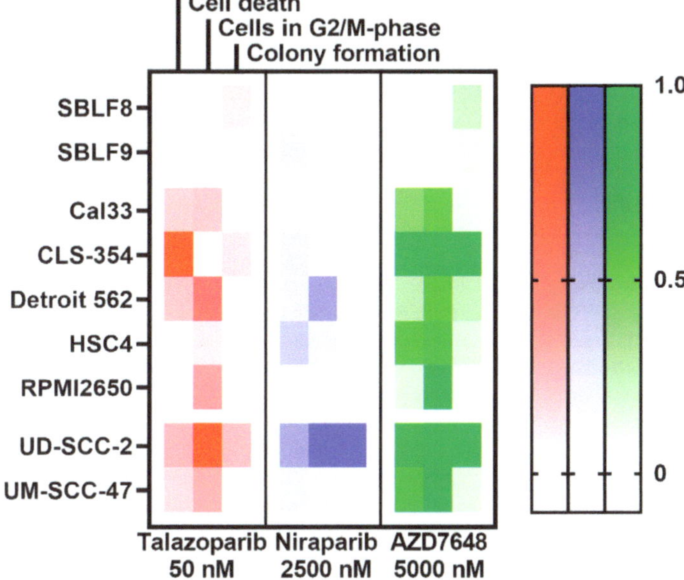

Figure 5. Overview of all effect sizes of combined treatments compared to IR alone. All combined treatments across all methods and all cell lines are aggregated in this heat map. Values were calculated

using the mean values either by subtracting the effect of IR alone from the effect of combined treatment (cell death; cells in G2/M-phase) or by dividing the SF of IR alone by the SF of the combined treatment (colony formation). The highest resulting value within one method, but across all inhibitors and cell lines, was normalized to the value "1". Inhibitors are color-coded, and color intensity correlates with the normalized value for the effect of combined treatment compared to IR alone.

3. Discussion

We aimed to assess if talazoparib, niraparib and AZD7648 radiosensitize HNSCC cell lines in vitro, and whether they are less toxic for healthy fibroblast cells and should therefore be further considered for clinical testing in HNSCC. While most of the comparable studies focus on clonogenic survival alone, we also measured apoptotic and necrotic cell death and G2/M-phase arrest, in order to get a more comprehensive picture of the effects resulting from the combined treatment with KI plus IR.

Unrepairable DNA damage leads to the activation of various cellular processes. For example, it can lead to cell death through apoptosis or necrosis induction or inhibit proliferation through senescence, where a cell grows in size and loses its ability to proliferate [71,72]. Tumor cells survive because these mechanisms are compromised in them [27]. Therefore, finding a treatment that induces cell death or senescence in tumor cells is desirable. We directly measured apoptotic and necrotic cell death induction and found AZD7648 plus IR to increase cell death in all HNSCC cell lines, except RPMI2650, which did not respond with a significant increase in cell death to any treatment. These effects were mostly supra-additive. Talazoparib and niraparib with IR prompted rather heterogeneous effects, being effective in only five and four of seven HNSCC cell lines, respectively. In the healthy fibroblast cell lines, the combined treatments did not relevantly increase cell death.

Next, the G2/M-phase arrest is of interest because G2/M is the most radiosensitive phase of the cell cycle [69,70]. In a clinical setting with fractionated radiotherapy, a patient receives radiotherapy every 24 h. If a treatment results in an increased G2/M-phase arrest for at least 48 h, the next cycle of IR may target many cells that are more vulnerable to irradiation, resulting in further increased DNA damage. Contrary to tumor cells, healthy cells still have a functioning and effective G1 block. The G1 phase is much more radioresistant [69,70]. In theory, this circumstance should facilitate the tumor-selectivity of the treatment. Such an identified mechanism of tumor-selectivity is demanded by a 2018 report, claiming that initiating clinical trials without it was ill-advised [73]. To gain information on that, we measured the cell cycle distribution. As with cell death induction, AZD7648 had the strongest effects overall, radiosensitizing all HNSCC cell lines in a supra-additive manner, while talazoparib and niraparib did not do so consistently and did not have any effect in CLS-354. The combined treatments did not increase the G2/M-phase arrest in healthy fibroblasts. AZD7648 especially had no such effect on healthy fibroblasts at all.

The colony formation assay is the gold standard for evaluating in vitro radiosensitivity. It displays the ability of the cells for long-term proliferation after treatment. We found that all combined treatments were effective in suppressing colony formation, with AZD7648 being by far the most effective overall, with supra-additive effects in all cell lines. The combined treatments also had relevant effects in the two healthy fibroblast cell lines, with AZD7648 being the most effective again. However, while talazoparib and niraparib alone were toxic, AZD7648 alone had very little effect on healthy fibroblasts.

Overall, we found all three KIs to be promising, with AZD7648 being the most potent, while irradiation-induced damage to healthy tissue is a concern. In the following, we first focus on the PARPis talazoparib and niraparib before moving on to the DNA-PKcsi AZD7648. Both are discussed, beginning with their effects on HNSCC cell lines and continuing with an evaluation of their effects in healthy fibroblasts.

3.1. Talazoparib and Niraparib Radiosensitize HNSCC Cell Lines Heterogeneously

PARP is involved in SSB repair. The inhibition of PARP combined with an inductor of DNA damage such as IR should therefore lead to the accumulation of SSBs, which can then transform into the much more toxic DSBs in the S-phase [14,17]. The PARPis talazoparib and niraparib are already market-approved in some gynecological tumor entities [42–47]. While multiple preclinical studies have found olaparib, another PARPi, to successfully radiosensitize HNSCC cell lines in vitro [48,51], data on the effects of talazoparib and niraparib combined with IR in HNSCC are scarce. The potential of a combined treatment with IR in HNSCC cell lines in vitro has been shown by one study for talazoparib [52] and by two studies for niraparib [49,50].

In our study, we are furthering this knowledge by examining additional cell lines, including HPV-negative and HPV-positive ones, as well as healthy fibroblasts, and by measuring cell death induction and G2/M-phase arrest, in addition to clonogenic survival. We found both talazoparib and niraparib to have radiosensitizing effects in HNSCC cell lines. Overall, neither PARPi had a clear advantage over the other. The HPV-positive cell lines seemed to be more consistently affected by the combination of PARPis plus IR compared to the HPV-negative cell lines, where the effects were more heterogeneous. This was particularly prominent in apoptosis and necrosis induction. The G2/M-phase arrest was detectable for the combined treatment with PARPis plus IR for all cell lines, except for the HPV-negative cell line CLS-354. While CLS-354 stands out in this way, it is important to note that the effects in the other cell lines were heterogeneous in size, and in HSC4, for example, the effects were very small. In clonogenic survival, almost all cell lines were affected by both combinations, however, in the HPV-positive cell lines, the effects were supra-additive, while the decrease in SF was additive in most HPV-negative ones. Further in vivo studies are warranted, and establishing a predictive marker for the effectiveness of PARPis plus IR in HNSCC would be desirable.

The higher radiosensitivity of the HPV-positive HNSCC has been known for a long time [9,11,12]. We show that the HPV-positive HNSCC is not only more radiosensitive per se, but also exhibits supra-additive effects when treated with talazoparib or niraparib in combination with IR. These effects represent radiosensitization capabilities, which were not consistently observed in the case in HPV-negative cell lines. Comparable data on this are conflicting, and the results depend on the experimental design and the specific inhibitor used. While the results of one study on talazoparib are similar to our findings [52], two studies on niraparib found a slightly stronger effect in HPV-negative cell lines [49,50]. However, these differences seem negligible since all the mentioned studies found effects in both HPV-negative and HPV-positive cell lines. It is worth mentioning that, at 50 nM, we used a lower concentration of talazoparib compared to the 100 nM used in the aforementioned study [52]. The two studies on niraparib showed these promising results with a concentration of 1000 nM [49,50], while we used 2500 nM. Despite that, the results are similar. Zhou et al. concluded that a combined treatment of PARPis and IR should be further considered in HNSCC, especially in the more radioresistant HPV-negative cell lines [52]. Our findings not only support but extend beyond their conclusion. All in all, in our experiments, despite the fact that the HPV-positive cell lines were overall more strongly affected, PARPis with IR affected all HNSCC cell lines, both HPV-negative and HPV-positive, in a relevant fashion.

Therefore, we conclude that for the potential treatment of HNSCC tumors, regardless of HPV status, the combination of IR with PARPi, specifically talazoparib and niraparib, should be considered for further studies. As the response to the combined treatment was to some extent heterogeneous between cell lines, individual testing prior to treatment with PARPis and IR seems reasonable. It would therefore be desirable to explore the underlying molecular mechanisms involved in the heterogeneity of the effects in order to find a predictive marker for therapeutic success.

One possible approach to understanding the heterogeneity in response to treatment with PARPis and IR, especially among the HPV-negative HNSCC cell lines, is the influence of p53. The p53 protein is a major component in the regulation of cellular responses to stress, including apoptosis, necrosis, cell cycle, cell proliferation, DNA repair and senescence [74,75]. While HPV-positive HNSCC cells usually encode a wild-type (wt) p53, which is then inactivated and degraded by the HPV oncoproteins E6, E7 and others, rendering the HPV-positive HNSCC functionally p53-deficient [76–78], the HPV-negative HNSCC can be divided into two groups, based on their p53 status: p53 wt and p53 mutated (mut). In a clinical sample of 243 patients with HPV-negative HNSCC, 84% were p53 mut [79]. The p53 status is of interest in the context of our study, since PARP is involved in p53 regulation [80,81]. On the one hand, there are studies that found a functional p53 to promote a radiosensitizing effect of PARPis in different tumor entities [82,83], while, on the other hand, there is evidence that the loss-of-function mutations of p53 lead to increased radiosensitivity by PARPis through the activation of oxidative stress pathways [84]. The accumulated evidence suggests that PARPis can have radiosensitizing effects in p53 wt, as well as in p53 mut tumors, through various different mechanisms [85].

Out of the five HPV-negative cell lines we investigated, Cal33, HSC4 and Detroit 562 are p53 mut [86,87], while RPMI2650 is a p53 wt [88]. The p53 status of CLS-354 is unclear, although one study found p53 expression in CLS-354 after treatment with IFNγ, which can trigger cell cycle arrest and apoptosis, among other functions [89]. In our data, no clear association can be made between the p53 status of an HPV-negative HNSCC cell line and the radiosensitization capabilities of either of the two PARPis. This is stressed by the fact that even within one cell line and method, oftentimes, the radiosensitizing effects of talazoparib and niraparib are incongruent. In our experiments, the influence of the p53 status of the HPV-negative HNSCC cell lines on the radiosensitizing capabilities of PARPis was not great enough to be noticeable among the effects of various other differences between the cell lines that may influence treatment effectiveness, and it does not explain the heterogeneity found.

3.2. Effects of Talazoparib and Niraparib on Healthy Fibroblast Cells

In addition to the therapeutically desirable effects of a treatment, it is important to consider the effects on healthy tissue as well. Since talazoparib and niraparib are already market-approved, their safety and tolerability have been proven before [90,91]. Nevertheless, contrary to the above-mentioned studies on PARPis plus IR in HNSCC [49,50,52], we included two healthy fibroblast cell lines that were treated and evaluated exactly the same as the HNSCC cell lines for comparison. We focused on how the PARPis could amplify the potential irradiation-induced side effects.

On the one hand, neither combination of PARPi and IR increased cell death or G2/M-phase arrest compared to IR alone in SBLF8 or SBLF9. On the other hand, in the colony formation assay, both combined treatments showed toxicity, resulting in a decreased SF compared to IR alone, for talazoparib in both fibroblast cell lines and for niraparib in SBLF8. This constellation leads us to hypothesize that the irradiation-associated effects of talazoparib and niraparib on healthy fibroblasts may be mediated by the induction of senescence. The possible induction of senescence by KIs with IR and the implications for clinical development are discussed in more detail below in the context of the DNA-PKcsi AZD7648.

Additionally, the high G2/M-phase arrest by mono-treatment with talazoparib and especially niraparib in SBLF8 is noteworthy. Interestingly, this effect vanished in combination with IR. We do not have a conclusive explanation for this unusual behavior and attribute it to individual cellular composition of the donor of the fibroblast cell line. Unfortunately, we cannot investigate the context of this phenomenon further due to the protection of the donor's privacy.

To summarize, our experiments suggest that increased irradiation-induced side effects are likely but the anti-tumor potential of talazoparib and niraparib with IR justifies further consideration of these drugs for treatment of HNSCC.

3.3. AZD7648 Is a Potent Radiosensitizer in HNSCC Cell Lines

AZD7648 is a potent and selective inhibitor of the DNA-PKcs with radiosensitizing capabilities [92] first discovered in 2020 [93]. In HR-deficient tumor cells, inhibition of NHEJ through DNA-PKcsis combined with DNA damage inductors like IR leads to accumulation of DSBs in the tumor cells which results in cell death [14–16]. Meanwhile, healthy cells are still HR-proficient, which should facilitate tumor-selectivity.

We found AZD7648 to be a highly effective radiosensitizer of HNSCC cell lines in vitro, more effective than the PARPis talazoparib and niraparib. The combination of AZD7648 and IR induced cell death in all the HNSCC cell lines but RPMI2650, which was resistant to all the combined treatments. For the induction of G2/M-phase arrest and for clonogenic survival, the picture gets even clearer. AZD7648 plus IR stood out as by far the most powerful radiosensitizer, having supra-additive effects in all cell lines, irrespective of HPV status. Our data support the idea that AZD7648 has the potential to relevantly improve radiation therapy in HNSCC. Because AZD7648 has only recently been discovered, not much data exist on HNSCC. One study was conducted, which found AZD7648 to be a potent radiosensitizer, additionally, in two HNSCC cell lines in a mouse model, in both oxic and anoxic conditions [64]. Another study found promising effects of AZD7648 plus IR in vitro, especially in the two HPV-negative HNSCC cell lines investigated [65].

DNA-PK inhibitors (DNA-PKi) other than AZD7648 have been found to radiosensitize HNSCC cell lines. Comparable to our results, DNA-PKi KU57788 and IC87361 did so in a more potent way than the PARPis olaparib and veliparib [61]. Studies on other DNA-PKi, and also inhibitors of ATM, ATR and PARP, did not find any therapeutically exploitable differences between photon and proton irradiation in HNSCC cell lines [94,95]. Additionally, the combination of DNA-PKis with other substances can amplify the radiosensitizing effects in HNSCC. This has been shown for a combination with ATR inhibition [62], as well as with the PARPi olaparib, where the effect was comparable to cisplatin plus IR, while having considerably fewer toxic effects [63]. However, the effects of AZD7648 alone on healthy cells should be considered first, before proposing its combination with another radiosensitizer to enhance the radiosensitizing effects.

3.4. Healthy Fibroblast Cells Tolerate AZD7648 Well While Combining AZD7648 with IR Impairs Clonogenicity without Increased G2/M Arrest, Apoptosis or Necrosis

The high radiosensitization capability of AZD7648 in HNSCC cell lines begs the question to what degree healthy tissue is affected by it as well. We found that AZD7648 alone was tolerated very well in healthy fibroblasts despite the relatively high concentration of 5000 nM. There was no increase in the G2/M-phase arrest compared to the control, while one of the two healthy fibroblast cell lines experienced a slight increase in cell death. Clonogenicity was only marginally affected in SBLF8 (SF = 0.83) and not affected in SBLF9 (SF = 0.85) compared to the control which was normalized to the value of 1. For comparison, SF for IR alone was 0.34 and 0.30, respectively.

Next to the tolerance of the substance alone, potential effects in the irradiated area are a concern. One study found AZD7648 to relevantly radiosensitize intestinal and mucosal stem cells in mouse models [64]. The effects of AZD7648 plus IR in healthy fibroblast cell lines observed in another study are very similar to our observations [65]. In our analyses, AZD7648 had no radiosensitizing effects on healthy fibroblasts regarding cell death induction and G2/M-phase arrest. However, we could observe a strong decrease in the survival fraction compared to the IR alone. This leads us to hypothesize that AZD7648 with IR induces senescence in healthy fibroblast cells. Noticeably, the data on the dual DNA-PK/mTOR inhibitor CC-115 in similar fibroblast cell lines, where senescence was directly measured, do not support this hypothesis, since the CC-115 with IR decreased senescence compared to the IR alone, rather than inducing it [29]. Not having measured senescence directly but only indirectly, through clonogenic survival, is a clear limitation of our study. To test the hypothesis of senescence induction in healthy fibroblasts after treatment with AZD7648 and IR, we propose that further studies directly measure senescence after treatment using beta-galactosidase.

When healthy tissue experiences senescence, the senescent cells maintain tissue integrity. In contrast, after cell death, tissue integrity is lost. Side effects caused by induction of senescence in healthy tissue are likely less burdening for patients than those mediated by direct cell death and, therefore, cell loss, for example through apoptosis or necrosis [96]. Nevertheless, the irradiation-induced side effects of AZD7648 remain a concern and can only be adequately evaluated in clinical trials. There is one currently in progress, which evaluates AZD7648 alone and in combination with other anti-cancer agents, but not including IR (NCT03907969).

It is an obvious limitation of an in vitro study like this one that the results cannot be extrapolated into more complex systems, such as in vivo models, or into a clinical setting. There are a lot of factors that influence how the treatment affects the tumor cells in such settings, such as the distribution of the inhibitor, the tumor microenvironment and the immune system. Additional studies are therefore essential, and we believe that further in vivo studies and, as a consequence of these studies, potentially clinical trials are warranted given the high potential of AZD7648 to radiosensitize HNSCC cell lines. These studies can address the complexity of the tumor microenvironment in vivo and systemic effects that cannot be assessed in vitro.

Additionally, further in vitro analyses that could shed light on ways to avoid normal tissue damage include combining PARPis and DNA-PKcsis, while reducing the respective concentrations, as well as lowering the dose of ionizing radiation. Assessing cell death and cell cycle arrest at multiple time points after treatment to see differences in the response over time can also be of interest. Replicating and validating our results in organoids would be particularly compelling.

4. Materials and Methods

4.1. Cell Lines and Cell Culture

This study examined two healthy fibroblast cell lines, SBLF8 and SBLF9, and seven HNSCC cell lines. SBLF8 and SBLF9 were derived from healthy donors, while Cal33 (CVCL_1108), HSC4 (CVCL_1289), UD-SCC-2 (CVCL_E325) and UM-SCC-47 (CVCL_7759) were obtained from Dr. Thorsten Rieckmann (University of Medical Centre Hamburg-Eppendorf, Germany); CLS-354 (Cytion Catalog Number 300152), Detroit 562 (300399) and RPMI2650 (300323) were obtained from Cytion (formerly CLS, Eppelheim, Germany). UD-SCC-2 and UM-SCC-47 are HPV-positive, while the remaining five HNSCC cell lines are HPV-negative. Cal33, HSC4 and UM-SCC-47 were derived from tumors of the oral tongue, and CLS-354 from a tumor of the oral cavity. RPMI2650, Detroit 562 and UD-SCC-2 were derived from tumors of the nasal septum, the pharynx and the hypopharynx, respectively. All cells were incubated in a humidified environment at 37 °C and 5% CO_2 and kept in cell culture flasks. The medium for SBLF8 and SBLF9 consisted of an F-12 medium supplemented with 15% fetal bovine serum (FBS, Sigma-Aldrich, St. Louis, MO, USA), 2% non-essential amino acids (NEA, Gibco, Waltham, MA, USA) and 1% penicillin-streptomycin (Gibco, Waltham, MA, USA), while HNSCC cells were kept in Dulbecco's Modified Eagle Medium (DMEM, Gibco, Waltham, MA, USA), supplemented with 10% FBS and 1% penicillin–streptomycin. Cells were passaged twice a week, or when they reached a confluence of 90%. The medium was removed, cells were washed with phosphate-buffered saline (PBS, Gibco, Waltham, MA, USA), trypsinated (Gibco, Waltham, MA, USA) and incubated for 2 to 5 min until they were detached from the flask. The trypsination was terminated with a fresh medium. The average number of passages the cell lines experienced before seeding was between 5 and 23 for all cell lines, except for Detroit 562 with 44, which had already been exposed to a high passage number before storage.

4.2. Kinase Inhibitors and Radiation Treatment

Every analysis consisted of eight settings per cell line. We analyzed three inhibitors, and each substance was evaluated alone and with 2 Gy IR. The remaining two settings contained dimethyl sulfoxide (DMSO, Roth, Karlsruhe, Germany) to compensate for its use as the carrier medium for the inhibitors, so that every setting contained identical volumes of DMSO. The setting treated only with DMSO will be referred to as control, and the setting treated with DMSO plus 2 Gy will be referred to as 2 Gy or IR alone. The radiation dose of 2 Gy was chosen, since it is well established in research and is a standard single dose in clinical practice. It is high enough to induce measurable effects, but low enough for a sufficient number of cells to survive the treatment with IR alone.

The inhibitors used were the PARPis talazoparib (50 nM) and niraparib (2500 nM) and the DNA-PKcsi AZD7648 (5000 nM), all obtained from Selleck Chemicals (Houston, TX, USA). The concentrations used for talazoparib and niraparib were determined by a calculation based on the maximum plasma concentration and the molecular weight by Jonuscheit et al. [97]. The concentration for AZD7648 was based on internal data by Klieber et al. [65]. For radiation treatment, an ISOVOLT Titan X-ray generator (GE, Ahrensburg, Germany) was used. It was operated at 120 kV and a dose rate of 6 Gy per minute.

4.3. Flow Cytometry Analysis of Apoptosis and Necrosis

Apoptosis and necrosis were assessed using flow cytometry. Cells were seeded in T25 flasks. After 24 to 96 h of incubation, when a confluence of about 80% was reached, the medium was changed to a medium with a reduced content of 2% FBS, and the inhibitor was added. After 3 h of incubation post-treatment, half of the flasks were irradiated with 2 Gy. After 48 h, cells and supernatants were harvested using trypsin and PBS. Centrifuge tubes containing 14 mL of cell suspension were centrifuged at $180 \times g$ for 8 min at room-temperature, the supernatant was disposed, and 300 µL cell solution remained. An amount of 150 µL was transferred into Eppendorf tubes.

An amount of 1 mL starvation medium (2% FBS) and 10 mL 70% ethanol (Fischer, Saarbruecken, Germany) were added to the centrifuge tubes containing the remaining 150 µL of cell suspension, which were then stored at 4 °C for a minimum of 24 h to be used for cell cycle analysis.

For apoptosis/necrosis analysis, 200 µL of cold Ringer´s solution (Fresenius, Bad Homburg, Germany), as well as 10 µL of an equal mixture of APC Annexin V (BD Biosciences, Franklin Lakes, NJ, USA) and 7AAD (BD Biosciences, Franklin Lakes, NJ, USA), was added to 150 µL cell suspension. After 30 min of incubation on ice without light exposure, the tubes were centrifuged ($180 \times g$, 8 min, RT), the supernatant was removed, and cells were resuspended. Another 150 µL of cold Ringer's solution was added, and 200 µL of each tube was transferred onto a 96-well plate. A CytoFLEX S flow cytometer (Beckmann Coulter, Brea, CA, USA) was used for measurements. Data were analyzed with Kaluza Analysis software (Version 2.1, Beckmann Coulter, Brea, CA, USA). Representative gating strategies are displayed in Supplementary Figure S1.

Annexin V and 7AAD double negative (Ann7AAD − −) were defined as living, double positive (Ann7AAD + +) was defined as necrotic and Annexin V positive only (Ann7AAD + −) as apoptotic.

4.4. Flow Cytometry Analysis of Cell Cycle

Cell cycle analysis was conducted using flow cytometry. Seeding and harvesting were performed as described for apoptosis–necrosis analysis. The centrifuge tubes with cells fixed in ethanol were centrifuged ($180 \times g$, 8 min, RT), supernatants were removed, and the pellet was resuspended. An amount of 1 mL of cold Ringer´s solution and 3 µL of 10-fold soluted Hoechst 33342 (Molecular Probes, Eugene, OR, USA) were added per sample, after which they were incubated on ice for 1 h. After another centrifugation cycle, disposing of the supernatant and resuspending, 150 µL of cold Ringer´s solution was added, resulting in a total of about 200–250 µL of cell suspension. For each condition, 200 µL of the resulting cell suspension was transferred into a 96-well plate.

CytoFLEX S flow cytometer (Beckmann Coulter, Brea, CA, USA) was used for measurements. Data were analyzed with Kaluza Analysis software (Version 2.1, Beckmann Coulter, Brea, CA, USA). Representative gating strategies are displayed in Supplementary Figure S1. Gating excluded doublets, and G1/G0-, S- and G2M-phase were differentiated by the intensity of the Hoechst signal.

4.5. Colony Formation Assay

Following 48 h of passaging, when confluence equated to about 80%, cells were harvested and seeded into Petri dishes. Depending on cell line growth characteristics and intended treatment, between 250–1000 cells were added to 3 mL of cell line-specific medium. They were then evenly distributed throughout the Petri dish by swirling the dish so that cells were separated by a sufficient distance. Cells were incubated overnight, but not longer than for a maximum of 24 h, until they were attached to the bottom of the Petri dish, but before they had the opportunity to proliferate, as this would have corrupted the statistical analysis. If a single cell had enough time to proliferate, the probability of this aggregate forming a colony would be disproportionately higher than that for cells that did not proliferate, since both cells would have to be damaged for them not to form a colony. This can be avoided by limiting the incubation time to a maximum of 24 h or less, which is usually not enough time for a relevant number of individual cells to divide. Then, the inhibitor was added, and 3 h later, half of the samples were irradiated.

After another 48 h of incubation, the medium was replaced, and cells were then incubated for the next 10–14 d, depending on their growth rate. The colony formation process was terminated by disposing of the medium and adding Wright's eosin methylene blue solution (Carl Roth, Karlsruhe, Germany) until the bottom of the Petri dish was completely covered. After 30 min, the methylene blue staining was removed, and the Petri dishes were washed with water and deionized water. All samples were allowed to dry overnight, and images were captured using a stationary camera under specific lighting conditions. Colony counting was performed by using the images of the Petri dishes, and only those colonies with at least 50 cells were counted. Representative Petri dish images are displayed in Supplementary Figure S2. Plating efficiency (PE) describes the fraction of seeded cells able to form a colony in the control:

$$PE = \text{colonies [control]/seeded cells [control].} \tag{1}$$

Survival fraction (SF) is calculated after treatment using the PE calculated from the corresponding control:

$$SF = \text{colonies [treatment]/(seeded cells [treatment]} \times PE). \tag{2}$$

4.6. Statistical Analysis

GraphPad Prism 9 (GraphPad Software, Boston, MA, USA) was used for statistical analysis. Each experiment was repeated at least four times to allow for statistical analysis using an unpaired, two-tailed Mann–Whitney U-test. A p-value of 0.05 determined statistical significance. A minimum of four replicates with two technical replicates were performed in the colony formation assay.

To study radiosensitization, supra-additivity was assessed. We added the effects of mono-treatment with IR and mono-treatment with an inhibitor together, and compared the additive effect to the effect of the combined treatment. Supra-additivity means that the effect of the combined treatment was greater than the added isolated effects. This means that there could be a clinical advantage of the combined treatment because there is a synergistic effect that extends beyond the added effects. Such supra-additivity points to the fact that the inhibitor is capable of radiosensitizing the cell line, rather than just having a toxic effect that adds to the effect of the IR. Therefore, we show if there is a radiosensitizing mechanism involved through the concept of supra-additivity.

Supra-additivity was assessed using the aforementioned unpaired, two-tailed Mann–Whitney U-test, where, again, a p-value of 0.05 determined statistical significance. For induction of apoptosis and necrosis and of G2/M-phase arrest using values in percent, the comparison was between

$$[IR] - [Co] + [KI] - [Co] \text{ vs. } [KI+IR] - [Co]. \tag{3}$$

For colony formation assay, SF was used, and the comparison was between

$$[IR] \text{ vs. } [KI + IR] / [KI]. \tag{4}$$

In each case, the effect of the corresponding approach is shown in square brackets, where Co = control without treatment, IR = ionizing radiation, KI = kinase inhibitor.

5. Conclusions

In conclusion, we found that the PARPis talazoparib and niraparib and the DNA-PKcsi AZD7648 relevantly radiosensitize HNSCC cell lines in vitro, with AZD7648 being the most potent and effective across all cell lines. Talazoparib and niraparib in combination with IR resulted in more heterogeneous effects, depending on the cell line. Therefore, it appears that for PARPis, individual patient testing would be necessary. AZD7648 alone had little to no effect on healthy fibroblast cells, making toxicity unlikely. The combination of AZD7648 with IR did not relevantly affect cell death or G2/M arrest in healthy fibroblast cells, but did decrease clonogenic survival relevantly, rendering radiation-induced side effects a concern. In summary, while talazoparib and niraparib are promising as well, AZD7648, especially, has the potential to improve radiation therapy in HNSCC in the future. Our findings justify in vivo studies, especially on the DNA-PKcsi AZD7648 combined with IR in HNSCC.

Supplementary Materials: The following supporting information can be downloaded at: https://www.mdpi.com/article/10.3390/ijms25115629/s1.

Author Contributions: Conceptualization, L.V.D. and R.F.; methodology, J.M., L.S.H. and L.K.; software, L.V.D.; validation, L.S.H. and L.V.D.; formal analysis, J.M.; investigation, J.M.; resources, R.F.; data curation, J.M.; writing—original draft preparation, J.M.; writing—review and editing, L.V.D. and L.S.H.; visualization, J.M.; supervision, L.V.D.; project administration, R.F. All authors have read and agreed to the published version of the manuscript.

Funding: We acknowledge financial support by Deutsche Forschungsgemeinschaft and Friedrich-Alexander-Universität Erlangen-Nürnberg within the funding program "Open Access Publication Funding".

Data Availability Statement: The data used and analyzed during the current study are available from the corresponding author on reasonable request.

Acknowledgments: The authors would like to thank Doris Mehler for her excellent technical support of this study. The present work was performed by Jacob Mentzel in fulfillment of the requirements for obtaining the degree "Dr. med." at the Friedrich-Alexander-Universität Erlangen-Nürnberg (FAU).

Conflicts of Interest: The authors declare no conflicts of interest.

References

1. Sung, H.; Ferlay, J.; Siegel, R.L.; Laversanne, M.; Soerjomataram, I.; Jemal, A.; Bray, F. Global Cancer Statistics 2020: GLOBOCAN Estimates of Incidence and Mortality Worldwide for 36 Cancers in 185 Countries. *CA Cancer J. Clin.* **2021**, *71*, 209–249. [CrossRef]
2. Huang, J.; Chan, S.C.; Ko, S.; Lok, V.; Zhang, L.; Lin, X.; Lucero-Prisno, D.E., 3rd; Xu, W.; Zheng, Z.J.; Elcarte, E.; et al. Disease burden, risk factors, and trends of lip, oral cavity, pharyngeal cancers: A global analysis. *Cancer Med.* **2023**, *12*, 18153–18164. [CrossRef] [PubMed]
3. Johnson, D.E.; Burtness, B.; Leemans, C.R.; Lui, V.W.Y.; Bauman, J.E.; Grandis, J.R. Head and neck squamous cell carcinoma. *Nat. Rev. Dis. Primers* **2020**, *6*, 92. [CrossRef]

4. McDermott, J.D.; Bowles, D.W. Epidemiology of Head and Neck Squamous Cell Carcinomas: Impact on Staging and Prevention Strategies. *Curr. Treat. Options Oncol.* **2019**, *20*, 43. [CrossRef]
5. Warnakulasuriya, S. Global epidemiology of oral and oropharyngeal cancer. *Oral. Oncol.* **2009**, *45*, 309–316. [CrossRef]
6. Blot, W.J.; McLaughlin, J.K.; Winn, D.M.; Austin, D.F.; Greenberg, R.S.; Preston-Martin, S.; Bernstein, L.; Schoenberg, J.B. Stemhagen, A.; Fraumeni, J.F., Jr. Smoking and drinking in relation to oral and pharyngeal cancer. *Cancer Res.* **1988**, *48*, 3282–3287
7. Zhang, L.W.; Li, J.; Cong, X.; Hu, X.S.; Li, D.; Wu, L.L.; Hua, H.; Yu, G.Y.; Kerr, A.R. Incidence and mortality trends in oral and oropharyngeal cancers in China, 2005–2013. *Cancer Epidemiol.* **2018**, *57*, 120–126. [CrossRef] [PubMed]
8. O'Sullivan, B.; Huang, S.H.; Su, J.; Garden, A.S.; Sturgis, E.M.; Dahlstrom, K.; Lee, N.; Riaz, N.; Pei, X.; Koyfman, S.A.; et al. Development and validation of a staging system for HPV-related oropharyngeal cancer by the International Collaboration on Oropharyngeal cancer Network for Staging (ICON-S): A multicentre cohort study. *Lancet Oncol.* **2016**, *17*, 440–451. [CrossRef]
9. Vokes, E.E.; Agrawal, N.; Seiwert, T.Y. HPV-Associated Head and Neck Cancer. *J. Natl. Cancer Inst.* **2015**, *107*, djv344. [CrossRef]
10. zur Hausen, H. Papillomaviruses and cancer: From basic studies to clinical application. *Nat. Rev. Cancer* **2002**, *2*, 342–350 [CrossRef]
11. Fakhry, C.; Westra, W.H.; Li, S.; Cmelak, A.; Ridge, J.A.; Pinto, H.; Forastiere, A.; Gillison, M.L. Improved survival of patients with human papillomavirus-positive head and neck squamous cell carcinoma in a prospective clinical trial. *J. Natl. Cancer Inst.* **2008**, *100*, 261–269. [CrossRef]
12. Zhou, C.; Parsons, J.L. The radiobiology of HPV-positive and HPV-negative head and neck squamous cell carcinoma. *Expert Rev. Mol. Med.* **2020**, *22*, e3. [CrossRef] [PubMed]
13. Machiels, J.P.; René Leemans, C.; Golusinski, W.; Grau, C.; Licitra, L.; Gregoire, V. Squamous cell carcinoma of the oral cavity, larynx, oropharynx and hypopharynx: EHNS–ESMO–ESTRO Clinical Practice Guidelines for diagnosis, treatment and follow-up. *Ann. Oncol.* **2020**, *31*, 1462–1475. [CrossRef]
14. Monge-Cadet, J.; Moyal, E.; Supiot, S.; Guimas, V. DNA repair inhibitors and radiotherapy. *Cancer Radiother.* **2022**, *26*, 947–954 [CrossRef]
15. Zhao, Y.; Thomas, H.D.; Batey, M.A.; Cowell, I.G.; Richardson, C.J.; Griffin, R.J.; Calvert, A.H.; Newell, D.R.; Smith, G.C.; Curtin, N.J. Preclinical evaluation of a potent novel DNA-dependent protein kinase inhibitor NU7441. *Cancer Res.* **2006**, *66*, 5354–5362 [CrossRef] [PubMed]
16. Liang, S.; Thomas, S.E.; Chaplin, A.K.; Hardwick, S.W.; Chirgadze, D.Y.; Blundell, T.L. Structural insights into inhibitor regulation of the DNA repair protein DNA-PKcs. *Nature* **2022**, *601*, 643–648. [CrossRef] [PubMed]
17. Jackson, S.P.; Bartek, J. The DNA-damage response in human biology and disease. *Nature* **2009**, *461*, 1071–1078. [CrossRef] [PubMed]
18. Scheper, J.; Hildebrand, L.S.; Faulhaber, E.-M.; Deloch, L.; Gaipl, U.S.; Symank, J.; Fietkau, R.; Distel, L.V.; Hecht, M.; Jost, T. Tumor-specific radiosensitizing effect of the ATM inhibitor AZD0156 in melanoma cells with low toxicity to healthy fibroblasts. *Strahlenther. Onkol.* **2022**, *199*, 1128–1139. [CrossRef]
19. Kopa, P.; Macieja, A.; Galita, G.; Witczak, Z.J.; Poplawski, T. DNA Double Strand Breaks Repair Inhibitors: Relevance as Potential New Anticancer Therapeutics. *Curr. Med. Chem.* **2019**, *26*, 1483–1493. [CrossRef]
20. Khanna, K.K.; Jackson, S.P. DNA double-strand breaks: Signaling, repair and the cancer connection. *Nat. Genet.* **2001**, *27*, 247–254 [CrossRef]
21. Brown, J.S.; O'Carrigan, B.; Jackson, S.P.; Yap, T.A. Targeting DNA Repair in Cancer: Beyond PARP Inhibitors. *Cancer Discov.* **2017**, *7*, 20–37. [CrossRef]
22. Classen, S.; Petersen, C.; Borgmann, K. Crosstalk between immune checkpoint and DNA damage response inhibitors for radiosensitization of tumors. *Strahlenther. Onkol.* **2023**, *199*, 1152–1163. [CrossRef]
23. Bétermier, M.; Bertrand, P.; Lopez, B.S. Is non-homologous end-joining really an inherently error-prone process? *PLoS Genet.* **2014**, *10*, e1004086. [CrossRef] [PubMed]
24. San Filippo, J.; Sung, P.; Klein, H. Mechanism of eukaryotic homologous recombination. *Annu. Rev. Biochem.* **2008**, *77*, 229–257 [CrossRef] [PubMed]
25. Lieber, M.R. The mechanism of human nonhomologous DNA end joining. *J. Biol. Chem.* **2008**, *283*, 1–5. [CrossRef]
26. Faulhaber, E.-M.; Jost, T.; Symank, J.; Scheper, J.; Bürkel, F.; Fietkau, R.; Hecht, M.; Distel, L.V. Kinase Inhibitors of DNA-PK, ATM and ATR in Combination with Ionizing Radiation Can Increase Tumor Cell Death in HNSCC Cells While Sparing Normal Tissue Cells. *Genes* **2021**, *12*, 925. [CrossRef]
27. Hanahan, D. Hallmarks of Cancer: New Dimensions. *Cancer Discov.* **2022**, *12*, 31–46. [CrossRef] [PubMed]
28. Srivastava, M.; Raghavan, S.C. DNA Double-Strand Break Repair Inhibitors as Cancer Therapeutics. *Chem. Biol.* **2015**, *22*, 17–29 [CrossRef]
29. Dobler, C.; Jost, T.; Hecht, M.; Fietkau, R.; Distel, L. Senescence Induction by Combined Ionizing Radiation and DNA Damage Response Inhibitors in Head and Neck Squamous Cell Carcinoma Cells. *Cells* **2020**, *9*, 2012. [CrossRef]
30. Yamada, M.; Miwa, M.; Sugimura, T. Studies on poly (adenosine diphosphate-ribose). X. Properties of a partially purified poly (adenosine diphosphate-ribose) polymerase. *Arch. Biochem. Biophys.* **1971**, *146*, 579–586. [CrossRef]
31. Curtin, N.J.; Szabo, C. Poly(ADP-ribose) polymerase inhibition: Past, present and future. *Nat. Rev. Drug Discov.* **2020**, *19*, 711–736 [CrossRef] [PubMed]
32. Caldecott, K.W. DNA single-strand break repair. *Exp. Cell Res.* **2014**, *329*, 2–8. [CrossRef] [PubMed]

33. Vodenicharov, M.D.; Sallmann, F.R.; Satoh, M.S.; Poirier, G.G. Base excision repair is efficient in cells lacking poly(ADP-ribose) polymerase 1. *Nucleic Acids Res.* **2000**, *28*, 3887–3896. [CrossRef] [PubMed]

34. Trucco, C.; Oliver, F.J.; De Murcia, G.; Ménissier-De Murcia, J. DNA repair defect in poly(ADP-ribose) polymerase-deficient cell lines. *Nucleic Acids Res.* **1998**, *26*, 2644–2649. [CrossRef] [PubMed]

35. Ström, C.E.; Johansson, F.; Uhlén, M.; Szigyarto, C.A.-K.; Erixon, K.; Helleday, T. Poly (ADP-ribose) polymerase (PARP) is not involved in base excision repair but PARP inhibition traps a single-strand intermediate. *Nucleic Acids Res.* **2010**, *39*, 3166–3175. [CrossRef] [PubMed]

36. Ding, R.; Pommier, Y.; Kang, V.H.; Smulson, M. Depletion of poly(ADP-ribose) polymerase by antisense RNA expression results in a delay in DNA strand break rejoining. *J. Biol. Chem.* **1992**, *267*, 12804–12812. [CrossRef] [PubMed]

37. Dantzer, F.; De La Rubia, G.; Ménissier-De Murcia, J.; Hostomsky, Z.; De Murcia, G.; Schreiber, V. Base excision repair is impaired in mammalian cells lacking poly(ADP- ribose) polymerase-1. *Biochemistry* **2000**, *39*, 7559–7569. [CrossRef] [PubMed]

38. Murai, J.; Huang, S.Y.; Das, B.B.; Renaud, A.; Zhang, Y.; Doroshow, J.H.; Ji, J.; Takeda, S.; Pommier, Y. Trapping of PARP1 and PARP2 by Clinical PARP Inhibitors. *Cancer Res.* **2012**, *72*, 5588–5599. [CrossRef] [PubMed]

39. Helleday, T. The underlying mechanism for the PARP and BRCA synthetic lethality: Clearing up the misunderstandings. *Mol. Oncol.* **2011**, *5*, 387–393. [CrossRef]

40. Nambiar, D.K.; Mishra, D.; Singh, R.P. Targeting DNA repair for cancer treatment: Lessons from PARP inhibitor trials. *Oncol. Res.* **2023**, *31*, 405–421. [CrossRef]

41. Ceccaldi, R.; Rondinelli, B.; D'Andrea, A.D. Repair Pathway Choices and Consequences at the Double-Strand Break. *Trends Cell Biol.* **2016**, *26*, 52–64. [CrossRef]

42. Deeks, E.D. Olaparib: First global approval. *Drugs* **2015**, *75*, 231–240. [CrossRef]

43. Hoy, S.M. Talazoparib: First Global Approval. *Drugs* **2018**, *78*, 1939–1946. [CrossRef] [PubMed]

44. Ledermann, J.; Harter, P.; Gourley, C.; Friedlander, M.; Vergote, I.; Rustin, G.; Scott, C.; Meier, W.; Shapira-Frommer, R.; Safra, T.; et al. Olaparib Maintenance Therapy in Platinum-Sensitive Relapsed Ovarian Cancer. *N. Engl. J. Med.* **2012**, *366*, 1382–1392. [CrossRef]

45. Litton, J.K.; Rugo, H.S.; Ettl, J.; Hurvitz, S.A.; Gonçalves, A.; Lee, K.-H.; Fehrenbacher, L.; Yerushalmi, R.; Mina, L.A.; Martin, M.; et al. Talazoparib in Patients with Advanced Breast Cancer and a Germline *BRCA* Mutation. *N. Engl. J. Med.* **2018**, *379*, 753–763. [CrossRef]

46. Mirza, M.R.; Monk, B.J.; Herrstedt, J.; Oza, A.M.; Mahner, S.; Redondo, A.; Fabbro, M.; Ledermann, J.A.; Lorusso, D.; Vergote, I.; et al. Niraparib Maintenance Therapy in Platinum-Sensitive, Recurrent Ovarian Cancer. *N. Engl. J. Med.* **2016**, *375*, 2154–2164. [CrossRef]

47. Scott, L.J. Niraparib: First Global Approval. *Drugs* **2017**, *77*, 1029–1034. [CrossRef] [PubMed]

48. Güster, J.D.; Weissleder, S.V.; Busch, C.J.; Kriegs, M.; Petersen, C.; Knecht, R.; Dikomey, E.; Rieckmann, T. The inhibition of PARP but not EGFR results in the radiosensitization of HPV/p16-positive HNSCC cell lines. *Radiother. Oncol.* **2014**, *113*, 345–351. [CrossRef]

49. Molkentine, J.M.; Molkentine, D.P.; Bridges, K.A.; Xie, T.; Yang, L.; Sheth, A.; Heffernan, T.P.; Clump, D.A.; Faust, A.Z.; Ferris, R.L.; et al. Targeting DNA damage response in head and neck cancers through abrogation of cell cycle checkpoints. *Int. J. Radiat. Biol.* **2021**, *97*, 1121–1128. [CrossRef]

50. Wang, L.; Cao, J.; Wang, X.; Lin, E.; Wang, Z.; Li, Y.; Li, Y.; Chen, M.; Wang, X.; Jiang, B.; et al. Proton and photon radiosensitization effects of niraparib, a PARP-1/-2 inhibitor, on human head and neck cancer cells. *Head Neck* **2020**, *42*, 2244–2256. [CrossRef]

51. Wurster, S.; Hennes, F.; Parplys, A.C.; Seelbach, J.I.; Mansour, W.Y.; Zielinski, A.; Petersen, C.; Clauditz, T.S.; Münscher, A.; Friedl, A.A.; et al. PARP1 inhibition radiosensitizes HNSCC cells deficient in homologous recombination by disabling the DNA replication fork elongation response. *Oncotarget* **2016**, *7*, 9732–9741. [CrossRef] [PubMed]

52. Zhou, C.; Fabbrizi, M.R.; Hughes, J.R.; Grundy, G.J.; Parsons, J.L. Effectiveness of PARP inhibition in enhancing the radiosensitivity of 3D spheroids of head and neck squamous cell carcinoma. *Front. Oncol.* **2022**, *12*, 940377. [CrossRef] [PubMed]

53. Glorieux, M.; Dok, R.; Nuyts, S. Novel DNA targeted therapies for head and neck cancers: Clinical potential and biomarkers. *Oncotarget* **2017**, *8*, 81662. [CrossRef] [PubMed]

54. Harnor, S.J.; Brennan, A.; Cano, C. Targeting DNA-Dependent Protein Kinase for Cancer Therapy. *ChemMedChem* **2017**, *12*, 895–900. [CrossRef] [PubMed]

55. Mohiuddin, I.S.; Kang, M.H. DNA-PK as an Emerging Therapeutic Target in Cancer. *Front. Oncol.* **2019**, *9*, 635. [CrossRef] [PubMed]

56. Matsumoto, Y.; Asa, A.; Modak, C.; Shimada, M. DNA-Dependent Protein Kinase Catalytic Subunit: The Sensor for DNA Double-Strand Breaks Structurally and Functionally Related to Ataxia Telangiectasia Mutated. *Genes* **2021**, *12*, 1143. [CrossRef] [PubMed]

57. Jette, N.; Lees-Miller, S.P. The DNA-dependent protein kinase: A multifunctional protein kinase with roles in DNA double strand break repair and mitosis. *Prog. Biophys. Mol. Biol.* **2015**, *117*, 194–205. [CrossRef]

58. Lovejoy, C.A.; Cortez, D. Common mechanisms of PIKK regulation. *DNA Repair* **2009**, *8*, 1004–1008. [CrossRef]

59. Zhao, Y.; Ye, X.; Xiong, Z.; Ihsan, A.; Ares, I.; Martínez, M.; Lopez-Torres, B.; Martínez-Larrañaga, M.R.; Anadón, A.; Wang, X.; et al. Cancer Metabolism: The Role of ROS in DNA Damage and Induction of Apoptosis in Cancer Cells. *Metabolites* **2023**, *13*, 796. [CrossRef]

60. Chughtai, A.A.; Pannhausen, J.; Dinger, P.; Wirtz, J.; Knüchel, R.; Gaisa, N.T.; Eble, M.J.; Rose, M. Effective Radiosensitization of Bladder Cancer Cells by Pharmacological Inhibition of DNA-PK and ATR. *Biomedicines* **2022**, *10*, 1277. [CrossRef]

61. Lee, T.W.; Wong, W.W.; Dickson, B.D.; Lipert, B.; Cheng, G.J.; Hunter, F.W.; Hay, M.P.; Wilson, W.R. Radiosensitization of head and neck squamous cell carcinoma lines by DNA-PK inhibitors is more effective than PARP-1 inhibition and is enhanced by SLFN11 and hypoxia. *Int. J. Radiat. Biol.* **2019**, *95*, 1597–1612. [CrossRef] [PubMed]

62. Hafsi, H.; Dillon, M.T.; Barker, H.E.; Kyula, J.N.; Schick, U.; Paget, J.T.; Smith, H.G.; Pedersen, M.; McLaughlin, M.; Harrington, K.J. Combined ATR and DNA-PK Inhibition Radiosensitizes Tumor Cells Independently of Their p53 Status. *Front. Oncol.* **2018**, *8*, 245. [CrossRef] [PubMed]

63. Zeng, L.; Boggs, D.H.; Xing, C.; Zhang, Z.; Anderson, J.C.; Wajapeyee, N.; Veale, C.; Bredel, M.; Shi, L.Z.; Bonner, J.A.; et al. Combining PARP and DNA-PK Inhibitors With Irradiation Inhibits HPV-Negative Head and Neck Cancer Squamous Carcinoma Growth. *Front. Genet.* **2020**, *11*, 1036. [CrossRef] [PubMed]

64. Hong, C.R.; Buckley, C.D.; Wong, W.W.; Anekal, P.V.; Dickson, B.D.; Bogle, G.; Hicks, K.O.; Hay, M.P.; Wilson, W.R. Radiosensitisation of SCCVII tumours and normal tissues in mice by the DNA-dependent protein kinase inhibitor AZD7648. *Radiother. Oncol.* **2022**, *166*, 162–170. [CrossRef] [PubMed]

65. Klieber, N.; Hildebrand, L.S.; Faulhaber, E.; Symank, J.; Häck, N.; Härtl, A.; Fietkau, R.; Distel, L.V. Different Impacts of DNA-PK and mTOR Kinase Inhibitors in Combination with Ionizing Radiation on HNSCC and Normal Tissue Cells. *Cells* **2024**, *13*, 304. [CrossRef]

66. Munster, P.; Mita, M.; Mahipal, A.; Nemunaitis, J.; Massard, C.; Mikkelsen, T.; Cruz, C.; Paz-Ares, L.; Hidalgo, M.; Rathkopf, D.; et al. First-In-Human Phase I Study Of A Dual mTOR Kinase And DNA-PK Inhibitor (CC-115) In Advanced Malignancy. *Cancer Manag. Res.* **2019**, *11*, 10463–10476. [CrossRef]

67. Franken, N.A.P.; Rodermond, H.M.; Stap, J.; Haveman, J.; van Bree, C. Clonogenic assay of cells in vitro. *Nat. Protoc.* **2006**, *1*, 2315–2319. [CrossRef] [PubMed]

68. Wang, L.; Zhang, P.; Molkentine, D.P.; Chen, C.; Molkentine, J.M.; Piao, H.; Raju, U.; Zhang, J.; Valdecanas, D.R.; Tailor, R.C.; et al. TRIP12 as a mediator of human papillomavirus/p16-related radiation enhancement effects. *Oncogene* **2017**, *36*, 820–828. [CrossRef]

69. Pawlik, T.M.; Keyomarsi, K. Role of cell cycle in mediating sensitivity to radiotherapy. *Int. J. Radiat. Oncol. Biol. Phys.* **2004**, *59*, 928–942. [CrossRef]

70. Sinclair, W.K.; Morton, R.A. X-ray sensitivity during the cell generation cycle of cultured Chinese hamster cells. 1966. *Radiat. Res.* **2012**, *178*, Av88-101. [CrossRef]

71. Sharpless, N.E.; Sherr, C.J. Forging a signature of in vivo senescence. *Nat. Rev. Cancer* **2015**, *15*, 397–408. [CrossRef] [PubMed]

72. Hernandez-Segura, A.; Nehme, J.; Demaria, M. Hallmarks of Cellular Senescence. *Trends Cell Biol.* **2018**, *28*, 436–453. [CrossRef] [PubMed]

73. Brown, J.M. Beware of Clinical Trials of DNA Repair Inhibitors. *Int. J. Radiat. Oncol. Biol. Phys.* **2019**, *103*, 1182–1183. [CrossRef] [PubMed]

74. Wang, H.; Guo, M.; Wei, H.; Chen, Y. Targeting p53 pathways: Mechanisms, structures, and advances in therapy. *Signal Transduct. Target. Ther.* **2023**, *8*, 92. [CrossRef] [PubMed]

75. Hafner, A.; Bulyk, M.L.; Jambhekar, A.; Lahav, G. The multiple mechanisms that regulate p53 activity and cell fate. *Nat. Rev. Mol. Cell Biol.* **2019**, *20*, 199–210. [CrossRef] [PubMed]

76. Busch, C.-J.; Kriegs, M.; Laban, S.; Tribius, S.; Knecht, R.; Petersen, C.; Dikomey, E.; Rieckmann, T. HPV-positive HNSCC cell lines but not primary human fibroblasts are radiosensitized by the inhibition of Chk1. *Radiother. Oncol.* **2013**, *108*, 495–499. [CrossRef]

77. Moody, C.A.; Laimins, L.A. Human papillomavirus oncoproteins: Pathways to transformation. *Nat. Rev. Cancer* **2010**, *10*, 550–560. [CrossRef] [PubMed]

78. Caicedo-Granados, E.; Lin, R.; Fujisawa, C.; Yueh, B.; Sangwan, V.; Saluja, A. Wild-type p53 reactivation by small-molecule Minnelide™ in human papillomavirus (HPV)-positive head and neck squamous cell carcinoma. *Oral. Oncol.* **2014**, *50*, 1149–1156. [CrossRef] [PubMed]

79. Lawrence, M.S.; Sougnez, C.; Lichtenstein, L.; Cibulskis, K.; Lander, E.; Gabriel, S.B.; Getz, G.; Ally, A.; Balasundaram, M.; Birol, I.; et al. Comprehensive genomic characterization of head and neck squamous cell carcinomas. *Nature* **2015**, *517*, 576–582. [CrossRef]

80. Wieler, S.; Gagné, J.-P.; Vaziri, H.; Poirier, G.G.; Benchimol, S. Poly(ADP-ribose) Polymerase-1 Is a Positive Regulator of the p53-mediated G1 Arrest Response following Ionizing Radiation. *J. Biol. Chem.* **2003**, *278*, 18914–18921. [CrossRef]

81. Li, W.H.; Wang, F.; Song, G.Y.; Yu, Q.H.; Du, R.P.; Xu, P. PARP-1: A critical regulator in radioprotection and radiotherapy-mechanisms, challenges, and therapeutic opportunities. *Front. Pharmacol.* **2023**, *14*, 1198948. [CrossRef] [PubMed]

82. Sizemore, S.T.; Mohammad, R.; Sizemore, G.M.; Nowsheen, S.; Yu, H.; Ostrowski, M.C.; Chakravarti, A.; Xia, F. Synthetic Lethality of PARP Inhibition and Ionizing Radiation is p53-dependent. *Mol. Cancer Res.* **2018**, *16*, 1092–1102. [CrossRef] [PubMed]

83. Xiao, G.; Lundine, D.; Annor, G.K.; Canar, J.; Ellison, V.; Polotskaia, A.; Donabedian, P.L.; Reiner, T.; Khramtsova, G.F.; Olopade, O.I.; et al. Gain-of-Function Mutant p53 R273H Interacts with Replicating DNA and PARP1 in Breast Cancer. *Cancer Res.* **2020**, *80*, 394–405. [CrossRef] [PubMed]

84. Liu, Q.; Gheorghiu, L.; Drumm, M.; Clayman, R.; Eidelman, A.; Wszolek, M.F.; Olumi, A.; Feldman, A.; Wang, M.; Marcar, L.; et al. PARP-1 inhibition with or without ionizing radiation confers reactive oxygen species-mediated cytotoxicity preferentially to cancer cells with mutant TP53. *Oncogene* **2018**, *37*, 2793–2805. [CrossRef] [PubMed]

35. Li, S.; Wang, L.; Wang, Y.; Zhang, C.; Hong, Z.; Han, Z. The synthetic lethality of targeting cell cycle checkpoints and PARPs in cancer treatment. *J. Hematol. Oncol.* **2022**, *15*, 147. [CrossRef] [PubMed]

36. Kasten-Pisula, U.; Saker, J.; Eicheler, W.; Krause, M.; Yaromina, A.; Meyer-Staeckling, S.; Scherkl, B.; Kriegs, M.; Brandt, B.; Grénman, R.; et al. Cellular and Tumor Radiosensitivity is Correlated to Epidermal Growth Factor Receptor Protein Expression Level in Tumors Without EGFR Amplification. *Int. J. Radiat. Oncol. Biol. Phys.* **2011**, *80*, 1181–1188. [CrossRef]

37. Ganci, F.; Pulito, C.; Valsoni, S.; Sacconi, A.; Turco, C.; Vahabi, M.; Manciocco, V.; Mazza, E.M.C.; Meens, J.; Karamboulas, C.; et al. PI3K Inhibitors Curtail MYC-Dependent Mutant p53 Gain-of-Function in Head and Neck Squamous Cell Carcinoma. *Clin. Cancer Res.* **2020**, *26*, 2956–2971. [CrossRef]

38. Zeng, Y.; Prabhu, N.; Meng, R.; Eldeiry, W. Adenovirus-mediated p53 gene therapy in nasopharyngeal cancer. *Int. J. Oncol.* **1997**, *11*, 221–226. [CrossRef]

39. El Jamal, S.M.; Taylor, E.B.; Abd Elmageed, Z.Y.; Alamodi, A.A.; Selimovic, D.; Alkhateeb, A.; Hannig, M.; Hassan, S.Y.; Santourlidis, S.; Friedlander, P.L.; et al. Interferon gamma-induced apoptosis of head and neck squamous cell carcinoma is connected to indoleamine-2,3-dioxygenase via mitochondrial and ER stress-associated pathways. *Cell Div.* **2016**, *11*, 11. [CrossRef]

40. Hennes, E.R.; Dow-Hillgartner, E.N.; Bergsbaken, J.J.; Piccolo, J.K. PARP-inhibitor potpourri: A comparative review of class safety, efficacy, and cost. *J. Oncol. Pharm. Pract.* **2020**, *26*, 718–729. [CrossRef]

41. Cai, Z.; Liu, C.; Chang, C.; Shen, C.; Yin, Y.; Yin, X.; Jiang, Z.; Zhao, Z.; Mu, M.; Cao, D.; et al. Comparative safety and tolerability of approved PARP inhibitors in cancer: A systematic review and network meta-analysis. *Pharmacol. Res.* **2021**, *172*, 105808. [CrossRef] [PubMed]

42. Fok, J.H.L.; Ramos-Montoya, A.; Vazquez-Chantada, M.; Wijnhoven, P.W.G.; Follia, V.; James, N.; Farrington, P.M.; Karmokar, A.; Willis, S.E.; Cairns, J.; et al. AZD7648 is a potent and selective DNA-PK inhibitor that enhances radiation, chemotherapy and olaparib activity. *Nat. Commun.* **2019**, *10*, 5065. [CrossRef] [PubMed]

43. Goldberg, F.W.; Finlay, M.R.V.; Ting, A.K.T.; Beattie, D.; Lamont, G.M.; Fallan, C.; Wrigley, G.L.; Schimpl, M.; Howard, M.R.; Williamson, B.; et al. The Discovery of 7-Methyl-2-[(7-methyl[1,2,4]triazolo[1,5-a]pyridin-6-yl)amino]-9-(tetrahydro-2H-pyran-4-yl)-7,9-dihydro-8H-purin-8-one (AZD7648), a Potent and Selective DNA-Dependent Protein Kinase (DNA-PK) Inhibitor. *J. Med. Chem.* **2020**, *63*, 3461–3471. [CrossRef] [PubMed]

44. Meerz, A.; Deville, S.S.; Müller, J.; Cordes, N. Comparative Therapeutic Exploitability of Acute Adaptation Mechanisms to Photon and Proton Irradiation in 3D Head and Neck Squamous Cell Carcinoma Cell Cultures. *Cancers* **2021**, *13*, 1190. [CrossRef] [PubMed]

45. Vitti, E.T.; Kacperek, A.; Parsons, J.L. Targeting DNA Double-Strand Break Repair Enhances Radiosensitivity of HPV-Positive and HPV-Negative Head and Neck Squamous Cell Carcinoma to Photons and Protons. *Cancers* **2020**, *12*, 1490. [CrossRef] [PubMed]

46. Ewald, J.A.; Desotelle, J.A.; Wilding, G.; Jarrard, D.F. Therapy-induced senescence in cancer. *J. Natl. Cancer Inst.* **2010**, *102*, 1536–1546. [CrossRef]

47. Jonuscheit, S.; Jost, T.; Gajdošová, F.; Wrobel, M.; Hecht, M.; Fietkau, R.; Distel, L. PARP Inhibitors Talazoparib and Niraparib Sensitize Melanoma Cells to Ionizing Radiation. *Genes* **2021**, *12*, 849. [CrossRef]

International Journal of
Molecular Sciences

Article

Aspartate β-Hydroxylase Is Upregulated in Head and Neck Squamous Cell Carcinoma and Regulates Invasiveness in Cancer Cell Models

Pritha Mukherjee [1,2,†], Xin Zhou [1,2,3,†], Susana Galli [4], Bruce Davidson [5], Lihua Zhang [1,2], Jaeil Ahn [6], Reem Aljuhani [2,4], Julius Benicky [1,2], Laurie Ailles [7], Vitor H. Pomin [8,9], Mark Olsen [10,11] and Radoslav Goldman [1,2,4,*]

1   Department of Oncology, Lombardi Comprehensive Cancer Center, Georgetown University, Washington, DC 20057, USA
2   Clinical and Translational Glycoscience Research Center, Georgetown University, Washington, DC 20057, USA
3   Biotechnology Program, Northern Virginia Community College, Manassas, VA 20109, USA
4   Department of Biochemistry and Molecular & Cellular Biology, Georgetown University, Washington, DC 20057, USA
5   Department of Otolaryngology-Head and Neck Surgery, MedStar Georgetown University Hospital, Washington, DC 20057, USA
6   Department of Biostatistics, Bioinformatics and Biomathematics, Georgetown University, Washington, DC 20057, USA
7   Department of Medical Biophysics, University of Toronto, Toronto, ON M5G 1L7, Canada
8   Department of BioMolecular Sciences, University of Mississippi, Oxford, MS 38677, USA; vpomin@olemiss.edu
9   Research Institute of Pharmaceutical Sciences, School of Pharmacy, University of Mississippi, Oxford, MS 38677, USA
10   Department of Pharmaceutical Sciences, College of Pharmacy Glendale Campus, Midwestern University, Glendale, AZ 85308, USA
11   Pharmacometrics Center of Excellence, Midwestern University, Downers Grove, IL 60515, USA
*   Correspondence: rg26@georgetown.edu; Tel.: +1-202-687-9868
†   These authors contributed equally to this work.

Abstract: Aspartate β-hydroxylase (ASPH) is a protein associated with malignancy in a wide range of tumors. We hypothesize that inhibition of ASPH activity could have anti-tumor properties in patients with head and neck cancer. In this study, we screened tumor tissues of 155 head and neck squamous cell carcinoma (HNSCC) patients for the expression of ASPH using immunohistochemistry. We used an ASPH inhibitor, MO-I-1151, known to inhibit the catalytic activity of ASPH in the endoplasmic reticulum, to show its inhibitory effect on the migration of SCC35 head and neck cancer cells in cell monolayers and in matrix-embedded spheroid co-cultures with primary cancer-associated fibroblast (CAF) CAF 61137 of head and neck origin. We also studied a combined effect of MO-I-1151 and HfFucCS, an inhibitor of invasion-blocking heparan 6-O-endosulfatase activity. We found ASPH was upregulated in HNSCC tumors compared to the adjacent normal tissues. ASPH was uniformly high in expression, irrespective of tumor stage. High expression of ASPH in tumors led us to consider it as a therapeutic target in cell line models. ASPH inhibitor MO-I-1151 had significant effects on reducing migration and invasion of head and neck cancer cells, both in monolayers and matrix-embedded spheroids. The combination of the two enzyme inhibitors showed an additive effect on restricting invasion in the HNSCC cell monolayers and in the CAF-containing co-culture spheroids. We identify ASPH as an abundant protein in HNSCC tumors. Targeting ASPH with inhibitor MO-I-1151 effectively reduces CAF-mediated cellular invasion in cancer cell models. We propose that the additive effect of MO-I-1151 with HfFucCS, an inhibitor of heparan 6-O-endosulfatases, on HNSCC cells could improve interventions and needs to be further explored.

Keywords: HNSCC; aspartate β-hydroxylase (ASPH); heparan 6-O-endosulfatase 2 (SULF2); MO-I-1151; HfFucCS; primary head and neck CAF; SCC35 co-culture spheroid

1. Introduction

Aspartate β-hydroxylase (ASPH) belongs to the alpha-ketoglutarate-dependent dioxygenase family of proteins. It has been identified as one of the cell surface proteins associated with malignant transformation of 70–90% of tumors, mainly in the cells of breast, hepatic, colon, pancreatic, and neural origin [1–9]. ASPH catalyzes hydroxylation of aspartic acid or asparagine residues in EGF-like domains of several proteins, including clotting factors, extracellular matrix proteins, low-density lipoprotein receptor, Notch homologues, and Notch ligand homologues [5,9–11]. It is a type II transmembrane protein with a highly conserved sequence and, besides cell-surface localization, it has been also detected in the endoplasmic and sarcoplasmic reticula. Recent reports describe its presence in mitochondria in specific cancer types [5]. The mature ASPH is transported from the endoplasmic reticulum to the plasma membrane, which exposes the C-terminal region to the extracellular environment. The enzymatic activity of cell surface ASPH has been associated with enhanced cell motility, migration, and invasion, metastatic spread, and drug resistance in solid tumors through its impact on a Notch, proto-oncogene tyrosine-protein kinase Src, phosphoinositide 3-kinases (PI3Ks), and mitogen-activated protein kinase (MAPK) signaling pathways [4,5,11,12].

Specific and selective small molecule inhibitors (SMIs) have been designed to target the hydroxylase activity of ASPH; these compounds inhibit tumor development and metastasis [1,3,5,13]. Antibody-drug conjugate systems, bio-nanoparticle-based therapeutic vaccines, and dendritic cells fused to the ASPH protein have yielded substantial antitumor effects in cell lines and animal models [14–17]. We and others have developed SMIs of ASPH based on the crystal structure of the ASPH catalytic site. The first ASPH SMIs were the tetronimides MO-I-500 and MO-I-1100 [18,19] followed by MO-I-1151 and MO-I-1182 [3,5], which are modified with trifluoromethyl and carboxymethyl groups, respectively. MO-I-1151 and MO-I-1182 show enhanced activity [3,20]; they not only inhibit catalytic activity but also penetrate cells [5,21]. MO-I-1100 prevented cell migration, invasion and metastasis in hepatocellular carcinoma by modulating the Notch pathway [1]. MO-I-1144 inhibited tumor development and metastasis in colorectal cancer by decreasing Notch expression [22].

ASPH is a documented oncogene in several cancers [5,11] but has not been studied in head and neck squamous cell carcinoma to our knowledge. Besides the inhibition of colorectal cancer [22], ASPH inhibition blocked cell invasion, EMT, and metastasis in pancreatic and hepatocellular carcinomas by interacting with vimentin [2,12]. In pancreatic cancer, ASPH has also been reported to interact with ADAM 12/15, thus activating SRC kinase pathway proteins that control MMP-mediated extracellular matrix degradation and tumor invasion [23]. Tumor cell invasion into the local tissue initiates a metastatic cascade important for the prognosis of cancer patients [24,25]. In the early part of the metastatic process, the tumor cells acquire the ability to penetrate the basement membrane and ECM. The cell–matrix and cell–cell interactions enable malignant tumor cells to invade the surrounding stroma [26–28]. Cancer-associated fibroblasts (CAF) are an important mediator of cancer cell invasiveness [29,30] and could contribute to the ASPH function in HNSCC. Single-cell transcriptomic analysis of HNSCC shows that ASPH is expressed in cancer cells but also in the cancer-associated fibroblasts (CAFs) and endothelial cells [31]. In addition, CAFs effectively deposit and remodel the ECM in the tumor micro-environment (TME) [32], and stromal alpha-SMA is an independent prognostic factor in OSCC patients [33]. In this study, we therefore examine the expression of ASPH in HNSCC tumors, and we use a CAF-cancer cell co-culture spheroid model [34] to evaluate the impact of ASPH inhibition on cancer cell invasion.

In summary, we examined the expression of ASPH in tissues of 155 HNSCC patients and in cell models of HNSCC invasion. We tested the ability of the MO-I-1151 ASPH inhibitor alone or in combination with a recently described SULF2 inhibitor HfFucCS [34] with regard to the invasion of HNSCC tumor cells in a spheroid co-culture model with a primary HNSCC cancer-associated fibroblast (CAF 61137). Our results show that inhibition of ASPH inhibits invasion of the HNSCC cell line into Matrigel and that the effect is additive with the SULF2 inhibitor HfFucCS. The results warrant further mechanistic and in vivo studies of ASPH in HNSCC.

2. Results

2.1. Expression of ASPH in HNSCC Tissues

ASPH staining of 142 patients with primary HNSCC tumors, 10 patients with lymph node metastasis, and 3 patients with CIS were evaluated on our TMA (Figure 1). A subset of the patients (n = 35) had adjacent normal cores represented on the TMA in addition to the primary tumor (Figure 1A). We present higher magnification images of tissue sections showing ASPH expression and distribution in the tumor and adjacent normal tissues in Figure S1. The mean ASPH score (Intensity + Distribution) in primary tumors (mean = 5.0, SD = 0.40) is significantly higher ($p < 0.0001$) than the mean score in the adjacent normal tissues (mean = 3.2, SD = 0.98) (Figure 1B). Analysis of the TCGA and CPTAC dataset for 43 and 64 HNSCC patients, respectively, corroborated our data, showing that ASPH is significantly higher in tumors compared to paired normal tissues at both mRNA ($p = 0.0004$) and protein ($p = 0.0026$) expression levels (Figure S2). The lymph node metastasis showed high expression of ASPH (mean = 5.0, SD = 0.50) similar to the primary tumors; the node scores were also significantly higher ($p < 0.0001$) compared to the normal tissue cores (Figure 1B). The CIS cores stain positive (mean = 5.0), which suggests that ASPH is elevated early, already at the pre-cancerous stage, and is uniformly high at all stages of the HNSCC tumors. However, the CIS cases are poorly represented on our TMA (n = 3), and further studies will need to confirm this result (Figure 1B). Patients with paired primary tumor and lymph node cores (n = 17) showed no difference in mean score (Figure 1C). The uniformly high expression of ASPH across all tumor stages prevented a meaningful analysis of the impact of ASPH on patient survival in our cohort of samples. In our patient cohort, we notice that the intensity of ASPH expression correlated with the grade of primary tumors. The mean scores increase from 1.9 (w), to 2.0 (m), and 2.2 (p), which shows a significant increase in poorly differentiated tumors ($p < 0.05$ for 'w' to 'm', and $p < 0.001$ for 'm' to 'p' grades), expected to be more aggressive. ASPH expression does not differ significantly between the tumor epithelium and stroma either in intensity or in distribution, which shows that CAF and other non-cancer cells are an important source of ASPH.

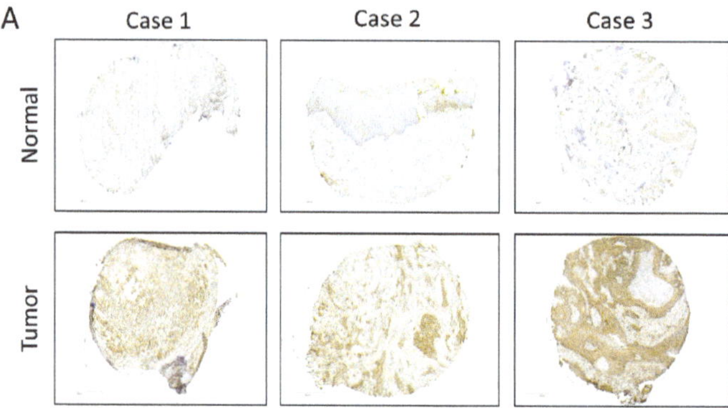

Figure 1. *Cont.*

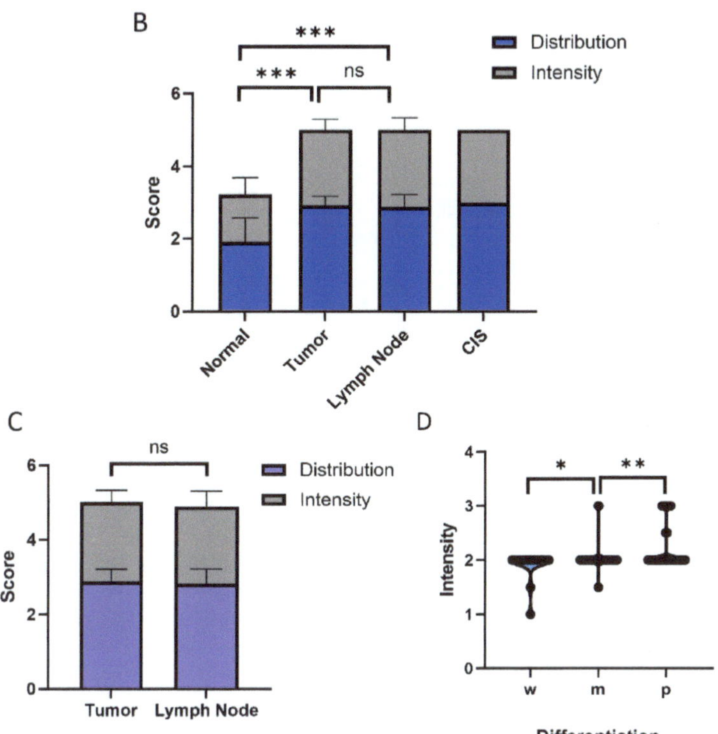

Figure 1. ASPH protein is elevated in the HNSCC tumor tissues. (**A**). Typical IHC staining for ASPH in HNSCC tumors and adjacent normal tissues. (**B**). ASPH staining scores (distribution and intensity) increase significantly from normal adjacent tissues (n = 35) to primary tumors (n = 142) and SCC-positive lymph nodes (n = 10). CIS scores appear equally high as tumors but low numbers of the cores (n = 3) prevent statistically powered evaluation. (**C**). ASPH score of paired tumors and SCC-positive lymph nodes (n = 17) show equally high expression of ASPH. (**D**). ASPH staining intensity in well- (w, n = 15), moderately (m, n = 92), and poorly (p, n = 48) differentiated tumors significantly increases. Statistical significance * p value < 0.05, ** p value < 0.001, *** p value < 0.0001, ns = not significant.

2.2. MO-1151 Inhibits Migration of HNSCC Cells

We quantified the migration of SCC35 cells in adherent cultures using a wound healing assay. Our results show that MO-I-1151 reduced the migration of cells and delayed wound closure compared to the untreated controls (Figure 2A). Two doses of the inhibitor were tested, and both doses significantly reduced the wound closure at 24 h of treatment, MO-I-1151$_{Low}$ (5μM) 54% (p < 0.0001) and MO-I-1151$_{High}$ (25 μM) 52% (p < 0.0001) (Figure 2B). The lower dose was sufficient to arrest migration of the SCC35 cells. The untreated controls closed the wound by 48 h in contrast to the treated cells at both doses.

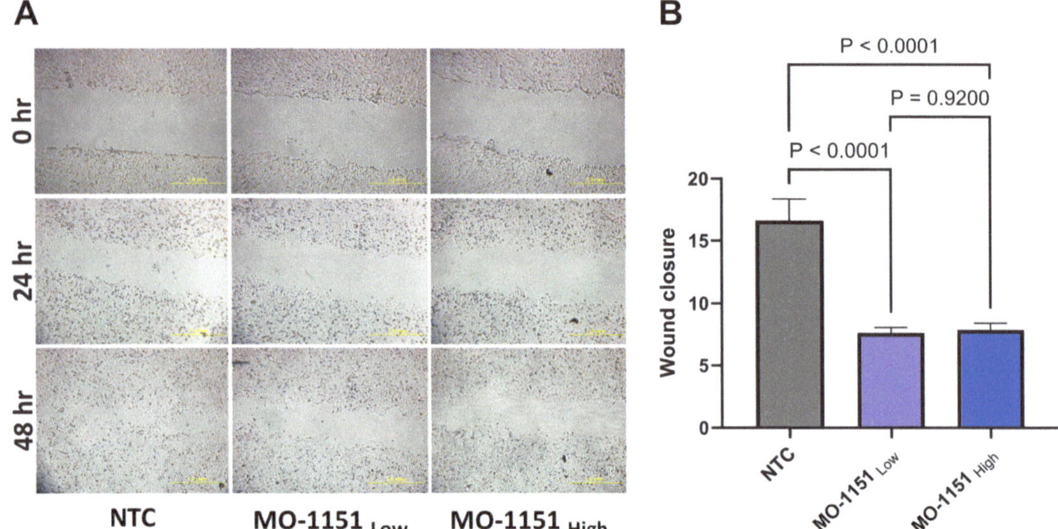

Figure 2. ASPH inhibitor MO-1151 arrests migration of SCC35 cells. (**A**). Representative images of a wound healing assay showing migration of cell at three timepoints (0, 24, 48 h) in non-treated control (NTC) and MO-1151 (low dose—5μM, high dose—25 μM) treated cells. Scale bar: 1 mm. (**B**). MO-1151 treatment shows significant difference in wound closure in the low and high doses compared to the control cells. Graphs represent mean value of each group with SD, n = 5 independent repeats. *p*-values for group-wise comparison are presented in panel (**B**).

2.3. MO-1151 Inhibits CAF-Mediated Invasion of Spheroids in Matrigel

We used a Matrigel-embedded co-culture spheroid model of SCC35 tumor cells with CAF 61137, optimized previously [34], to evaluate the impact of MO-I-1151 on cell invasion (Figure 3A). The CAF-supported tumor cell invasion was recorded on day 5, and our results show that MO-I-1151 reduced the area of the co-culture spheroids by 44% ($p < 0.0001$) for MO-I-1151$_{Low}$ and 45% ($p < 0.0001$) for MO-I-1151$_{High}$ (Figure 3B) compared to the untreated controls. Inverse circularity of the treated spheroids was also significantly reduced, by 61% at the MO-I-1151$_{Low}$ ($p < 0.0001$) and 70% ($p < 0.0001$) at the MO-I-1151$_{high}$ doses (Figure 3C). These results show that MO-I-1151 decreases cancer cell invasion at the CAF–cancer cell interface.

2.4. Combination of MO-I-1151 with HfFucCS Has an Additive Effect on Migration of HNSCC Cells

We have shown previously that the marine fucosylated chondroitin sulfate HfFucCS, a newly identified inhibitor of heparan 6-O-endosulfatase SULF2, is an inhibitor of HNSCC tumor cell invasion [34]. We treated the SCC35 cells with a combination of the MO-I-1151 and HfFucCS inhibitors to see if they together inhibit the invasion more efficiently (Figure 4A). We observed that, compared to the untreated controls, the migration of SCC35 cells in adherent cultures was reduced by 74% for the HfFucCS + MO-I-1151$_{Low}$ and 91% for the HfFucCS + MO-I-1151$_{High}$ treatment groups (Figure 4B). The wound closure rate was also 48% lower for the HfFucCS + MO-I-1151$_{Low}$ and 82% for the HfFucCS + MO-I-1151$_{High}$ combinations compared to the treatment with HfFucCS alone (Figure 4B). These data are also well supported by the migration of cells using transwell chambers, where treatment with inhibitors showed reduction in cellular migration (Figure S2). The additive effect of the combination treatment indicates independent mechanisms of action and could be potentially explored therapeutically.

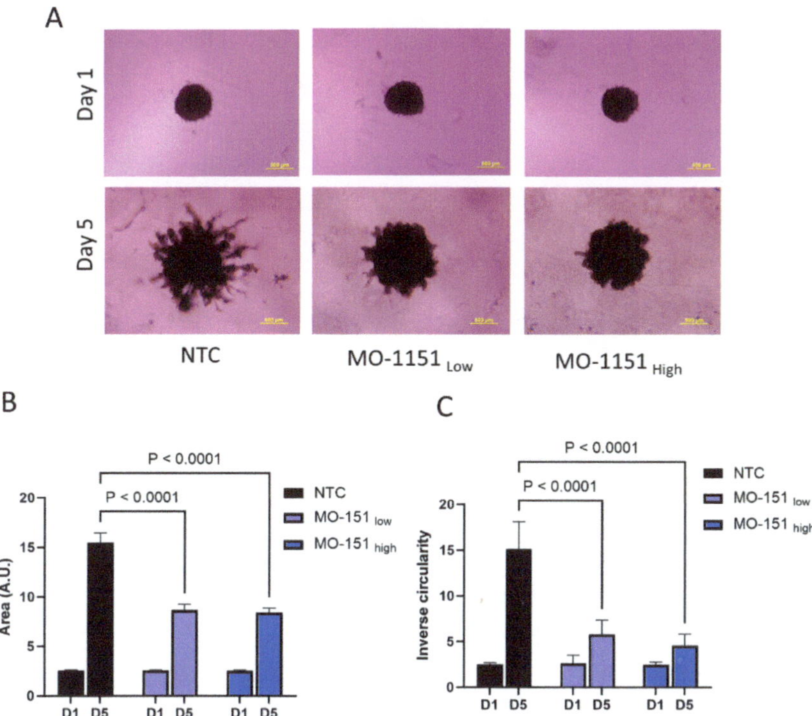

Figure 3. ASPH inhibitor MO-1151 reduces CAF-mediated invasion of SCC35 cells. (**A**). Representative images of a Matrigel invasion assay using co-culture spheroids (CAF 61137 + SCC35 cells) showing reduced cellular protrusions in MO-1151-treated spheroids (Low dose—5µM, high dose—25 µM) compared to NTC on day 5. Scale bar: 500um. (**B**). MO-1151 treatment leads to significantly reduced spheroid area on day 5 compared to the NTC. (**C**). MO-1151 treatment leads to significantly reduced inverse circularity of the spheroids on day 5 compared to the NTC. Graphs represent mean value of each group with SD, n = 5 independent repeats. *p*-values for group-wise comparison are presented in panels (**B**,**C**).

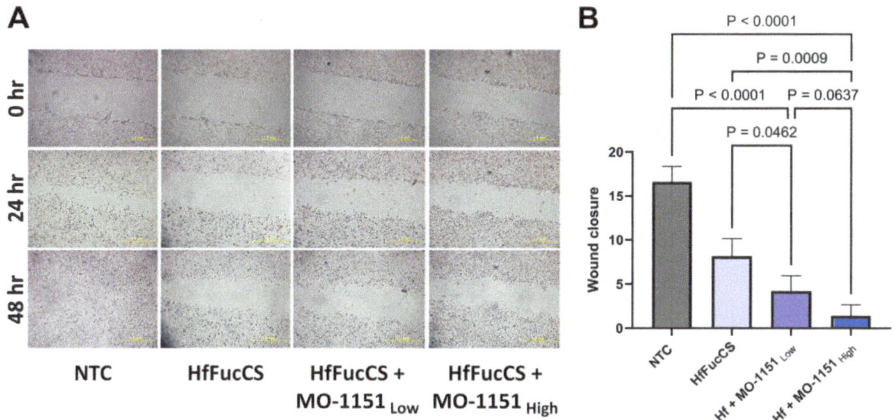

Figure 4. Treatment of SCC35 cells with a combination of the MO-1151 and HfFucCS inhibitors shows additive effect on migratory arrest. (**A**). Representative images of a wound healing assay

showing migration of SCC35 cells at three timepoints (0, 24, 48 h) in non-treated control (NTC), HfFucCS (10 µg/mL), and HfFucCS (10 µg/mL) with MO-1151 (Low dose—5 µM, and high dose—25 µM) treated cells. Scale bar: 1 mm (**B**). Combination treatments significantly reduced the wound closure compared to the NTC or HfFucCS treatment alone. Graphs represent mean value of each group with SD, n = 5 independent repeats. *p*-values for group-wise comparisons are presented in panel (**B**).

2.5. Combination of MO-I-1151 with HfFucCS Inhibitors Reduces the CAF-Mediated Invasion of Tumor Cells

The effect of the combination treatment of MO-I-1151 and HfFucCS was tested on the CAF 61137-mediated invasion of cancer cells in Matrigel-embedded spheroids. Compared to the untreated control, the combined treatment showed reduced invasion of spheroids (Figure 5A). HfFucCS reduced the area (20%, $p = 0.14$) and inverse circularity (52%, $p < 0.0001$) compared to the control (Figure 5B,C). The combined treatments reduced the area of the spheroids by 47% ($p < 0.0001$) for HfFucCS + MO-I-1151$_{Low}$ and 48% ($p < 0.0001$) for HfFucCS + MO-I-1151$_{High}$ (Figure 5B); inverse circularity was reduced by 68% ($p < 0.0001$) for both HfFucCS + MO-I-1151$_{Low}$ and HfFucCS + MO-I-1151$_{High}$ (Figure 5C). Thus, the combined treatment proved to be effective in blocking the invasion of the cancer cells supported by the CAF 61137.

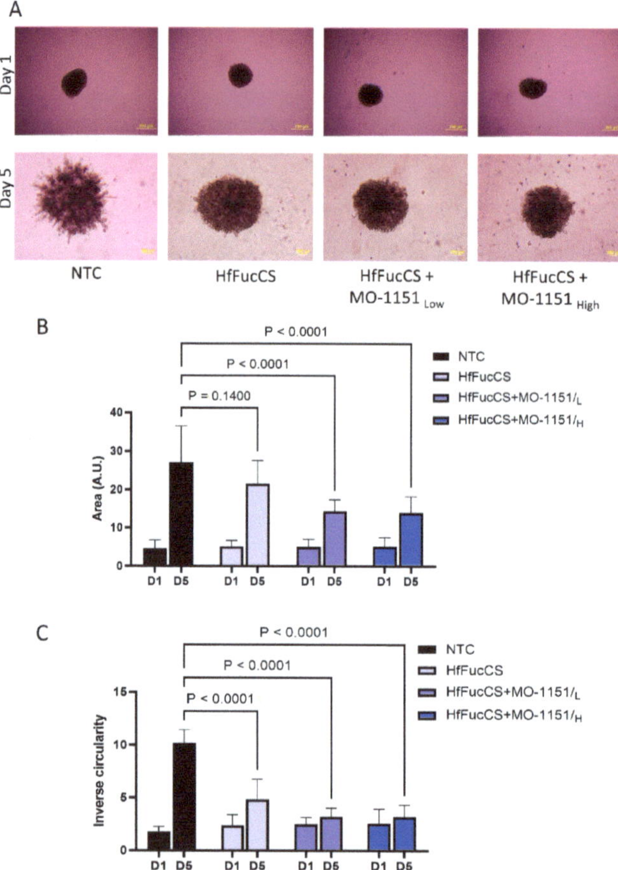

Figure 5. MO-1151 and HfFucCS additively reduce invasion of the co-culture spheroids into the Matrigel. (**A**). Representative images of Matrigel invasion of co-culture spheroids (CAF 61137 + SCC35)

show reduced cellular protrusions in spheroids treated with HfFucCS (10 mg/L) or HfFucCS (10 mg/L) + MO-1151 (Low dose—5 µM, high dose—25 µM) compared to non-treated control (NTC) on day 5. Scale bar: 200 µm. (**B**). MO-1151 + HfFucCS combined treatments significantly reduced spheroid area or (**C**). inverse circularity on day 5 compared to NTC. Graphs represent mean value of each group with SD, n = 5 independent repeats. *p*-values for group-wise comparison are presented in panels (**B**,**C**).

3. Discussion

HNSCC is an assembly of epithelial squamous malignancies of the oral cavity, larynx, and pharyngeal regions and the seventh most common cancer worldwide [35]. Approximately 50,000 new cases and 12,000 cancer deaths occur annually in the US [36]. Smoking and alcohol are the dominant etiology of this mostly preventable disease; however, the etiology of up to 80% of oropharyngeal cancers is related to human papillomavirus (HPV) [37]. This is important because the tumors of HPV origin have different molecular pathogenesis and a better prognosis. Our study emphasizes analysis of the oral cancers (OSCCs) because they are most frequent and have common etiology, disease characteristics, and treatment approaches. Surgery is the standard treatment, but retained functionality and comorbid conditions affect selection of the therapeutic regimens [36]. Introduction of Cetuximab, inhibiting the epidermal growth factor receptor (EGFR) pathways activated in a majority of HNSCCs, and inhibition of immune checkpoints added new therapeutic options [38–41]. However, the survival of HNSCC patients remains largely unchanged [42–44], and further advances in molecular diagnostics and treatment at every stage of the disease are needed [45].

ASPH is an oncogenic protein upregulated in many types of cancer. Our study documents the upregulation of ASPH in HNSCC and shows that inhibition of ASPH activity by SMI MO-I-1151 limits HNSCC cancer cell migration and CAF-supported invasion into Matrigel.

Our results from the IHC analysis of patient tissues show that ASPH scores (intensity and distribution) are high in primary tumors of the oral cavity at all stages of the disease. The expression of ASPH in positive lymph nodes is as high as the expression in primary tumors (Figure 1), and it appears that even CIS lesions have already elevated ASPH compared to the adjacent normal tissue. The CIS representation on our tumor array is admittedly low, and this observation needs to be verified; however, overall, the upregulation of ASPH in the tumors is clearly demonstrated. The high expression of ASPH, even in early-stage tumors and CIS lesions, suggests that ASPH should be further examined as an early detection marker of HNSCC.

The uniformly high expression of ASPH prevented a meaningful study of the impact of ASPH on patient survival, reported in other malignancies [5,10,11]. However, the expression of ASPH in the primary tumors of HNSCC increases with the grade of the tumor. Poorly differentiated tumors express the most ASPH, and these tumors tend to be most aggressive. The intensity and distribution of ASPH in the stroma of the tumors is also elevated compared to the adjacent normal tissues. The distribution of the staining in the stromal cells is quite uniform, and the stromal cells contribute significantly to the content of ASPH in the HNSCC tissues. CAFs represent an important stromal cell type, and the CAF cells are clearly expressing ASPH, in line with prior single-cell RNA sequencing studies. Thus, the inhibitors of ASPH could be affecting the tumors by acting on both the tumor cells and on the CAF.

Given the high expression of ASPH in HNSCC tumors, we evaluated if ASPH inhibitors affect cancer cell migration and invasion in line with previous studies showing that ASPH regulates cellular migration, invasion, epithelial–mesenchymal transition, apoptosis, and stemness in other cancers [1,2,5,21]. We selected for our study MO-I-1151, a well-tested inhibitor of the β-hydroxylase activity of ASPH, with anti-tumor effects in other malignancies. MO-I-1151 reduced the invasion of colorectal cancer cells [22], reduced in vivo growth of cholangiocarcinoma tumor models [3], altered cellular senescence, and inhibited hepatocellular carcinoma growth and progression in pre-clinical in vivo models [1]. We tested the MO-I-1151 in scratch-migration assay, transwell chamber assay, and a spheroid co-culture cancer cell invasion model combining SCC35 cancer cells with a CAF (CAF 61137) derived from a primary head and neck tumor. The co-culture spheroids generated from these lines create a suitable invasive microenvironment that mimics the local microenvironment of HNSCC tumors, typically rich in stroma. MO-I-1151 clearly inhibits the CAF 61137-supported invasion of cancer cells into Matrigel in our co-culture spheroid model. The inhibitor significantly reduces the area of the invading spheroids and their inverse circularity, a measure of invasive extensions at the periphery of the spheroids. MO-I-1151 also showed a significant effect on inhibiting the migration of SCC35 cells over a period of 24 to 48 h in a wound closure assay and transwell migration chambers. This indicates that the ASPH inhibitor MO-I-1151 arrests the migration and invasion of SCC35 cells, whether they are in a monolayer culture or in a 3D spheroid model, where CAF 61137 stimulates their invasion into the extracellular matrix (ECM). The results suggest that inhibition of ASPH with MO-I-1151 could be an effective anti-invasive strategy and should be further tested using in vivo models.

We established in a recent study that HfFucCS, a marine fucosylated chondroitin sulfate isolated from the sea cucumber species *H. floridana*, previously studied as an anti-SARS-CoV-2 and anticoagulant agent [46], blocks the activity of heparan 6-O-endosulfatase SULF2 and inhibits the invasion of HNSCC cancer cells into the ECM [34]. The heparan 6-O-endosulfatases are upregulated in HNSCC tumors and are a prognostic marker of poor HNSCC outcomes [47–49]. SCC35 is an HNSCC-derived cancer cell line with high invasive potential and high expression of SULF2 [34,50]. SULF2 knockout in SCC35 significantly reduced the invasion of cells in vitro, which was also achieved by treating wild-type cells with HfFucCS [34]. Thus, in this study, we wanted to test the combination of the HfFucCS and MO-I-1151 inhibitors as an efficient strategy to block cancer cell invasion. We found that the combination of the two compounds has an additive effect on the CAF-supported spheroid invasion and on cellular migration assays. This is a first report of the effect of MO-I-1151 on inhibiting ASPH in HNSCC cancer cells and arresting cellular migration and invasion. Testing of the inhibitors in additional HNSCC cell lines, including HPV-positive cell lines, will be needed to confirm the results. In addition, we observe an additive effect when combining MO-I-1151 with HfFucCS, an effective inhibitor of HNSCC cell invasion. Further mechanistic studies are needed to determine how the inhibition of these enzymes in cancer cells and/or the CAF affects their communication and the invasion into Matrigel. It will also be important to evaluate how these cytostatic modulators of invasiveness affect tumor growth in combination with standard cytotoxic drugs and treatments as neo-adjuvant therapies.

4. Materials and Methods

4.1. Materials

Cell culture media (DMEM/F12, IMDM) and Growth Factor Reduced Matrigel were from Corning Inc., Corning, NY, USA. Hydrocortisone was from MilliporeSigma, Burlington, MA, USA. Cell culture supplements were from Gibco™, Billings, MT, USA. Cell culture flasks and dishes were from Nunc, Thermo Fisher Scientific, Rochester, NY, USA; 96-well round-bottom low attachment plates were from Corning, Corning, NY, USA. Monoclonal anti-ASPH-antibody was from Santa Cruz (SC271391). DAB chromagen was from Agilent (Dako #K3468), and Hematoxylin was from Thermo Fischer, Memphis, TN, USA (Harris Modified Hematoxylin).

4.2. Study Subjects and Samples

Participants were enrolled between 1995 and 2020 in collaboration with clinicians at the Department of Otolaryngology—Head and Neck Surgery at Georgetown University Hospital under a protocol approved by the Georgetown University Institutional Review

Board. Patients with newly diagnosed HNSCC primarily from the oral cavity (n = 155) undergoing surgical resection were selected for our study. HNSCC diagnosis was based on medical examination and was confirmed by histopathological evaluation of the tissues. Cancer classification was based on the 7th Edition of the American Joint Committee on Cancer Staging manual. An overview of the demographic and clinical features of the cohort is described in Table 1. The majority of the patients had early stage tumors (stage 1 and 2), with 36% (T1), 33% (T2), 8% (T3), and 22% (T4), and 48% of the patients had positive lymph nodes (LN+) with an N-stage distribution of N1 (n = 28), N2 (n = 42), and N3 (n = 4). The tumor grades of the patients were classified as well- (w = 10%), moderately (m = 59%), and poorly (*p* = 31%) differentiated. Disease recurrence was reported in 41% of the patients. The median follow-up time of the patients was 42 months (range: 0.2 to 286 months), with the outcomes of alive (n = 97), deceased (n = 56), and unknown (n = 2). The population distribution of patients reflects the demographics seen at the Georgetown University Hospital, with median age of the population of 62 years (range: 24–93) years.

Table 1. Demographic and clinicopathological characteristics of the study population.

Variable	Parameter	No. of Cases	Percentage %
Gender	Male	92	59
	Female	63	41
Age	<60 y	66	43
	≥60 y	86	56
	Unknown	3	2
Race	CA	115	74
	AA	11	7
	Asian	2	1
	Other	11	7
	Unknown	16	10
T-Stage	Early (T1, T2)	107	69
	Late (T3, T4)	46	30
	Unknown	2	1
Node	N+	74	48
	N-	79	51
	Unknown	2	1
M-Stage	M+	1	1
	M-	152	98
	Unknown	2	1
Disease Recurrence	Yes	64	41
	No	72	47
	Unknown	19	12
Margin	Positive	18	11
	Negative	69	45
	Close	58	37
	Unknown	10	7
Grade	w	15	10
	m	92	59
	p	48	31

Gene expression and proteomic data for ASPH were analyzed in the TCGA (Cancer Genome Atlas Consortium) and CPTAC (Clinical Proteomic Tumor Analysis Consortium) HNSCC datasets as described previously [48,49]. Briefly, 43 HNSCC patients with RNA-seq data available for both tumor and adjacent benign tissues were analyzed for ASPH differential expression, and log-transformed fold-changes (log_2FC) were computed by difference of log_2(counts+1) of RNA-seq data. The proteomic data and clinical information of 64 HNSCC patients, including matched tumor and adjacent non-tumor tissue pairs, were obtained from CPTAC. The quantification of protein abundance was conducted using the CPTAC Common Data Analysis Pipeline, which determined the log2 ratio of individual proteins to an internal control using only peptides not shared between quantified proteins.

4.3. Tissue Microarray and Histological Sections

Tissue microarray and histological sections were generated by the Lombardi Comprehensive Cancer Center's Histopathology and Tissue Shared Resource, Georgetown University Medical Center, using a T-sue Microarray mold kit from Electron Microscopy Sciences [51]. Briefly, tissue cores were taken from paraffin-embedded blocks using a 1.5 mm punch tip from Electron Microscopy Sciences and manually inserted into the recipient paraffin blocks. Sections of 5 μm were cut from the array using a Leica tissue microtome and analyzed by immunohistochemistry (IHC). We evaluated 155 patient cases spread across 6 TMAs in 314 cores. The array distribution scored for ASPH expression was as follows: Primary lesion (92 cases, 144 cores); Primary + LN (15 cases, 46 cores); Primary + Adjacent normal (31 cases, 90 cores); Primary + Normal + LN (2 cases, 8 cores); Primary + CIS (Carcinoma in situ) (2 cases, 5 cores); CIS (1 case, 1 core); LN (10 cases, 18 cores); Normal tissue (2 cases, 2 cores).

4.4. Immunohistochemistry and Pathology Scoring

TMAs of patient tissue samples were stained using anti-ASPH monoclonal antibody (1/100 dilution) and the DAB chromogen kit following the manufacturer's instructions. Slides were counterstained with Hematoxylin (Fisher, Harris Modified Hematoxylin), blued in 1% ammonium hydroxide, dehydrated, and mounted with Acrymount. Consecutive sections with the primary antibody omitted were used as negative controls. All TMAs were scanned with the Aperio image scope (Aperio GT450) and scored by a pathologist. Staining intensities were classified as 0 = negative, 1 = weak, 2 = moderate, and 3 = strong. The proportion of positive tumor epithelial cells was assessed, and the staining distribution was classified as 0 = no cells, 1 = 1–25% of cells, 2 = 26–50% of cells, and 3 > 50% of cells. Positive fibroblast cells in the stromal compartments were evaluated separately using the same scoring criteria.

4.5. Cell Culture

The SCC35 cell line, a human squamous cell carcinoma cell line derived from a hypopharyngeal tumor, was kindly provided by Prof. Vicente Notario, Georgetown University. Primary head and neck fibroblast CAF 61137 was derived at the Princess Margaret Cancer Centre, Toronto, CA, from a post-surgery tumor sample of a patient (male/59 years of age) before radio/chemotherapy, in line with an approved Research Ethics Board protocol. The tumor site was the tongue, and the tumor was categorized as SCC, stage IVA, T3/N2b/M0. SCC35 cells were grown in DMEM/F12 supplemented with 400 ng/mL hydrocortisone. CAF 61137 cells were grown in IMDM. The cells used in this study were in passage <10 and in log-phase growth. All media were supplemented with 10% fetal bovine serum, non-essential amino acids, and 1 mM sodium pyruvate and were grown in a humidified 5% CO_2 atmosphere. Cells were subcultured at a 1:3 ratio at 90 to 100% confluence.

4.6. Wound Healing

SCC35 cells were cultured in 96-well plates (1×10^4 cells/well) to form a confluent monolayer. After 24 h of incubation, scratches were created by scraping the monolayer cells using a 96-well pin block in Woundmaker™ (Essen Bioscience, Sartorius, Göttingen, Germany). Subsequently, the cells were gently washed with serum-free medium to remove dislodged cells. The cells were treated with two doses of MO-1151 (5 μM, 25 μM) to inhibit invasion either alone or in combination with HfFucCS (10 μg/mL) and observed at defined time intervals (0 h, 24 h and 48 h) after scraping. The migration of cells was analyzed by the area of the wound and quantified as follows: percentage wound closure = (area of initial wound at time t0—area of wound at time t1)/area of initial wound at time t0 × 100.

4.7. Spheroid Invasion

Spheroids were generated as described previously [34]. For co-culture spheroids, an equal number of SCC35 and CAF 61137 cells (3×10^4 cells/mL for each cell type) were mixed and seeded in ultra-low attachment 96-well round bottom plates, and the cells were grown for 1 day in an incubator followed by Matrigel embedding. Spheroids were cultured in an incubator and imaged using an Olympus IX71 inverted microscope (Olympus, Tokyo, Japan). Digital images of spheroids were analyzed using Particle Analysis in ImageJ software version 1.51p (NIH, Bethesda, MD, USA) to obtain the values of area and inverse circularity. The area for spheroids is the count of pixels comprising the object. Inverse Circularity = $\frac{1}{Circularity}$, where circularity ($Circularity = 4\pi \times [area]/[perimeter]$) is a property of the spheroid calculated using automated image analysis with the analyze particles tool. A circularity value of 1 (maximum) arbitrary unit (A.U.) indicates that the spheroid is perfectly circular; decreasing values indicates deviations from a circular spheroid. All our data processing and analysis parameters are identical for samples in the same dataset.

4.8. Inhibitor Treatment of Spheroids

Co-culture spheroids were treated with two doses of MO-I-1151 ASPH inhibitor (5 μM, 25 μM) with/without HfFucCS (10 μg/mL) and compared to non-treated spheroids (NTC). The inhibitors were added to the complete medium prior to treatment, and the solutions were gently added to the Matrigel spheroids. The plates were incubated for 5 days and imaged on the day of harvest using a phase-contrast microscope (Olympus). The results were quantified using Image J analysis version 1.51p (NIH, Bethesda, MD, USA), as described above.

4.9. Statistical Analysis

GraphPad Prism version 10 for Windows (GraphPad Software, La Jolla, CA, USA) was used for statistical analysis, with summary outcomes represented as mean values ± SD. Differential expression between the tumor and paired normal tissues was assessed through a paired *t*-test. IHC results were compared among normal, tumor, and lymph node samples using pairwise *t*-tests. Wound closure, area, and inverse circularity of the spheroids at day 5 were evaluated using one-way analysis of variance (ANOVA) with post hoc Tukey's test. A two-sided *p* value < 0.05 was considered statistically significant unless specified otherwise.

5. Conclusions

Our study documents the upregulation of Aspartate β-Hydroxylase (ASPH) in head and neck squamous cell carcinoma (HNSCC) based on immunohistochemical examination of human tumor tissues. At the same time, we document the inhibition of cancer cell invasion in a spheroid co-culture model with cancer-associated fibroblasts (CAFs). We used the model to establish the ability of the ASPH inhibitor MO-I-1151 to inhibit the invasion of HNSCC tumor cells into Matrigel, and we showed an additive inhibitory effect of its combination with a recently described heparan 6-O-endosulfatase (SULF2) inhibitor, HfFucCS.

Supplementary Materials: The following supporting information can be downloaded at: https://www.mdpi.com/article/10.3390/ijms25094998/s1.

Author Contributions: Study design: J.B. and R.G.; patient sample procurement and primary cell lines: L.A. and B.D.; inhibitor synthesis and isolation: M.O. and V.H.P.; cell model experiments: P.M. and X.Z.; immunohistochemical analysis: S.G. and X.Z.; data analysis: J.A., P.M., R.A. and L.Z.; writing—original draft: P.M. and X.Z.; review and editing: J.A., L.A., J.B., B.D., R.G., M.O., V.H.P., R.A. and L.Z. All authors have read and agreed to the published version of the manuscript.

Funding: This work was supported in part by the National Institutes of Health grants R01CA238455 and S10OD023557 to RG and CCSG Grant P30 CA51008. The content is solely the responsibility of the authors and does not necessarily represent the official views of the National Institutes of Health.

Institutional Review Board Statement: The study was conducted in accordance with the Declaration of Helsinki, and approved by the Institutional Review Board of Georgetown University-Medstar Health (protocol 2001-277) and by the Research Ethics Board of the Princess Margaret Cancer Centre, Toronto, CA (protocol 08-0888-T) for studies involving humans.

Informed Consent Statement: Informed consent was obtained from all subjects involved in the study.

Data Availability Statement: The data presented in this study are available in this article.

Acknowledgments: We thank the Histopathology and Tissue Shared Resource, Lombardi Comprehensive Cancer Center, Georgetown University, Washington, D.C., for the immunohistochemical analysis.

Conflicts of Interest: The authors declare no conflicts of interest.

Abbreviations

ASPH: aspartate β-hydroxylase; AA, African American; CA, Caucasian American; CAF, cancer-associated fibroblast; CIS, carcinoma in situ; ECM, extracellular matrix; HNSCC, head and neck squamous cell carcinoma; HfFucCS, fucosylated chondroitin sulfate isolated from the sea cucumber *Holothuria floridana*; IHC, immunohistochemistry; m, moderately differentiated tumor; OSCC, oral squamous cell carcinoma; p, poorly differentiated tumor; SMI, small molecule inhibitor; SULF2, heparan 6-O-endosulfatase 2; TME, tumor microenvironment; TMA, tumor microarray; w, well-differentiated tumor.

References

1. Aihara, A.; Huang, C.-K.; Olsen, M.J.; Lin, Q.; Chung, W.; Tang, Q.; Dong, X.; Wands, J.R. A cell-surface β-hydroxylase is a biomarker and therapeutic target for hepatocellular carcinoma. *J. Hepatol.* **2014**, *60*, 1302–1313. [CrossRef] [PubMed]
2. Dong, X.; Lin, Q.; Aihara, A.; Li, Y.; Huang, C.-K.; Chung, W.; Tang, Q.; Chen, X.; Carlson, R.; Nadolny, C.; et al. Aspartate β-hydroxylase expression promotes a malignant pancreatic cellular phenotype. *Oncotarget* **2014**, *6*, 1231–1248. [CrossRef]
3. Huang, C.-K.; Iwagami, Y.; Aihara, A.; Chung, W.; de la Monte, S.; Thomas, J.-M.; Olsen, M.; Carlson, R.; Yu, T.; Dong, X.; et al. Anti-Tumor Effects of Second Generation β-Hydroxylase Inhibitors on Cholangiocarcinoma Development and Progression. *PLoS ONE* **2016**, *11*, e0150336. [CrossRef] [PubMed]
4. Ince, N.; de la Monte, S.M.; Wands, J.R. Overexpression of Human Aspartyl (Asparaginyl) β-Hydroxylase Is Associated with Malignant Transformation1. *Cancer Res.* **2000**, *60*, 1261–1266. [PubMed]
5. Kanwal, M.; Smahel, M.; Olsen, M.; Smahelova, J.; Tachezy, R. Aspartate β-hydroxylase as a target for cancer therapy. *J. Exp. Clin Cancer Res.* **2020**, *39*, 163. [CrossRef] [PubMed]
6. Lavaissiere, L.; Jia, S.; Nishiyama, M.; de la Monte, S.; Stern, A.M.; Wands, J.R.; Friedman, P.A. Overexpression of human aspartyl(asparaginyl)beta-hydroxylase in hepatocellular carcinoma and cholangiocarcinoma. *J. Clin. Investig.* **1996**, *98*, 1313–1323 [CrossRef]
7. Lin, Q.; Chen, X.; Meng, F.; Ogawa, K.; Li, M.; Song, R.; Zhang, S.; Zhang, Z.; Kong, X.; Xu, Q.; et al. ASPH-notch Axis guided Exosomal delivery of Prometastatic Secretome renders breast Cancer multi-organ metastasis. *Mol. Cancer* **2019**, *18*, 156. [CrossRef]
8. Silbermann, E.; Moskal, P.; Bowling, N.; Tong, M.; de la Monte, S.M. Role of aspartyl-(asparaginyl)-β-hydroxylase mediated notch signaling in cerebellar development and function. *Behav. Brain Funct.* **2010**, *6*, 68. [CrossRef]
9. Yang, H.; Song, K.; Xue, T.; Xue, X.-P.; Huyan, T.; Wang, W.; Wang, H. The distribution and expression profiles of human Aspartyl/Asparaginyl beta-hydroxylase in tumor cell lines and human tissues. *Oncol. Rep.* **2010**, *24*, 1257–1264. [CrossRef]
10. Yang, H.; Li, J.; Tang, R.; Li, J.; Liu, Y.; Ye, L.; Shao, D.; Jin, M.; Huang, Q.; Shi, J. The aspartyl asparaginyl beta-hydroxylase in carcinomas. *Front. Biosci.* **2015**, *20*, 902–909. [CrossRef]

11. Zheng, W.; Wang, X.; Hu, J.; Bai, B.; Zhu, H. Diverse molecular functions of aspartate β-hydroxylase in cancer (Review). *Oncol. Rep.* **2020**, *44*, 2364–2372. [CrossRef] [PubMed]

12. Zou, Q.; Hou, Y.; Wang, H.; Wang, K.; Xing, X.; Xia, Y.; Wan, X.; Li, J.; Jiao, B.; Liu, J.; et al. Hydroxylase Activity of ASPH Promotes Hepatocellular Carcinoma Metastasis Through Epithelial-to-Mesenchymal Transition Pathway. *EBioMedicine* **2018**, *31*, 287–298. [CrossRef] [PubMed]

13. Brewitz, L.; Tumber, A.; Pfeffer, I.; McDonough, M.A.; Schofield, C.J. Aspartate/asparagine-β-hydroxylase: A high-throughput mass spectrometric assay for discovery of small molecule inhibitors. *Sci. Rep.* **2020**, *10*, 8650. [CrossRef] [PubMed]

14. Bai, X.; Zhou, Y.; Lin, Q.; Huang, C.-K.; Zhang, S.; Carlson, R.I.; Ghanbari, H.; Sun, B.; Wands, J.R.; Dong, X. Bio-nanoparticle based therapeutic vaccine induces immunogenic response against triple negative breast cancer. *Am. J. Cancer Res.* **2021**, *11*, 4141–4174. [PubMed]

15. Nagaoka, K.; Bai, X.; Ogawa, K.; Dong, X.; Zhang, S.; Zhou, Y.; Carlson, R.I.; Jiang, Z.-G.; Fuller, S.; Lebowitz, M.S.; et al. Anti-tumor activity of antibody drug conjugate targeting aspartate-β-hydroxylase in pancreatic ductal adenocarcinoma. *Cancer Lett.* **2019**, *449*, 87–98. [CrossRef] [PubMed]

16. Ragothaman, M.; Yoo, S.Y. Engineered Phage-Based Cancer Vaccines: Current Advances and Future Directions. *Vaccines* **2023**, *11*, 919. [CrossRef] [PubMed]

17. Zhou, Y.; Li, C.; Shi, G.; Xu, X.; Luo, X.; Zhang, Y.; Fu, J.; Chen, L.; Zeng, A. Dendritic cell-based vaccine targeting aspartate-β-hydroxylas represents a promising therapeutic strategy for HCC. *Immunotherapy* **2019**, *11*, 1399–1407. [CrossRef]

18. Dahn, H.; Lawendel, J.S.; Hoegger, E.F.; Fischer, R.; Schenker, E. Über eine neue Herstellung aromatisch substituierter Reduktone. *Cell. Mol. Life Sci.* **1954**, *10*, 245–246. [CrossRef]

19. Zheng, G.; Cox, T.; Tribbey, L.; Wang, G.Z.; Iacoban, P.; Booher, M.E.; Gabriel, G.J.; Zhou, L.; Bae, N.; Rowles, J.; et al. Synthesis of a FTO Inhibitor with Anticonvulsant Activity. *ACS Chem. Neurosci.* **2014**, *5*, 658–665. [CrossRef]

20. Ogawa, K.; Lin, Q.; Li, L.; Bai, X.; Chen, X.; Chen, H.; Kong, R.; Wang, Y.; Zhu, H.; He, F.; et al. Aspartate β-hydroxylase promotes pancreatic ductal adenocarcinoma metastasis through activation of SRC signaling pathway. *J. Hematol. Oncol.* **2019**, *12*, 144. [CrossRef]

21. Huang, C.-K.; Iwagami, Y.; Zou, J.; Casulli, S.; Lu, S.; Nagaoka, K.; Ji, C.; Ogawa, K.; Cao, K.Y.; Gao, J.-S.; et al. Aspartate beta-hydroxylase promotes cholangiocarcinoma progression by modulating RB1 phosphorylation. *Cancer Lett.* **2018**, *429*, 1–10. [CrossRef] [PubMed]

22. Benelli, R.; Costa, D.; Mastracci, L.; Grillo, F.; Olsen, M.J.; Barboro, P.; Poggi, A.; Ferrari, N. Aspartate-β-Hydroxylase: A Promising Target to Limit the Local Invasiveness of Colorectal Cancer. *Cancers* **2020**, *12*, 971. [CrossRef] [PubMed]

23. Ogawa, K.; Lin, Q.; Li, L.; Bai, X.; Chen, X.; Chen, H.; Kong, R.; Wang, Y.; Zhu, H.; He, F.; et al. Prometastatic secretome trafficking via exosomes initiates pancreatic cancer pulmonary metastasis. *Cancer Lett.* **2020**, *481*, 63–75. [CrossRef] [PubMed]

24. Swarnali, A.; Matrisian, L.; Welch, D.R.; Massagué, J. 18—Invasion and Metastasis. In *The Mo-Lecular Basis of Cancer*, 4th ed.; Mendelsohn, J., Gray, J.W., Howley, P.M., Israel, M.A., Thompson, C.B., Eds.; W.B. Saunders: Philadelphia, PA, USA, 2015; pp. 269–284.e2. [CrossRef]

25. Friedl, P.; Wolf, K. Tumour-cell invasion and migration: Diversity and escape mechanisms. *Nat. Rev. Cancer* **2003**, *3*, 362–374. [CrossRef]

26. Asif, P.J.; Longobardi, C.; Hahne, M.; Medema, J.P. The Role of Cancer-Associated Fibroblasts in Cancer Invasion and Metastasis. *Cancers* **2021**, *13*, 4720. [CrossRef] [PubMed]

27. Attieh, Y.; Vignjevic, D.M. The hallmarks of CAFs in cancer invasion. *Eur. J. Cell Biol.* **2016**, *95*, 493–502. [CrossRef]

28. Sun, H.; Wang, X.; Wang, X.; Xu, M.; Sheng, W. The role of cancer-associated fibroblasts in tumorigenesis of gastric cancer. *Cell Death Dis.* **2022**, *13*, 874. [CrossRef] [PubMed]

29. Bienkowska, K.J.; Hanley, C.J.; Thomas, G.J. Cancer-Associated Fibroblasts in Oral Cancer: A Current Perspective on Function and Potential for Therapeutic Targeting. *Front. Oral Health* **2021**, *2*, 686337. [CrossRef] [PubMed]

30. Marsh, D.; Suchak, K.; Moutasim, K.A.; Vallath, S.; Hopper, C.; Jerjes, W.; Upile, T.; Kalavrezos, N.; Violette, S.M.; Weinreb, P.H.; et al. Stromal features are predictive of disease mortality in oral cancer patients. *J. Pathol.* **2011**, *223*, 470–481. [CrossRef]

31. Puram, S.V.; Tirosh, I.; Parikh, A.S.; Patel, A.P.; Yizhak, K.; Gillespie, S.; Rodman, C.; Luo, C.L.; Mroz, E.A.; Emerick, K.S.; et al. Single-Cell Transcriptomic Analysis of Primary and Metastatic Tumor Ecosystems in Head and Neck Cancer. *Cell* **2017**, *171*, 1611–1624.e24. [CrossRef]

32. Sahai, E.; Astsaturov, I.; Cukierman, E.; DeNardo, D.G.; Egeblad, M.; Evans, R.M.; Fearon, D.; Greten, F.R.; Hingorani, S.R.; Hunter, T.; et al. A framework for advancing our understanding of cancer-associated fibroblasts. *Nat. Rev. Cancer* **2020**, *20*, 174–186. [CrossRef] [PubMed]

33. Patel, A.K.; Vipparthi, K.; Thatikonda, V.; Arun, I.; Bhattacharjee, S.; Sharan, R.; Arun, P.; Singh, S. A subtype of cancer-associated fibroblasts with lower expression of alpha-smooth muscle actin suppresses stemness through BMP4 in oral carcinoma. *Oncogenesis* **2018**, *7*, 78. [CrossRef] [PubMed]

34. Mukherjee, P.; Zhou, X.; Benicky, J.; Panigrahi, A.; Aljuhani, R.; Liu, J.; Ailles, L.; Pomin, V.H.; Wang, Z.; Goldman, R. Heparan-6-O-Endosulfatase 2 Promotes Invasiveness of Head and Neck Squamous Carcinoma Cell Lines in Co-Cultures with Cancer-Associated Fibroblasts. *Cancers* **2023**, *15*, 5168. [CrossRef] [PubMed]

35. Chow, L.Q.M. Head and Neck Cancer. *N. Engl. J. Med.* **2020**, *382*, 60–72. [CrossRef] [PubMed]

36. Argiris, A.; Karamouzis, M.V.; Raben, D.; Ferris, R.L. Head and neck cancer. *Lancet* **2008**, *371*, 1695–1709. [CrossRef] [PubMed]

37. Marur, S.; D'Souza, G.; Westra, W.H.; Forastiere, A.A. HPV-associated head and neck cancer: A virus-related cancer epidemic *Lancet Oncol.* **2010**, *11*, 781–789. [CrossRef] [PubMed]

38. Botticelli, A.; Zizzari, I.G.; Scagnoli, S.; Pomati, G.; Strigari, L.; Cirillo, A.; Cerbelli, B.; Di Filippo, A.; Napoletano, C.; Scirocchi, F.; et al. The Role of Soluble LAG3 and Soluble Immune Checkpoints Profile in Advanced Head and Neck Cancer: A Pilot Study. *J. Pers. Med.* **2021**, *11*, 651. [CrossRef] [PubMed]

39. Ferris, R.L.; Blumenschein, G., Jr.; Fayette, J.; Guigay, J.; Colevas, A.D.; Licitra, L.; Harrington, K.; Kasper, S.; Vokes, E.E.; Even, C.; et al. Nivolumab for Recurrent Squamous-Cell Carcinoma of the Head and Neck. *N. Engl. J. Med.* **2016**, *375*, 1856–1867 [CrossRef] [PubMed]

40. Miyauchi, S.; Kim, S.S.; Pang, J.; Gold, K.A.; Gutkind, J.S.; Califano, J.A.; Mell, L.K.; Cohen, E.E.; Sharabi, A.B. Immune Modulation of Head and Neck Squamous Cell Carcinoma and the Tumor Microenvironment by Conventional Therapeutics. *Clin. Cancer Res.* **2019**, *25*, 4211–4223. [CrossRef]

41. Shin, D.M.; Khuri, F.R. Advances in the management of recurrent or metastatic squamous cell carcinoma of the head and neck *Head Neck* **2011**, *35*, 443–453. [CrossRef]

42. Alsahafi, E.; Begg, K.; Amelio, I.; Raulf, N.; Lucarelli, P.; Sauter, T.; Tavassoli, M. Clinical update on head and neck cancer: Molecular biology and ongoing challenges. *Cell Death Dis.* **2019**, *10*, 540. [CrossRef] [PubMed]

43. Carlisle, J.W.; Steuer, C.E.; Owonikoko, T.K.; Saba, N.F. An update on the immune landscape in lung and head and neck cancers *CA A Cancer J. Clin.* **2020**, *70*, 505–517. [CrossRef] [PubMed]

44. Cohen, E.E.W.; Bell, R.B.; Bifulco, C.B.; Burtness, B.; Gillison, M.L.; Harrington, K.J.; Le, Q.-T.; Lee, N.Y.; Leidner, R.; Lewis, R.L.; et al. The Society for Immunotherapy of Cancer consensus statement on immunotherapy for the treatment of squamous cell carcinoma of the head and neck (HNSCC). J. Immunother. *Cancer* **2019**, *7*, 184. [CrossRef] [PubMed]

45. Ho, A.S.; Kraus, D.H.; Ganly, I.; Lee, N.Y.; Shah, J.P.; Morris, L.G.T. Decision making in the management of recurrent head and neck cancer. *Head Neck* **2013**, *36*, 144–151. [CrossRef] [PubMed]

46. Farrag, M.; Dwivedi, R.; Sharma, P.; Kumar, D.; Tandon, R.; Pomin, V.H. Structural requirements of Holothuria floridana fucosylated chondroitin sulfate oligosaccharides in anti-SARS-CoV-2 and anticoagulant activities. *PLoS ONE* **2023**, *18*, e0285539 [CrossRef] [PubMed]

47. Flowers, S.A.; Zhou, X.; Wu, J.; Wang, Y.; Makambi, K.; Kallakury, B.V.; Singer, M.S.; Rosen, S.D.; Davidson, B.; Goldman, R Expression of the extracellular sulfatase SULF2 is associated with squamous cell carcinoma of the head and neck. *Oncotarget* **2016**, *7*, 43177–43187. [CrossRef] [PubMed]

48. Yang, Y.; Ahn, J.; Edwards, N.J.; Benicky, J.; Rozeboom, A.M.; Davidson, B.; Karamboulas, C.; Nixon, K.C.J.; Ailles, L.; Goldman, R. Extracellular Heparan 6-O-Endosulfatases SULF1 and SULF2 in Head and Neck Squamous Cell Carcinoma and Other Malignancies. *Cancers* **2022**, *14*, 5553. [CrossRef] [PubMed]

49. Yang, Y.; Ahn, J.; Raghunathan, R.; Kallakury, B.V.; Davidson, B.; Kennedy, Z.B.; Zaia, J.; Goldman, R. Expression of the Extracellular Sulfatase SULF2 Affects Survival of Head and Neck Squamous Cell Carcinoma Patients. *Front. Oncol.* **2021**, *10*, 582827. [CrossRef] [PubMed]

50. Benicky, J.; Sanda, M.; Panigrahi, A.; Liu, J.; Wang, Z.; Pagadala, V.; Su, G.; Goldman, R. A 6-O-endosulfatase activity assay based on synthetic heparan sulfate oligomers. *Glycobiology* **2023**, *33*, 384–395. [CrossRef]

51. Ihemelandu, C.; Naeem, A.; Parasido, E.; Berry, D.; Chaldekas, K.; Harris, B.T.; Rodriguez, O.; Albanese, C. Clinicopathologic and prognostic significance of LGR5, a cancer stem cell marker in patients with colorectal cancer. *Color. Cancer* **2019**, *8*, CRC11. [CrossRef]

Article

Impact of T Cell Exhaustion and Stroma Senescence on Tumor Cell Biology and Clinical Outcome of Head and Neck Squamous Cell Carcinomas

Lukas A. Brust [1,*], Meike Vorschel [1], Sandrina Körner [1], Moritz Knebel [1], Jan Philipp Kühn [1], Silke Wemmert [1], Sigrun Smola [2], Mathias Wagner [3], Bernhard Schick [1] and Maximilian Linxweiler [1]

[1] Department of Otorhinolaryngology, Head and Neck Surgery, Saarland University, 66421 Homburg, Germany; vorschelm@gmail.com (M.V.); sandrina.koerner@uks.eu (S.K.); moritz.knebel@uks.eu (M.K.); jan.kuehn@uks.eu (J.P.K.); silke.wemmert@uks.eu (S.W.); bernhard.schick@icloud.com (B.S.); maximilian.linxweiler@uks.eu (M.L.)

[2] Institute of Virology, Saarland University, 66421 Homburg, Germany; sigrun.smola@uks.eu

[3] Department of General and Surgical Pathology, Saarland University, 66421 Homburg, Germany; mathias.wagner@uks.eu

* Correspondence: lukas.brust@uks.eu

Abstract: Head and neck squamous cell carcinomas (HNSCC) have an overall poor prognosis, especially in locally advanced and metastatic stages. In most cases, multimodal therapeutic approaches are required and show only limited cure rates with a high risk of tumor recurrence. Anti-PD-1 antibody treatment was recently approved for recurrent and metastatic cases but to date, response rates remain lower than 25%. Therefore, the investigation of the immunological tumor microenvironment and the identification of novel immunotherapeutic targets in HNSCC is of paramount importance. In our study, we used tissue samples of n = 116 HNSCC patients for the immunohistochemical detection of the intratumoral and peritumoral expression of T cell exhaustion markers (PD-1, LAG-3, TIM-3) on tumor infiltration leukocytes (TIL), as well as the expression level of stromal senescence markers (IL-8, MMP-3) on tumor-associated fibroblasts. The clinical parameter of the vitamin D serum status as well as the histopathological HPV infection status of the tumor was correlated with the expression rates of the biomarkers and the overall patient survival. An increased peritumoral and intratumoral expression of the biomarkers PD-1 and TIM-3 significantly correlated with improved overall patient survival. A high peritumoral expression of LAG-3 correlated with better overall survival. A positive HPV tumor status correlated with a significantly elevated expression of PD-1 and TIM-3. Biomarkers of stromal senescence showed no influence on the patient outcome. However, the vitamin D serum status showed no influence on patient outcomes or biomarker expressions. Our study identified PD-1, LAG-3, and TIM-3 as promising targets of a therapeutic strategy targeting the tumor microenvironment in HNSCC, particularly among HPV-positive patients, where a higher expression of these checkpoints correlated with an improved overall survival. These findings support the potential of antibodies targeting these immune checkpoints to enhance treatment efficacy, especially in the context of bispecific targeting.

Keywords: HNSCC; biomarker; TME; HPV; PD-1; LAG-3; TIM-3; IL-8; MMP-3

1. Introduction

Head and neck squamous cell carcinomas (HNSCC) are the sixth-most common cancer type globally, with 895,000 new cases and 457,000 deaths in 2022 [1]. Major risk factors include chronic nicotine and alcohol consumption, with a growing contribution from high-risk HPV infection, particularly in oropharyngeal cancers [2]. Locally advanced and metastatic HNSCC stages maintain a poor prognosis, with five-year survival rates of approximately 60% [3].

The majority of HNSCC patients require multimodal treatment, including surgery, radiation, and chemotherapy. However, tumor recurrence is common [4] with highly limited therapeutic approaches for this challenging clinical setting. Recently, two anti-PD-1 immune checkpoint inhibitors (ICIs), pembrolizumab and nivolumab, have been approved for the treatment of recurrent and metastatic HNSCC [5] for first-line [6] and second-line treatment either as a monotherapy or in combination with platinum-based chemotherapy [7,8]. However, response rates remain at a low level, around 25%, and resistance to immune-checkpoint inhibition is frequently observed over time. The limited clinical success of the currently approved immunotherapeutic strategies underscore the need for further research in head and neck cancer immuno-oncology, particularly as HNSCC is among the most immune-infiltrated human cancers [9,10]. In the context of possible adjuvants, vitamin D has been shown in preclinical and retrospective clinical studies to be a positive prognostic marker, as well as a possible agent for increasing response rates due to its immunomodulatory properties [11,12]. Furthermore, the molecular processes of T cell exhaustion and stroma senescence as potential mechanisms of evading immune surveillance and an antitumoral immune response came into the spotlight over the past years with, however, only limited data in head and neck cancer so far [13,14].

T cell exhaustion represents a state of T cell dysfunction that arises during chronic infections and cancer development. It is defined by poor T cell effector function, a high expression of inhibitory checkpoint receptors including CTLA-4, PD-1, TIM-3, BTLA, VISTA, and LAG-3 [15], and an altered transcriptional program. LAG-3, a key checkpoint receptor alongside PD-1 and CTLA-4, promotes tumor growth by inhibiting the immune response at high expression levels [15]. Similarly, TIM-3, which is highly expressed in tumor-infiltrating lymphocytes, suppresses anti-tumor immunity through its interaction with Galectin 9 in various cancers [16]. Overall, T cell exhaustion was shown to be associated with the ineffective immunological control of chronic infections and several cancer types [17,18] including melanoma [19], chronic myeloid leukemia [20], ovarian cancer [21], and non-small cell lung cancer [22].

In addition to tumor-infiltrating leucocytes, peritumoral stromal tissue represents the major component of the tumor microenvironment (TME). Peritumoral stroma primarily consists of cancer-associated fibroblasts (CAFs) and an extracellular matrix (ECM) and was shown to have a relevant role in cancer progression through cell–cell and cell–matrix interactions in different cancer types [23]. Cancer development and progression as well as tumor treatment can induce stromal changes and lead to the accumulation of senescent stromal cells that are characterized by the so-called senescence-associated secretory phenotype (SASP). Those senescent stromal cells produce and secret a multitude of small molecules including cytokines, growth factors, and ECM components, which can create an immunosuppressive, inflammatory TME that may promote tumor growth and metastasis [24]. Among those secreted small molecules, matrix metalloproteinases (MMPs) modify the ECM, contributing to premature stroma aging, and are involved in angiogenesis, which promotes cancer cell growth and migration [25]. Another important factor in TME is IL-8, which is a chemokine with various pro-tumorigenic functions within the TME. It promotes tumor cell proliferation and transformation into migratory or mesenchymal phenotypes, angiogenesis, and the recruitment of immunosuppressive cells [26].

Against this background, we investigated the expression of five surrogate markers of T cell exhaustion (PD-1, TIM-3, LAG-3) and stromal senescence (IL-8, MMP-3) in a cohort of $n = 116$ HNSCC patients. By correlating the biomarker expression with the patients' clinical and histopathological data as well as vitamin D status, we aimed to gain new insights into head and neck cancer tumor and immunological biology as well as potentially identify new therapeutic strategies targeting the tumor microenvironment.

2. Results

2.1. Expression of T Cell Exhaustion and Stromal Senescence Markers Correlates with HPV Tumor Status but Not Vitamin D Serum Level

In an initial step, a semiquantitative analysis of intratumoral immunohistochemical staining targeting the five biomarkers PD-1, TIM-3, LAG-3, IL8, and MMP-3 was performed, followed by an assessment of peritumoral staining. The analysis revealed distinct differences in expression levels, ranging from negative or minimal staining (score 0–1) to high levels of staining (score up to 12). A representative illustration is provided in Figure 1, where, on the left, low expression levels are demonstrated by a corresponding low number of positively stained cells and a weak staining intensity. On the right, tumors and the TME exhibit a significantly stronger staining intensity and a higher number of positively stained cells. A subsequent analysis of immune cell infiltration patterns with respective IHC biomarker staining in the peritumoral and intratumoral region is presented in Figure 2, highlighting the variability in the degree of infiltration across the tumor microenvironment. The intratumoral and peritumoral biomarker expression was correlated with the HPV tumor status (positivity defined as p16+ and HPV-DNA+), as well as the 25-OH-vitamin D serum level (VitD low ≤ 10 ng/mL; VitD high > 10 ng/mL).

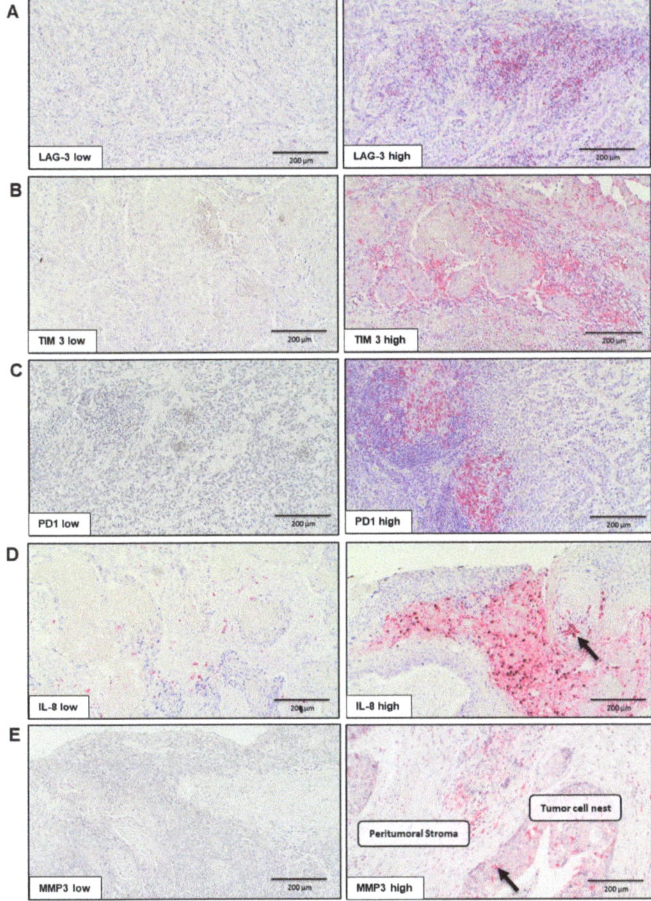

Figure 1. IHC representation of immune markers LAG-3, TIM-3, PD-1, IL-8, and MMP-3 (**A–E**). The images illustrate representative sections of the tumor as well as the tumor microenvironment. On the

left side, a low immune reactive score (IRS) is depicted, characterized by a small number of positively stained cells and a weak staining intensity. On the right, a correspondingly high IRS is shown. In panels D and E, arrows indicate positively stained cells. The intratumoral regions are identified by tumor cell nests, while the peritumoral stroma is shown accordingly. Magnification: 10×.

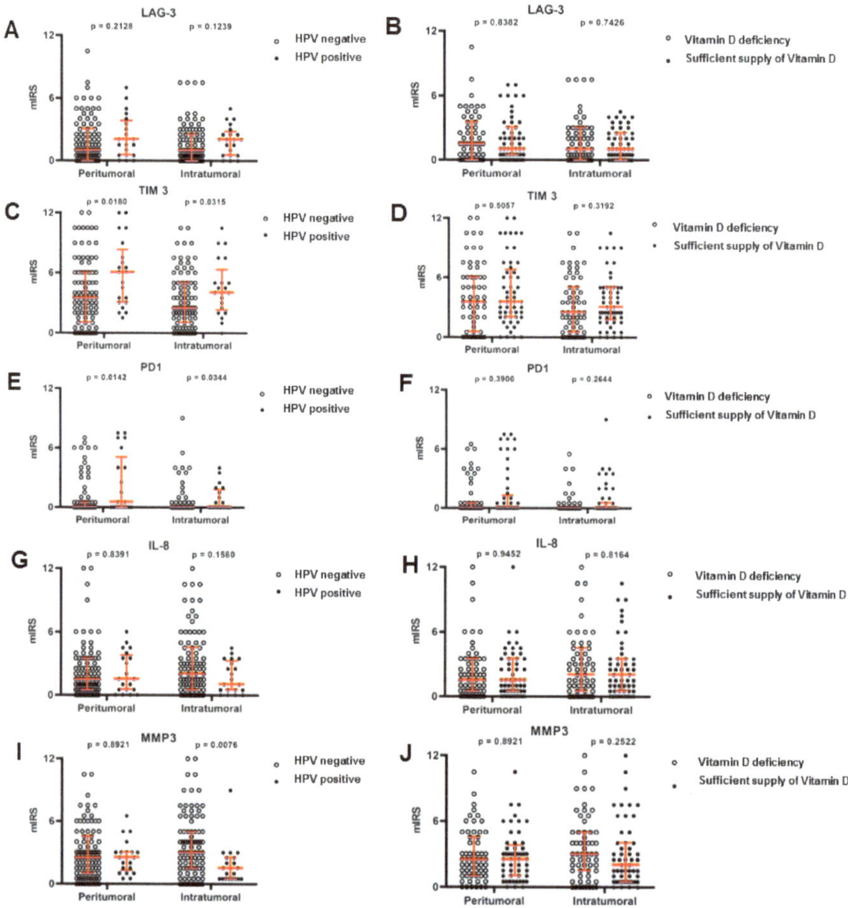

Figure 2. Correlation of the expression of T cell exhaustion and stroma senescence biomarkers with HPV tumor status and vitamin D status. (**A,B**) IRS of LAG-3 depending on HPV and vitamin D status. (**C,D**) IRS of TIM-3 depending on HPV and vitamin D status. (**E,F**) IRS of PD-1 depending on HPV and vitamin D status. (**G,H**) IRS of IL-8 depending on HPV and vitamin D status. (**I,J**) IRS of MMP-3 depending on HPV and vitamin D status. Statistical analysis was performed using the Mann–Whitney U test in all cases. The black dots symbolize one patient each, and the red lines show the median with the interquartile range.

Here, HPV-positive tumors showed a significant increase in both the peritumoral and intratumoral expression of the T cell exhaustion markers PD-1 ($p = 0.0142$, peritumoral median IRS of 0.0 vs. 0.5; $p = 0.0344$, intratumoral median IRS of 0.0 vs. 0.0) and TIM-3 ($p = 0.0180$, peritumoral median IRS of 6.0 vs. 3.5; $p = 0.0315$, intratumoral median IRS of 4.0 vs. 2.5) as shown in Figure 2. The LAG-3 expression demonstrates a clear tendency toward increased levels in association with HPV infection; however, statistical significance was not reached, with $p = 0.21$ for peritumoral and $p = 0.12$ for intratumoral regions (median IRS of 2.0 vs. 1.0 peritumoral and 2.0 vs. 1.0 intratumoral). By contrast, HPV infection resulted in a highly significant suppression (p = 0.0076) of MMP-3 expression on intratumoral TILs (median IRS of 1.5 vs. 3.0), while the expression on peritumoral TILs remained unchanged. No influence of the HPV tumor status on the fibroblast expression of stroma senescence surrogate markers could be observed. Regarding the patients' vitamin D status, none of the analyzed biomarkers showed a significant correlation with vitamin D supply. However, there was a trend toward increased expression levels of PD-1, TIM-3, and LAG-3, and a decreased expression level of MMP-3 in VitD-high patients compared to VitD-low patients. For LAG-3, the intratumoral median IRS was 1.0 in both groups, while the peritumoral median IRS was 1.5 in VitD-low and 1.0 in VitD-high patients. TIM-3 showed intratumoral median IRS values of 2.5 for VitD-low and 3.0 for VitD-high patients, with a consistent peritumoral IRS of 3.5 in both groups. For PD-1, the intratumoral median IRS was 2.0 in VitD-low patients and 3.0 in VitD-high patients, while the peritumoral IRS was 1.5 in both groups. The IL-8 expression showed no variation, with intratumoral and peritumoral IRS values of 2.0 and 1.5, respectively, for both groups. Lastly, MMP-3 had a higher intratumoral IRS in VitD-low patients (3.0) compared to VitD-high patients (2.0), with identical peritumoral IRS values of 2.5 for both groups. No significant differences were observed across all biomarkers analyzed.

2.2. High Levels of PD-1, LAG-3, and TIM-3 Expression Predict Improved Overall Patient Survival

Looking at the potentially prognostic relevance of the analyzed surrogate markers of T cell exhaustion and stroma senescence, we found significant correlations only for the T cell exhaustion markers PD-1, TIM-3, and LAG-3. Here, a high expression of PD-1 (defined by a PD-1 expression above the statistical mean of all samples) correlated with a significantly improved overall survival. PD-1-high patients showed a two-year overall survival of 87% in contrast to 58% in the PD-1-low group. This correlation was significant for PD-1 expression on both peritumoral (p = 0.0101) and intratumoral immune cells (p = 0.0266, Figure 3E,F).

A similar trend was observed for TIM-3: a high peritumoral TIM-3 expression showed a highly significant (p < 0.0001) overall survival benefit compared to low peritumoral TIM-3 expression. The two-year overall survival within these groups was 86% (TIM-3 peritumoral high) vs. 51% (TIM-3 peritumoral low). For TIM-3 expression on intratumoral immune cells, comparable effects were shown. Moreover, a significant survival advantage was observed for highly expressed LAG-3 on peritumoral TILs with a two-year survival of 77% (LAG-3 peritumoral high) vs. 56% (LAG-3 peritumoral low). However, no significant overall survival advantage was observed for a high intratumoral LAG-3 expression (p = 0.34).

By contrast, there was no significant correlation with the patients' overall survival regarding the expression of the stromal senescent markers MMP-3 and IL-8 on intra- and peritumoral fibroblasts (Figure 3G–J).

2.3. Positive HPV Tumor Status but Not Vitamin D Serum Level Predicts Improved Overall Patient Survival

The overall survival of the included HNSCC patients was correlated with the HPV tumor status and vitamin D status. We could show that HPV-positive HNSCCs showed a significantly better prognosis, with an overall survival rate of 85% after 2 years, compared to 55% in HPV-negative cases. For vitamin D, a trend towards improved overall survival in VitD-high patients compared to VitD-low patients was observed, particularly within the first 24 months after diagnosis (Figure 4). However, no statistical significance could be achieved.

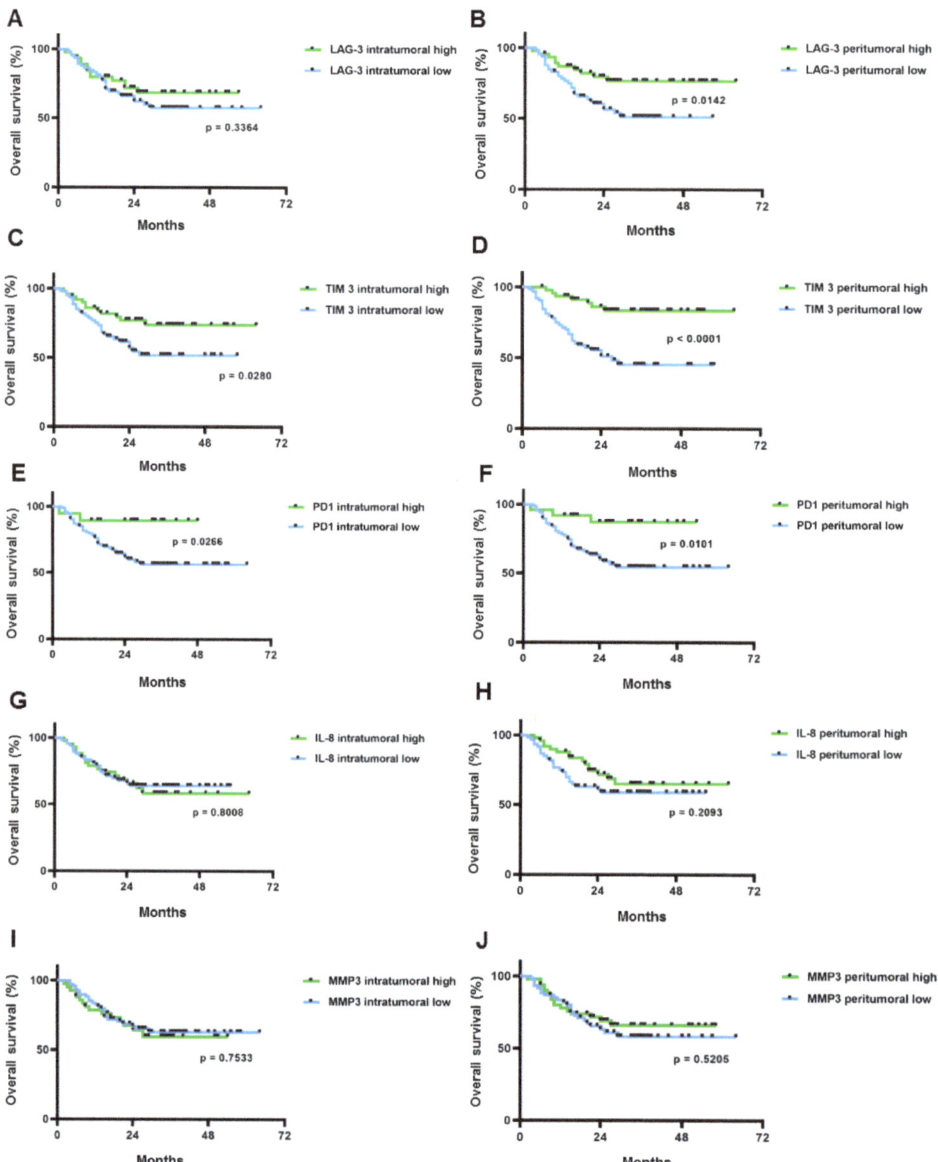

Figure 3. Overall survival of HNSCC patients depending on intratumoral and peritumoral expression of T cell exhaustion and stroma senescent biomarkers. Patient overall survival depending on LAG-3 expression (**A**,**B**), TIM-3 expression (**C**,**D**), and PD-1 expression (**E**,**F**) on intratumoral (left image) and peritumoral TILs (right image), respectively. From (**G**–**J**), the influence of intratumoral (left image) and peritumoral (right image) expression of stroma senescence markers on tumor-associated fibroblasts on overall survival is shown (**G**,**H**) for IL-8; (**I**,**J**) for MMP-3. The log-rank test (Mantel–Cox) was used for the statistical analysis in each case. Censored data are indicated as black dots on the Kaplan–Meier curves.

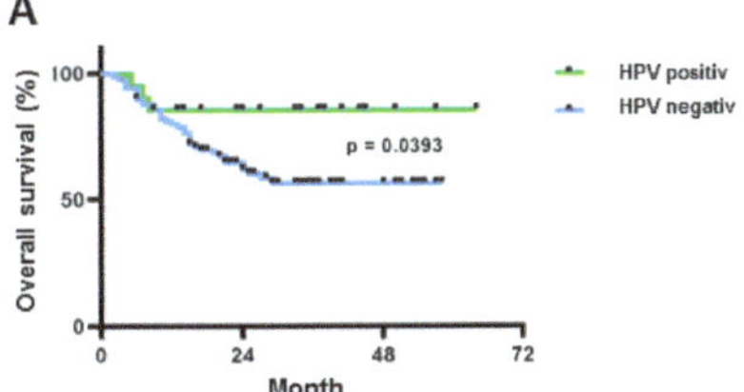
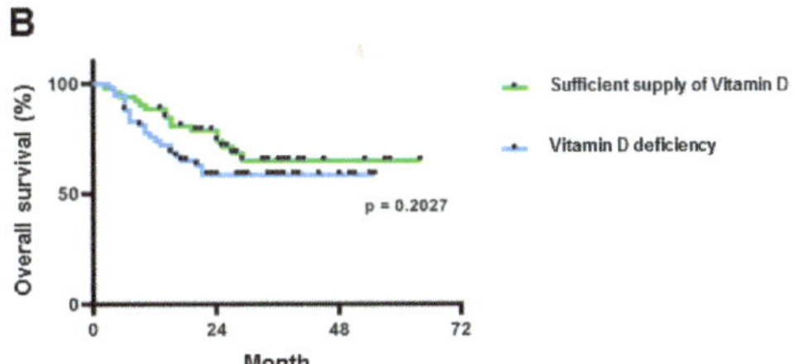

Figure 4. Patients' overall survival depending on HPV tumor status and vitamin D status. (**A**) Illustration of overall survival as a function of HPV tumor status. (**B**) Illustration of overall survival as a function of vitamin D status. The log-rank test (Mantel–Cox) was used for the statistical analysis in each case. Censored data are indicated as black dots on the Kaplan–Meier curves.

3. Discussion

HNSCCs are among the most common cancers worldwide, presenting a substantial social and economic burden [1]. Advanced-stage HNSCCs have limited treatment options and are associated with a poor prognosis [3]. Immune checkpoint inhibitors (ICIs) such as the PD-1/PD-L1 axis blockade provide new therapeutic options for recurrent or metastatic HNSCCs, but response rates remain modest with clinically relevant remissions being observed in less than 25% of cases [9,27]. Hence, new immunotherapeutic strategies are urgently needed and have been intensively studied over the past years with an increasing focus on the immunological and non-immunological tumor microenvironment [28]. The PD-1 pathway enables tumor cells to evade immune surveillance and resist treatment [29]. Anti-PD-1/PD-L1 antibodies have shown promise as checkpoint inhibitors, with overall low response rates, and adverse events have been noted, underscoring the need for a better understanding of the PD-1-mediated immunosuppression of cancer [29].

In this context, we investigated the expression of the T cell exhaustion markers PD-1, TIM-3, and LAG-3 on intra- and peritumoral TILs as well as the expression of the stroma senescence markers IL-8 and MMP-3 on intra- and peritumoral tumor-associated fibroblasts in a cohort of n = 116 HNSCC patients. We found a significant survival benefit for patients with an increased expression of the T cell exhaustion markers PD-1, LAG-3, and TIM-3 on intra- and peritumoral immune cells while the expression of the stroma senescence markers MMP-3 and IL-8 on intra- and peritumoral fibroblasts showed no influence on the patient outcome.

Considering the prognostic value of the aforementioned biomarkers in head and neck cancer, current literature evidence remains sparse with only a few studies including in most cases only a limited subset of patients.

The immune markers LAG-3, TIM-3, and PD-1 were examined in a multicenter study by Zou et al. in head and neck lymphoepithelioma-like carcinomas [30]. TIM-3 and LAG-3 were co-expressed with markers like PD-L1, B7H3, IDO-1, and VISTA, indicating a role in immune regulation within the tumor microenvironment. However, high expressions of these biomarkers were linked to worse disease-free and overall survival. The contrast to our findings may be linked to the relationship between checkpoint inhibitor expression and HPV infection. In HPV-positive HNSCCs, a higher expression of markers like TIM-3 and LAG-3 suggests a more active immune environment, potentially leading to a better response to checkpoint blockade therapy. HPV-positive tumors typically show greater immune cell infiltration, including TILs, which express these markers [31]. Conversely, HPV-negative tumors often have a less active immune landscape and respond poorly to immunotherapy. Thus, in HPV-positive cases, elevated checkpoint expression may signal a better therapeutic prognosis due to increased immunogenicity.

Another study from Yang et al. found that TIM-3 was highly expressed on intratumoral and/or stromal TILs in 91.3% of HNSCC cases [32]. TIM-3 TIL expression correlated with the tumor size, lymph node metastasis, and TNM stage, with lower TIM-3+ TIL levels linked to significantly better survival and prognosis. Here, too, there are discrepant results to our trials, but these can be attributed to the positivity of the HPV status. The results suggest that TIM-3 is a potential oncologic target in HNSCCs.

When looking at the stromal senescence markers IL-8 and MMP-3, no significant effect on overall survival was observed, but there was a significant association of MMP-3 expression with HPV positivity. Liu et al. could show that MMP-3 mRNA expression was elevated in HNSCCs compared to normal tissue and was significantly correlated with the pathological stage of HNSCC patients [33]. Additionally, MMP-3 expression correlated with immune cell infiltration, and as significant predictors of clinical outcomes in HNSCCs.

With respect to the HPV tumor status, the significantly improved overall survival in HPV-positive compared to HPV-negative HNSCC patients in our cohort is in line with numerous prospective and retrospective large-scale clinical studies of the past years and underlines the outstanding relevance of HPV as a prognostic and predictive biomarker in head and neck cancer, especially in oropharyngeal SCCs [34,35]. Considering the relevance of HPV for the response to immune checkpoint inhibition, phase III clinical trials that led to the FDA and EMA approval of pembrolizumab and nivolumab for RM-HNSCC treatment found no predictive value of HPV. However, several studies have shown better outcomes of HPV+ HNSCC patients undergoing a PD-1/PD-L1 axis blockade in contrast to HPV- HNSCC patients [36]. Exemplarily, Wang et al. demonstrated that the HPV status can predict the efficacy of PD-1 inhibition in HNSCC patients independent of PD-L1 expression, likely due to an HPV-induced inflamed immune microenvironment [37]. As our study showed that a positive HPV tumor status is associated with an increased expression of the T cell exhaustion biomarkers PD-1, TIM-3, and LAG-3, our results provide a potential explanation for their observation. ICI could thus positively influence and reactivate antitumoral T cell response, which seems to be driven into an exhausted stage in a much stronger manner than in non-HPV associated cases.

Apart from their relevance as potential prognostic biomarkers, the proteins investigated in our study may also serve as potential targets for new TME-directed immunotherapeutic strategies. Wuerdemann et al. demonstrated that intratumoral CD8+ T cells in oral HNSCCs showed a significantly upregulated expression of LAG-3, TIM-3, and VISTA, and concluded that those proteins could be used as targets for new immunotherapeutic strategies [38]. Indeed, numerous ongoing clinical trials are investigating the efficacy of ICIs targeting LAG-3 and TIM-3 in various cancer types including head and neck cancer (e.g., NCT04811027, NCT05287113), especially in combination with PD-1 antibodies. In melanoma, the dual LAG-3 and PD-1 inhibitor Opdualag was already approved by the FDA for treating unresectable or metastatic disease in adults and children [39,40]. In addition, combining ICIs or using bispecific antibodies (BsAbs) that target two ICPs at the same time is a promising approach to overcoming resistance to single-agent therapy as proven by the recently approved BsAbs targeting anti-LAG-3/TIGIT [40]. However, no LAG-3 and/or TIM-3 inhibitors are approved for the treatment of head and neck cancer so far, so further clinical studies are needed.

In addition to the HPV tumor status, we also investigated a potential correlation of the patients' 25-OH-vitamin D serum level with T cell exhaustion and stroma senescence biomarkers. In previous projects, vitamin D was shown to stimulate the infiltration of TME in head and neck cancer with various immune cells subtypes and additionally enhance their anti-tumor effector function, resulting in improved patient survival [11]. In the present study, we only found a tendency towards improved overall survival in VitD-high patients, which is in line with numerous previous studies of our own and other groups [11,12,41–43]. However, we did not find any significant correlation of 25-OH-vitamin D serum levels with the expression levels of the T cell exhaustion and stroma senescence biomarkers investigated, suggesting that vitamin D has no major role in those molecular processes.

From a critical point of view, there are some limitations that need to be considered when interpreting the study results. Our study highlights the complex interplay of T cell exhaustion and stromal senescence markers within the tumor microenvironment and their potential prognostic value in HNSCC. While we observed associations between certain checkpoint markers and patient survival, the limited size and heterogeneity of our patient cohort constrain the generalizability of these findings. Additionally, the surrogate markers we used only partially represent the biology of T cell exhaustion and stromal senescence, as these markers are also involved in other cellular functions.

Future studies should leverage RNA sequencing and pathway analyses to provide a more comprehensive and specific assessment of these biological processes. Moreover, further analyses with a larger and more homogeneous cohort, ideally with multivariate models, would be crucial to confirm the independent prognostic relevance of these markers. Expanding upon these findings could guide the development of therapeutic approaches that more effectively target the unique immune landscape in HNSCCs.

Another limitation of this study is potential interobserver variability in IHC interpretation, despite involving three independent investigators, including a board-certified pathologist. While using the mean IRS reduced variability, standardized protocols or automated tools are needed for greater consistency in future studies.

In our study, we excluded salivary gland tumors, but their differential diagnosis is important. Salivary gland tumors, such as pleomorphic adenomas, Warthin's tumors, and malignancies like mucoepidermoid carcinoma and adenoidcystic carcinoma, share clinical features with HNSCCs, complicating diagnosis. Accurate differentiation relies on clinical presentation, imaging, histopathology, and molecular profiling. Immunohistochemical markers like p63, CK7, and EGFR distinguish these tumors [44]. Molecular characteristics, such as MAML2 fusions in mucoepidermoid carcinoma, differentiate salivary gland tumors from HNSCCs, which often involve TP53 mutations [44]. Although salivary gland tumors were not included, further research into their molecular profiles and comparisons to HNSCC will enhance diagnostic accuracy and treatment strategies, improving patient outcomes.

4. Materials and Methods

A total of n = 116 patients with histologically proven HNSCC were included in our study. All patients were diagnosed and treated between 2006 and 2021 at the Department of Otorhinolaryngology, Head and Neck Surgery at the Saarland University Medical Center (Homburg/Saar, Germany). The patient cohort consisted of 97 male and 19 female patients with a median age of 64.2 years. Tumor node metastasis (TNM) and American Joint Committee on Cancer (AJCC) stages were defined according to the seventh version of the AJCC/Union for International Cancer Control (UICC) head and neck cancer staging system. Further epidemiological and clinical characteristics are shown in Table 1. The findings in this study are based on treatment modalities, specifically surgery, radiotherapy (RT), and radiochemotherapy (R(C)T).

For all patients, the pre-therapeutic 25-OH-vitamin D serum level was available and therefore included in our analyses. Here, a distinction was made between patients with a sufficient vitamin D supply (57 patients, 25-OH-vitamin D, $\geq$10 ng/mL, VitD-high) and insufficient vitamin D supply (59 patients, <10 ng/mL, VitD-low). All patients gave their written informed consent for the scientific use of their tissue samples and clinical data. All experiments were performed in accordance with the Declaration of Helsinki and its later amendments as well as the relevant guidelines and regulations. The study was approved by the Saarland Ethics Review Committee (reference number 280/10). For the experiments in our study, tumor tissue samples either taken during diagnostic panendoscopy for the histological verification of tumor diagnosis or during therapeutic tumor resection were used.

Table 1. Demographic and clinical patient data. * The 7th version of the TNM/UICC classification was used to categorize the carcinomas.

		HNSCC Patients
No. of Patients		**116**
Sex	male	97 (84%)
	female	19 (16%)
Median age [years]	male	64.5
	female	62.8
HPV Status	positive	22 (19%)
	negative	94 (81%)
Vitamin D Status	high	57 (49%)
	low	59 (51%)
Primary tumor	oral cavity	36 (31%)
	larynx	32 (28%)
	oropharynx	31 (27%)
	hypopharynx	11 (9%)
	multiple localizations	6 (5%)
T * stage	1	18 (16%)
	2	48 (41%)
	3	26 (22%)
	4	24 (21%)
N * stage	0	38 (32%)
	1	18 (16%)
	2	52 (45%)
	3	8 (7%)
M * stage	0	108 (93%)
	1	8 (7%)
UICC * Stage	I	14 (12%)
	II	14 (12%)
	III	24 (21%)
	IVa	49 (42%)
	IVb	7 (6%)
	IVc	8 (7%)

4.1. HPV Tumor Status

The HPV tumor status was determined using a combination of p16 immunohistochemical (IHC) staining and HPV-DNA-PCR analysis. Among the 116 HNSCCs patients, 81% were found to be HPV-negative, while 19% tested positive for HPV. Only those patients who showed both positive p16 IHC staining and positive HPV-DNA-PCR results were classified as having HPV-positive tumors. Due to the notably poorer prognosis and distinct tumor biology observed in discordant cases (where patients tested as p16-negative/HPV-positive or p16-positive/HPV-negative), it was predefined that both tests need to be positive to assign an HPV-positive tumor status.

For HPV-DNA-PCR testing, DNA was extracted from fresh-frozen tumor samples using the QIAamp DNA Blood Mini Kit (Qiagen, Hilden, Germany) according to the manufacturer's protocol. The HPV-DNA-PCR was conducted on the LightCycler 2.0

(Roche Diagnostics, Mannheim, Germany) using GP5+/6+ primers, following previously established procedures. Detection of the PCR amplification products was achieved with SYBR Green and gel electrophoresis. The PCR process included an initial denaturation step at 95 °C for 15 min, followed by 45 cycles of denaturation at 95 °C for 10 s, annealing at 45 °C for 5 s, and elongation at 72 °C for 18 s. After amplification, a melting curve analysis was performed over a temperature range of 45 °C to 95 °C, with an increase of 0.2 °C per second. Each PCR run included HPV16- and HPV18-positive controls, with melting temperatures (Tm) of 79 °C and 82 °C, respectively. The gene for glyceraldehyde-3-phosphate dehydrogenase (GAPDH) was amplified as an internal control.

For the immunohistochemical detection of p16, the CINtec p16 histology kit (Roche Diagnostics) was used according to the manufacturer's guidelines on formalin-fixed paraffin-embedded tissue samples obtained as described below. Epitope retrieval was achieved by heat-induced unmasking after deparaffinization in a rice cooker for 20 min, using the provided retrieval buffer. The p16 antibody was then applied, and the detection of staining was performed as recommended. Each batch of staining included both positive and negative controls.

4.2. Immunohistochemistry

Formalin-fixed paraffin-embedded (FFPE) tissue samples of the included patients were used for histopathological and immunohistochemical analyses of the tumor microenvironment (TME). Therefore, fresh tissue samples were first placed in PBS-buffered 4% formalin for 24 h and then embedded in paraffin using Tissue-Tek®VIPTM5 JR (Olympus, Tokyo, Japan). Next, FFPE tissue sections were prepared to perform immunohistochemical staining targeting PD-1, TIM-3, IL-8, LAG-3, and MMP-3. Initially, 3 tissue sections of 10 μm thickness were discarded to subsequently generate 3 μm thick sections using a Leica RM2235 rotary microtome (Leica Microsystems, Wetzlar, Germany). Sections were then transferred onto Superfrost Ultra Plus microscope slides (Menzel-Gläser, Braunschweig, Germany) and dried at 37 °C overnight. Deparaffinization was carried out, followed by heat-induced epitope unmasking in a 10 mM citrate buffer (pH 6.0). Nonspecific binding sites were blocked by the subsequent incubation of the slides with 3% BSA (Sigma Aldrich, St. Louis, MO, USA) in PBS (Sigma Aldrich) at pH 7.2 for 30 min. Sections were then exposed to primary antibodies targeting IL-8 (1:2350, ab18672), PD-1 (1:750, ab52587), TIM-3 (1:5600, ab241332), LAG-3 (1:350, ab209236), and MMP-3 (1:1000, ab52915; all antibodies from abcam, Cambridge, UK) for 1 h at room temperature. Visualization was performed using the Dako REALTM Detection System, Alkaline Phosphatase/RED (Dako Agilent Technologies, Glostrup, Denmark) according to the manufacturer's instructions. Finally, counterstaining with hematoxylin (Sigma Aldrich) was performed before the slides were mounted with coverslips.

A semiquantitative analysis of immunohistochemically stained tumor samples was performed using the Immunoreactivity Score (IRS) according to Remmele and Stegner (1987). The IRS assigns numerical values from 0 to 4, depending on the staining intensity and the percentage of stained cells in relation to all cells. The grading includes no reaction (0), weak staining reaction (1), moderate staining reaction (2), and strong staining reaction (3). The percentage of stained cells in relation to all cells was quantified with 0% (0), <10% (1), 10–50% (2), 51–80% (3), and >80% (4). Both numerical values were then multiplied resulting in a final IRS between 0 (negative) and 12 (strongly positive). For PD-1, TIM-3, and LAG-3, only immunoreactivity on peri- and intratumoral leukocytes was analyzed, and for MMP-3 and IL-8, only immunoreactivity on intra- and peritumoral fibroblasts was analyzed. The boundary between intratumoral and peritumoral regions was defined based on histopathological landmarks. The intratumoral region refers to the area within the tumor mass, including tumor cell nests and the immediately surrounding stroma infiltrated by leukocytes. By contrast, the peritumoral region is defined as the stromal area within a close margin around the tumor mass, carefully avoiding overlap with adjacent non-tumor tissues. To distinguish CAFs microscopically, we relied on their characteristic spindle-shaped morphology, elongated nuclei, and spatial localization within the stromal compartment of the tumor. All IHC stainings were analyzed by three independent investigators including one board-certified pathologist. For statistical analyses, the arithmetic mean of the three IRS values per tissue sample was used. To define whether a high IHC expression was present, the mean IRS of all samples was determined. The individual samples were defined as having either high or low expression according to the mean value as a diagnostic threshold.

4.3. Statistical Analysis

For statistical analysis, Prism 9 software (GraphPad Software, Boston, MA, USA) was used. To check the acquired data for Gaussian distribution, the Anderson–Darling test, D'Agostino and Pearson test, Shapiro–Wilk test, and Kolmogorov–Smirnov test were used. If data passed ≥ 2 of the normality tests, parametric tests were used for statistical testing (unpaired *t* test with Welch's correction, one-way ANOVA test). If the data did not pass ≥ 2 of the aforementioned normality tests, non-parametric tests were used (Mann–Whitney U test, Kruskal–Wallis test). The overall survival rates of the patient collective were analyzed using the Mantel–Cox test (log-rank test) and presented in Kaplan–Meier curves. *p* values < 0.05 were considered statistically significant ($\alpha = 0.05$). The tests that were used for statistical testing are indicated in the figure legends or the text, respectively.

5. Conclusions

Taken together, we have shown that the increased expression of the T cell exhaustion markers PD-1, LAG-3, and TIM-3 is associated with a significantly improved overall survival in HNSCC patients, and that HPV-positive disease is associated with an increased expression of these biomarkers. Further studies are necessary to uncover the clinical relevance of these observations and evaluate a potential clinical use of T cell exhaustion markers as single or combinational immunotherapeutic targets in head and neck cancer therapy.

Author Contributions: L.A.B. was responsible for generating the manuscript and conducting the final analysis. M.V. executed the majority of the experiments and generated the initial statistical analysis. M.L. and S.K. served as supervisors, providing oversight throughout the project. M.K., M.L., S.K., J.P.K., S.W., M.W., B.S. and S.S. all contributed to the critical review of the manuscript, offering valuable feedback and ensuring the accuracy and clarity of the final version. All authors have read and agreed to the published version of the manuscript.

Funding: We acknowledge support by the Saarland University within the Open Access Publication Funding program.

Institutional Review Board Statement: All experiments were performed in accordance with the Declaration of Helsinki and its later amendments as well as the relevant guidelines and regulations. The study was approved by the Saarland Ethics Review Committee on 24 October 2017 (reference number 280/10).

Informed Consent Statement: Informed consent was obtained from all subjects involved in the study.

Data Availability Statement: The raw data supporting the conclusions of this article will be made available by the authors upon request.

Conflicts of Interest: The authors declare no conflicts of interest.

Abbreviations

AJCC	American Joint Committee on Cancer
B7H3	B7 Homolog 3
BsAbs	Bispecific Antibodies
CAFs	Cancer-Associated Fibroblasts
CK7	Cytokeratin 7
CTLA-4	Cytotoxic T-Lymphocyte-Associated Protein 4

ECM	Extracellular Matrix
EGFR	Epidermal Growth Factor Receptor
EMA	European Medicines Agency
FFPE	Formalin-Fixed Paraffin-Embedded
FDA	U.S. Food and Drug Administration
GAPDH	Glyceraldehyde-3-Phosphate Dehydrogenase
HNSCC	Head and Neck Squamous Cell Carcinoma
HPV	Human Papillomavirus
IHC	Immunohistochemistry
ICI	Immune Checkpoint Inhibitors
IL-8	Interleukin-8
IRS	Immunoreactivity Score
LAG-3	Lymphocyte Activation Gene-3
MAML2	Mastermind-like domain containing 2
MMP-3	Matrix Metalloproteinase 3
MMPs	Matrix Metalloproteinases
OS	Overall Survival
PCR	Polymerase Chain Reaction
PD-1	Programmed Cell Death Protein 1
PD-L1	Programmed Death-Ligand 1
p16	Cyclin-Dependent Kinase Inhibitor 2A (p16INK4a)
p63	Tumor protein p63
SASP	Senescence-Associated Secretory Phenotype
TILs	Tumor-Infiltrating Lymphocytes
TME	Tumor Microenvironment
TIM-3	T cell Immunoglobulin and Mucin-Domain Containing-3
TNM	Tumor, Node, Metastasis (Staging System)
TP53	Tumor Protein 53
UICC	Union for International Cancer Control
VitD	Vitamin D
VISTA	V-Domain Immunoglobulin Suppressor of T Cell Activation

References

1. Bray, F.; Laversanne, M.; Sung, H.; Ferlay, J.; Siegel, R.L.; Soerjomataram, I.; Jemal, A. Global cancer statistics 2022: GLOBOCAN estimates of incidence and mortality worldwide for 36 cancers in 185 countries. *CA A Cancer J. Clin.* **2024**, *74*, 229–263. [CrossRef]
2. Dalianis, T. Human papillomavirus and oropharyngeal cancer, the epidemics, and significance of additional clinical biomarkers for prediction of response to therapy (Review). *Int. J. Oncol.* **2014**, *44*, 1799–1805. [CrossRef] [PubMed]
3. Pulte, D.; Brenner, H. Changes in Survival in Head and Neck Cancers in the Late 20th and Early 21st Century: A Period Analysis. *Oncologist* **2010**, *15*, 994–1001. [CrossRef] [PubMed]
4. Bhatia, A.; Burtness, B. Treating Head and Neck Cancer in the Age of Immunotherapy: A 2023 Update. *Drugs* **2023**, *83*, 217–248. [CrossRef] [PubMed]
5. Cohen, E.E.W.; Bell, R.B.; Bifulco, C.B.; Burtness, B.; Gillison, M.L.; Harrington, K.J.; Le, Q.-T.; Lee, N.Y.; Leidner, R.; Lewis, R.L.; et al. The Society for Immunotherapy of Cancer consensus statement on immunotherapy for the treatment of squamous cell carcinoma of the head and neck (HNSCC). *J. Immunother. Cancer* **2019**, *7*, 184. [CrossRef] [PubMed]
6. Burtness, B.; Harrington, K.J.; Greil, R.; Soulières, D.; Tahara, M.; de Castro, G., Jr.; Psyrri, A.; Basté, N.; Neupane, P.; Bratland, A.; et al. Pembrolizumab alone or with chemotherapy versus cetuximab with chemotherapy for recurrent or metastatic squamous cell carcinoma of the head and neck (KEYNOTE-048): A randomised, open-label, phase 3 study. *Lancet* **2019**, *394*, 1915–1928. [CrossRef]
7. Cohen, E.E.W.; Le Tourneau, C.; Licitra, L.; Ahn, M.-J.; Soria, A.; Machiels, J.-P.; Mach, N.; Mehra, R.; Zhang, P.; Cheng, J.; et al. Pembrolizumab versus methotrexate, docetaxel, or cetuximab for recurrent or metastatic head-and-neck squamous cell carcinoma (KEYNOTE-040): A randomised, open-label, phase 3 study. *Lancet* **2019**, *393*, 156–167. [CrossRef]
8. Ferris, R.L.; Blumenschein, G., Jr.; Fayette, J.; Guigay, J.; Colevas, A.D.; Licitra, L.; Harrington, K.; Kasper, S.; Vokes, E.E.; Even, C.; et al. Nivolumab for Recurrent Squamous-Cell Carcinoma of the Head and Neck. *N. Engl. J. Med.* **2016**, *375*, 1856–1867. [CrossRef]
9. Poulose, J.V.; Kainickal, C.T. Immune checkpoint inhibitors in head and neck squamous cell carcinoma: A systematic review of phase-3 clinical trials. *World J. Clin. Oncol.* **2022**, *13*, 388–411. [CrossRef]
10. Mandal, R.; Şenbabaoğlu, Y.; Desrichard, A.; Havel, J.J.; Dalin, M.G.; Riaz, N.; Lee, K.-W.; Ganly, I.; Hakimi, A.A.; Chan, T.A.; et al. The head and neck cancer immune landscape and its immunotherapeutic implications. *JCI Insight* **2016**, *1*, e89829. [CrossRef]

11. Bochen, F.; Balensiefer, B.; Körner, S.; Bittenbring, J.T.; Neumann, F.; Koch, A.; Bumm, K.; Marx, A.; Wemmert, S.; Papaspyrou, G.; et al. Vitamin D deficiency in head and neck cancer patients—Prevalence, prognostic value and impact on immune function. *OncoImmunology* **2018**, *7*, e1476817. [CrossRef] [PubMed]

12. Brust, L.A.; Linxweiler, M.; Schnatmann, J.; Kühn, J.-P.; Knebel, M.; Braun, F.L.; Wemmert, S.; Menger, M.D.; Schick, B.; Holick, M.F.; et al. Effects of Vitamin D on tumor cell proliferation and migration, tumor initiation and anti-tumor immune response in head and neck squamous cell carcinomas. *Biomed. Pharmacother.* **2024**, *180*, 117497. [CrossRef] [PubMed]

13. Dolina, J.S.; Van Braeckel-Budimir, N.; Thomas, G.D.; Salek-Ardakani, S. CD8+ T Cell Exhaustion in Cancer. *Front. Immunol.* **2021**, *12*, 715234. [CrossRef]

14. Ruhland, M.K.; Alspach, E. Senescence and Immunoregulation in the Tumor Microenvironment. *Front. Cell Dev. Biol.* **2021**, *9*, 754069. [CrossRef]

15. Shi, A.-P.; Tang, X.-Y.; Xiong, Y.-L.; Zheng, K.-F.; Liu, Y.-J.; Shi, X.-G.; Lv, Y.; Jiang, T.; Ma, N.; Zhao, J.-B. Immune Checkpoint LAG3 and Its Ligand FGL1 in Cancer. *Front. Immunol.* **2022**, *12*, 785091. [CrossRef] [PubMed]

16. Kandel, S.; Adhikary, P.; Li, G.; Cheng, K. The TIM3/Gal9 signaling pathway: An emerging target for cancer immunotherapy. *Cancer Lett.* **2021**, *510*, 67–78. [CrossRef] [PubMed]

17. Wherry, E.J. T cell exhaustion. *Nat. Immunol.* **2011**, *12*, 492–499. [CrossRef] [PubMed]

18. Wherry, E.J.; Kurachi, M. Molecular and cellular insights into T cell exhaustion. *Nat. Rev. Immunol.* **2015**, *15*, 486–499. [CrossRef] [PubMed]

19. Baitsch, L.; Baumgaertner, P.; Devêvre, E.; Raghav, S.K.; Legat, A.; Barba, L.; Wieckowski, S.; Bouzourene, H.; Deplancke, B.; Romero, P.; et al. Exhaustion of tumor-specific CD8$^+$ T cells in metastases from melanoma patients. *J. Clin. Investig.* **2011**, *121*, 2350–2360. [CrossRef]

20. Mumprecht, S.; Schürch, C.; Schwaller, J.; Solenthaler, M.; Ochsenbein, A.F. Programmed death 1 signaling on chronic myeloid leukemia–specific T cells results in T-cell exhaustion and disease progression. *Blood* **2009**, *114*, 1528–1536. [CrossRef] [PubMed]

21. Matsuzaki, J.; Gnjatic, S.; Mhawech-Fauceglia, P.; Beck, A.; Miller, A.; Tsuji, T.; Eppolito, C.; Qian, F.; Lele, S.; Shrikant, P.; et al. Tumor-infiltrating NY-ESO-1–specific CD8$^+$ T cells are negatively regulated by LAG-3 and PD-1 in human ovarian cancer. *Proc. Natl. Acad. Sci. USA* **2010**, *107*, 7875–7880. [CrossRef] [PubMed]

22. Zhang, Y.; Huang, S.; Gong, D.; Qin, Y.; Shen, Q. Programmed death-1 upregulation is correlated with dysfunction of tumor-infiltrating CD8+ T lymphocytes in human non-small cell lung cancer. *Cell. Mol. Immunol.* **2010**, *7*, 389–395. [CrossRef]

23. de Visser, K.E.; Joyce, J.A. The evolving tumor microenvironment: From cancer initiation to metastatic outgrowth. *Cancer Cell* **2023**, *41*, 374–403. [CrossRef] [PubMed]

24. Mavrogonatou, E.; Pratsinis, H.; Kletsas, D. The role of senescence in cancer development. *Semin. Cancer Biol.* **2020**, *62*, 182–191. [CrossRef] [PubMed]

25. Pittayapruek, P.; Meephansan, J.; Prapapan, O.; Komine, M.; Ohtsuki, M. Role of Matrix Metalloproteinases in Photoaging and Photocarcinogenesis. *Int. J. Mol. Sci.* **2016**, *17*, 868. [CrossRef]

26. Fousek, K.; Horn, L.A.; Palena, C. Interleukin-8: A chemokine at the intersection of cancer plasticity, angiogenesis, and immune suppression. *Pharmacol. Ther.* **2021**, *219*, 107692. [CrossRef] [PubMed]

27. Sacco, A.G.; Chen, R.; Worden, F.P.; Wong, D.J.L.; Adkins, D.; Swiecicki, P.; Chai-Ho, W.; Oppelt, P.; Ghosh, D.; Bykowski, J.; et al. Pembrolizumab plus cetuximab in patients with recurrent or metastatic head and neck squamous cell carcinoma: An open-label multi-arm, non-randomised, multicentre, phase 2 trial. *Lancet Oncol.* **2021**, *22*, 883–892. [CrossRef] [PubMed]

28. Ruffin, A.T.; Li, H.; Vujanovic, L.; Zandberg, D.P.; Ferris, R.L.; Bruno, T.C. Improving head and neck cancer therapies by immunomodulation of the tumour microenvironment. *Nat. Rev. Cancer* **2023**, *23*, 173–188. [CrossRef] [PubMed]

29. Wu, X.; Gu, Z.; Chen, Y.; Chen, B.; Chen, W.; Weng, L.; Liu, X. Application of PD-1 Blockade in Cancer Immunotherapy. *Comput. Struct. Biotechnol. J.* **2019**, *17*, 661–674. [CrossRef] [PubMed]

30. Zou, W.-Q.; Luo, W.-J.; Feng, Y.-F.; Liu, F.; Liang, S.-B.; Fang, X.-L.; Liang, Y.-L.; Liu, N.; Wang, Y.-Q.; Mao, Y.-P. Expression Profiles and Prognostic Value of Multiple Inhibitory Checkpoints in Head and Neck Lymphoepithelioma-Like Carcinoma. *Front. Immunol.* **2022**, *13*, 818411. [CrossRef] [PubMed]

31. Lechner, M.; Liu, J.; Masterson, L.; Fenton, T.R. HPV-associated oropharyngeal cancer: Epidemiology, molecular biology and clinical management. *Nat. Rev. Clin. Oncol.* **2022**, *19*, 306–327. [CrossRef] [PubMed]

32. Yang, F.; Zeng, Z.; Li, J.; Ren, X.; Wei, F. TIM-3 and CEACAM1 are Prognostic Factors in Head and Neck Squamous Cell Carcinoma. *Front. Mol. Biosci.* **2021**, *8*, 619765. [CrossRef] [PubMed]

33. Liu, M.; Huang, L.; Liu, Y.; Yang, S.; Rao, Y.; Chen, X.; Nie, M.; Liu, X. Identification of the MMP family as therapeutic targets and prognostic biomarkers in the microenvironment of head and neck squamous cell carcinoma. *J. Transl. Med.* **2023**, *21*, 208. [CrossRef]

34. Ang, K.K.; Harris, J.; Wheeler, R.; Weber, R.; Rosenthal, D.I.; Nguyen-Tân, P.F.; Westra, W.H.; Chung, C.H.; Jordan, R.C.; Lu, C.; et al. Human Papillomavirus and Survival of Patients with Oropharyngeal Cancer. *N. Engl. J. Med.* **2010**, *363*, 24–35. [CrossRef] [PubMed]

35. Fung, N.; Faraji, F.; Kang, H.; Fakhry, C. The role of human papillomavirus on the prognosis and treatment of oropharyngeal carcinoma. *Cancer Metastasis Rev.* **2017**, *36*, 449–461. [CrossRef]

36. Xu, Y.; Zhu, G.; Maroun, C.A.; Wu, I.X.Y.; Huang, D.; Seiwert, T.Y.; Liu, Y.; Mandal, R.; Zhang, X. Programmed Death-1/Programmed Death-Ligand 1-Axis Blockade in Recurrent or Metastatic Head and Neck Squamous Cell Carcinoma Stratified by Human Papillomavirus Status: A Systematic Review and Meta-Analysis. *Front. Immunol.* **2021**, *12*, 645170. [CrossRef]
37. Wang, J.; Sun, H.; Zeng, Q.; Guo, X.-J.; Wang, H.; Liu, H.-H.; Dong, Z.-Y. HPV-positive status associated with inflamed immune microenvironment and improved response to anti-PD-1 therapy in head and neck squamous cell carcinoma. *Sci. Rep.* **2019**, *9*, 13404. [CrossRef]
38. Wuerdemann, N.; Pütz, K.; Eckel, H.; Jain, R.; Wittekindt, C.; Huebbers, C.U.; Sharma, S.J.; Langer, C.; Gattenlöhner, S.; Büttner, R.; et al. LAG-3, TIM-3 and VISTA Expression on Tumor-Infiltrating Lymphocytes in Oropharyngeal Squamous Cell Carcinoma—Potential Biomarkers for Targeted Therapy Concepts. *Int. J. Mol. Sci.* **2020**, *22*, 379. [CrossRef] [PubMed]
39. Tawbi, H.A.; Schadendorf, D.; Lipson, E.J.; Ascierto, P.A.; Matamala, L.; Gutiérrez, E.C.; Rutkowski, P.; Gogas, H.J.; Lao, C.D.; De Menezes, J.J.; et al. Relatlimab and Nivolumab versus Nivolumab in Untreated Advanced Melanoma. *N. Engl. J. Med.* **2022**, *386*, 24–34. [CrossRef]
40. Cai, L.; Li, Y.; Tan, J.; Xu, L.; Li, Y. Targeting LAG-3, TIM-3, and TIGIT for cancer immunotherapy. *J. Hematol. Oncol.* **2023**, *16*, 101. [CrossRef]
41. Vaughan-Shaw, P.G.; O'Sullivan, F.; Farrington, S.M.; Theodoratou, E.; Campbell, H.; Dunlop, M.G.; Zgaga, L. The impact of vitamin D pathway genetic variation and circulating 25-hydroxyvitamin D on cancer outcome: Systematic review and meta-analysis. *Br. J. Cancer* **2017**, *116*, 1092–1110. [CrossRef]
42. Yokosawa, E.B.; Arthur, A.E.; Rentschler, K.M.; Wolf, G.T.; Rozek, L.S.; Mondul, A.M. Vitamin D intake and survival and recurrence in head and neck cancer patients. *Laryngoscope* **2018**, *128*, E371–E376. [CrossRef] [PubMed]
43. Starska-Kowarska, K. The Role of Different Immunocompetent Cell Populations in the Pathogenesis of Head and Neck Cancer—Regulatory Mechanisms of Pro- and Anti-Cancer Activity and Their Impact on Immunotherapy. *Cancers* **2023**, *15*, 1642. [CrossRef] [PubMed]
44. Vrinceanu, D.; Dumitru, M.; Bratiloveanu, M.; Marinescu, A.; Serboiu, C.; Manole, F.; Palade, D.O.; Costache, A.; Costache, M.; Patrascu, O. Parotid Gland Tumors: Molecular Diagnostic Approaches. *Int. J. Mol. Sci.* **2024**, *25*, 7350. [CrossRef] [PubMed]

International Journal of
Molecular Sciences

Article

Assessment of Concentration KRT6 Proteins in Tumor and Matching Surgical Margin from Patients with Head and Neck Squamous Cell Carcinoma

Dariusz Nałęcz [1,*], Agata Świętek [2,3,*], Dorota Hudy [2], Karol Wiczkowski [2,4], Zofia Złotopolska [1] and Joanna Katarzyna Strzelczyk [2]

[1] Department of Otolaryngology and Maxillofacial Surgery, St. Vincent De Paul Hospital, 1 Wójta Radtkego St., 81-348 Gdynia, Poland; zzlotopolska@gmail.com

[2] Department of Medical and Molecular Biology, Faculty of Medical Sciences in Zabrze, Medical University of Silesia in Katowice, 19 Jordana St., 41-808 Zabrze, Poland; karolwicz@gmail.com (K.W.); jstrzelczyk@sum.edu.pl (J.K.S.)

[3] Silesia LabMed Research and Implementation Centre, Medical University of Silesia in Katowice, 19 Jordana St., 41-808 Zabrze, Poland

[4] Students' Scientific Association, Department of Medical and Molecular Biology, Medical University of Silesia, Katowice, 19 Jordana St., 41-808 Zabrze, Poland

* Correspondence: dnalecz@szpitalepomorskie.eu (D.N.); agata.swietek@sum.edu.pl (A.Ś.)

Abstract: Head and neck squamous cell carcinomas (HNSCCs) are one of the most frequently detected cancers in the world; not all mechanisms related to the expression of keratin in this type of cancer are known. The aim of this study was to evaluate type II cytokeratins (KRT): KRT6A, KRT6B, and KRT6C protein concentrations in 54 tumor and margin samples of head and neck squamous cell carcinoma (HNSCC). Moreover, we examined a possible association between protein concentration and the clinical and demographic variables. Protein concentrations were measured using enzyme-linked immunosorbent assay (ELISA). Significantly higher KRT6A protein concentration was found in HNSCC samples compared to surgical margins. An inverse relationship was observed for KRT6B and KRT6C proteins. We showed an association between the KRT6C protein level and clinical parameters T and N in tumor and margin samples. When analyzing the effect of smoking and drinking on KRT6A, KRT6B, and KRT6C levels, we demonstrated a statistically significant difference between regular or occasional tobacco and alcohol habits and patients who do not have any tobacco and alcohol habits in tumor and margin samples. Moreover, we found an association between KRT6B and KRT6C concentration and proliferative index Ki-67 and HPV status in tumor samples. Our results showed that concentrations of KRT6s were different in the tumor and the margin samples and varied in relation to clinical and demographic parameters. We add information to the current knowledge about the role of KRT6s isoforms in HNSCC. We speculate that variations in the studied isoforms of the KRT6 protein could be due to the presence and development of the tumor and its microenvironment. It is important to note that the analyses were performed in tumor and surgical margins and can provide more accurate information on the function in normal and cancer cells and regulation in response to various factors.

Keywords: HNSCC; cancerogenesis; protein level; tumor; surgical margin; KRT6A; KRT6B; KRT6C; cytokeratins; KRT6s

1. Introduction

Head and neck cancers (HNSCC) are significant clinical and social problems [1]. Their share of all malignant neoplasms in Poland has invariably ranged between 5.5 and 6.2% in recent years, which accounts for approximately 5500 to 6000 new cases each year [2,3]. They are characterized by an overall poor prognosis, and the long-term overall survival

(5 years) in this group of patients is about 50% for all stages of HNSCC [4]. The causes of head and neck cancer are not completely clear, but it is known that cigarette smokers and alcohol abusers are at risk of the disease [5,6]. Several molecular mechanisms are known to be involved in the development and metastasis of cancer [6–9]. Some studies indicate that type II cytokeratins may also influence the process of carcinogenesis.

KRT6 (keratin 6) protein is a member of type II cytokeratins, of which three isoforms are known (KRT6A, KRT6B, and KRT6C), encoded by three genes: *KRT6A*, *KRT6B*, and *KRT6C*, respectively. They are found in the epidermis of the palms and soles, the filiform papillae of the tongue, the epithelium lining the mucosa of the mouth and esophagus, the epithelial cells of the nail bed, and hair follicles [10]. Recent studies have demonstrated the likely involvement of keratins in the processes of carcinogenesis, including cancer cell invasion and the formation of metastasis [11,12]. Therefore, these molecules have found their application as diagnostic markers for cancer, especially in unclear clinical cases, which is particularly valuable for correct identification of the tumor and selection of the most appropriate treatments [13–17]. The relationship between changes in the concentration of KRT6s proteins and HNSCC is still unknown and appears to have dual roles as both protooncogenes and tumor suppressors [13,17–20]. The role of KRT6 isoforms could be associated with the type of tumors. To the best of our understanding, it remains unclear whether alterations in KRT6s isoform concentrations have any association with HNSCC. Due to the role of KRT6 proteins in cell growth, invasion, and migration processes, it is hypothesized that the levels of KRT6A, KRT6B, and KRT6C will be changed in tumor samples compared to margin samples in patients with HNSCC.

The aim of this study was to evaluate the levels of selected proteins in tumors and the matched samples of surgical margins, in a group of patients with primary HNSCC. The association of clinical-pathological and demographical variables with the concentration of the proteins studied was also analyzed.

2. Results

2.1. Concentration of the Selected KRTs in Tumour Samples and Margin

Significantly higher KRT6A protein concentration was found in the tumor samples than in the margin sample (0.01976 (0.00962–0.02825) vs. 0.00933 (0.00488–0.01938)); (p = 0.0003). However, in the analysis of KRT6B and KRT6C, we observed significantly higher concentration of protein in the margin than in the tumor specimens (for KRT6B 1.264 (0.63545–1.80653) vs. 0.518 (0.215–0.8901)); (p = 0.0009) and (for KRT6C 0.15724 (0.10418–0.30146) vs. 0.13441 (0.079336–0.18031)); (p = 0.0274). The results are given in Figure 1.

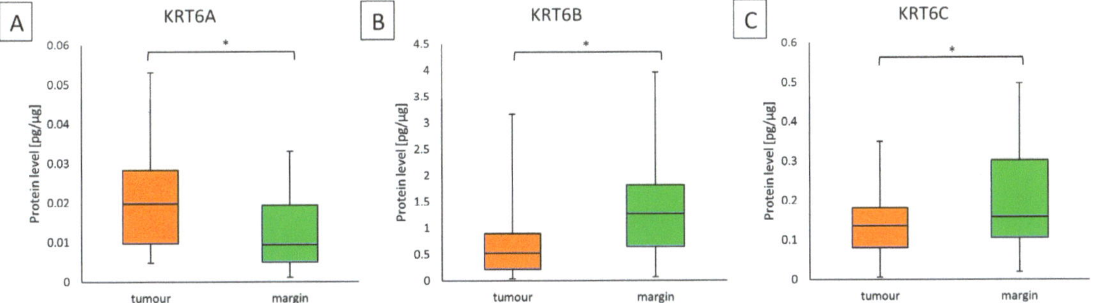

Figure 1. Results of the KRT6 proteins level analyses in tumor samples compared to margin samples. (**A**)—Protein KRT6A in tumor and margin samples, (**B**)—Protein KRT6B in tumor and margin samples, (**C**)—Protein KRT6C in tumor and margin samples. Statistical analysis was carried out with the Mann–Whitney U test where differences with $p \leq 0.05$ were considered statistically significant. The orange blocks indicate the tumor while the green ones represent the margin sample. * Indicate significant differences.

2.2. Protein Level of KRT6 and Cancer Classification, and Localization of Tumour

In the case of cancer classification due to a small number of some types only two groups were tested—OSCC (30 cases) and the combined group consisted of OPSCC, LSCC, and HPSCC subtypes (20 combined cases). There were no observed significant differences in KRT6 protein levels between those groups. The most of primary tumor regions used in this study were localized in the larynx (17 samples, 31.48%), the floor of the mouth (14, 25.93%), the tongue (10, 18.52%), the jaw (5, 9.26%). There were no observed differences in localization in KRT6 protein concentration. In both cases, there were no observed differences in the association of groups with sex, clinical parameters, tobacco, and alcohol habits (Supplementary S1, Tables S1 and S2).

2.3. Protein Level of KRT6 and Clinical Parameters

In the case of the nodal status, a combined group of patients with nodal status N2 and N3 was formed, due to a small number of N3 patients. The higher concentrations of KRT6B in margin samples showed a group of patients with higher nodal status, which was statistically significant (N0 vs. N2 + N3; 0.83449 (0.48671–1.5046) vs. 1.61342 (1.277–2.3565)); (p = 0.0200). The results are shown in Figure 2A.

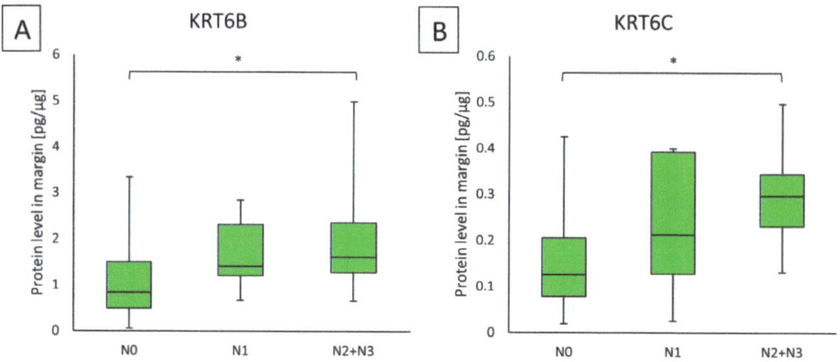

Figure 2. (A) The KRT6B protein level in the margin sample samples in the group of patients with N0, N1, and N2 + N3 nodal status; **(B)** the KRT6C protein level in the margin samples in the group of patients with N0, N1, and N2 + N3 nodal status. Statistical analysis was performed with Kruskal–Wallis and with Dunn–Sidak post hoc, differences of $p \leq 0.05$ are considered statistically significant. The orange blocks indicate the tumor while the green ones represent the margin sample. * Indicate significant differences.

KRT6C protein concentration was significantly higher in the margin samples in the group of patients with nodal status N2 or N3, as compared to samples in the group of patients with nodal status N0 (0.29846 (0.23041–0.34369) vs. 0.12591 (0.07801–0.20545)); (p = 0.0052). There was a significant difference in age between the groups of patients with N0 and N2 + N3 status, with higher nodal status related to younger age (p = 0.01958; 66.77 ± 11.75 vs. 56.17 ± 9.32). Tables with characteristics with clinical parameter groups are presented in Supplement S1 (Tables S3–S5). There were no observed associations with other parameters (T and G). The linkage of group T1 with T2, because of a small number of cases in the T1 group, did not obtain significant results vs. T3 and T4 cases as individual groups or joined groups.

2.4. Protein Level of KRT6 and Tobacco and Alcohol Habits

In the group of smokers, compared to non-smokers, KRT6A protein concentration was higher in the tumor (0.02126 (0.01055–0.03044) vs. 0.00858 (0.00669–0.01089)); (p = 0.0061). An opposite situation was observed for KRT6B in margins in the group of smokers compared to non-smokers (1.048 (0.59183–1.54050) vs. 2.357 (1.45890–4.5156)); (p = 0.0168).

These results are shown in Figure 3A,B. There was no observed correlation between KRT6 proteins and the amount of cigarettes per day or with years of smoking but there was a medium positive correlation of KRT6B protein in tumor samples with pack-years (0.32; $p = 0.0407$).

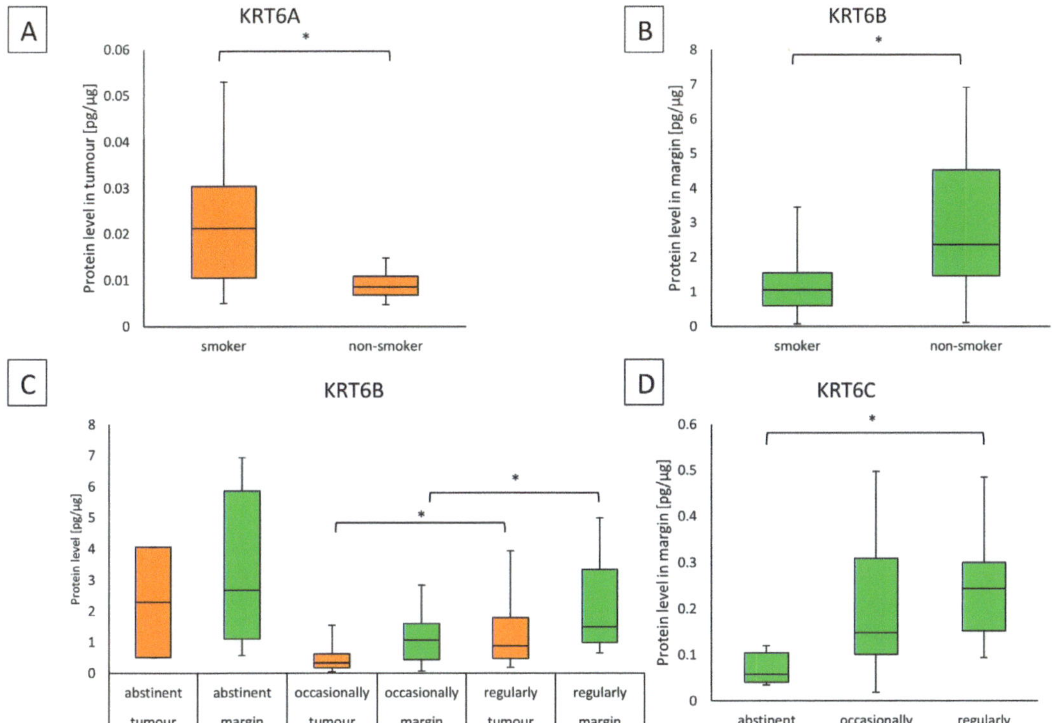

Figure 3. (**A**) The KRT6A protein level in the tumor samples according to smoking status; (**B**) the KRT6B protein level in the margin samples according to the smoking status; (**C**) the KRT6B protein level in the tumor and margin samples according to the drinking status; (**D**) the KRT6C protein level in the margin samples according to the drinking status. Statistical analysis for AB charts was performed with the Mann–Whitney U test, for CD charts Kruskal–Wallis was performed with Dunn–Sidak post hoc and differences with $p \leq 0.05$ are considered as statistically significant results. The orange blocks indicate the tumor while the green ones represent the margin sample. * Indicate significant differences.

Furthermore, regular alcohol drinkers had higher levels of KRT6B, compared to occasional drinkers 0.88750 (0.47487–1.78950) vs. 0.32989 (0.16330–0.62059); ($p = 0.0047$) in the tumor. Similar observations were found in the same groups in the surgical margin (1.49696 (0.98943–3.34760) vs. 1.05949 (0.44043–1.59470)); ($p = 0.0272$). The results are given in Figure 3C. Significantly higher KRT6C protein level was found in the margin samples collected from regular drinkers, in comparison to the abstinent patients (0.24470 (0.15152–0.30080) vs. 0.05722 (0.04014–0.10430)); ($p = 0.0059$) (Figure 3D). The group that was regularly drinking was significantly younger than the abstinent group ($p = 0.0069$; 53 (51–65) vs. 79 (71.5–82.75)) but this could be the effect of group size. Tables with characteristics of the tobacco and alcohol habits groups are presented in Supplement S1 (Tables S6 and S7). No association with other parameters was observed.

2.5. Protein Level of KRT6 and Proliferation Index Evaluated by Ki-67

The patients with proliferation index Ki-67 > 20 had a significantly higher KRT6C concentration than patients with Ki-67 $\leq$ 20 (0.13789 (0.08662–0.17819) vs. 0.05961 (0.04652–0.1062)); ($p = 0.0267$) in the tumor samples (Figure 4). A table with characteristics of the proliferative index evaluated $\leq$ 20 and Ki-67 > 20 is presented in Supplement S1 (Table S8). No association with other parameters was observed.

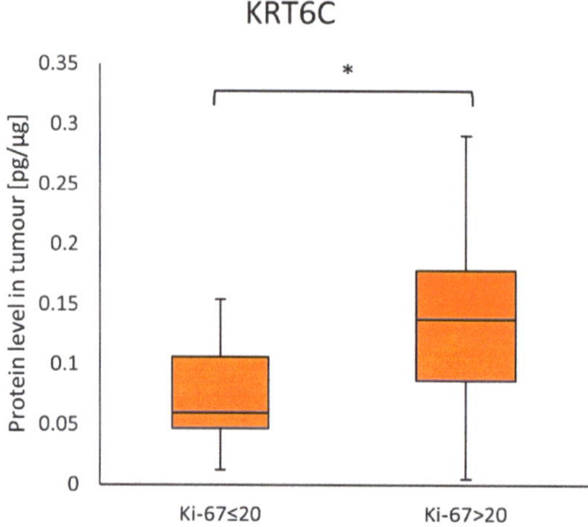

Figure 4. The KRT6C protein level in the tumor samples according to Ki-67 protein. Statistical analysis was performed with the Mann–Whitney U test and differences with $p \leq 0.05$ are considered statistically significant. The orange blocks indicate the tumor. * Indicate significant difference.

2.6. KRT6 Protein Concentration and HPV Status and p16 Status

The median protein concentration of KRT6B in the tumor was higher in HPV(+) patients than in the group of HPV(−) (2.43711 (0.59399–4.02520) vs. 0.36460 (0.16355–0.60257)); ($p = 0.0335$). Similarly, HPV(+) patients had a higher median level of KRT6C in the tumor samples (0.19478 (0.18870–0.26409) vs. 0.10846 (0.07466–0.16088)); ($p = 0.0199$), which is presented in Figure 5A,B.

The concentration of KRT6B protein was higher in the tumor samples of the p16(+) group than in the p16(−) group (2.43711 (0.59399–4.0252) vs. 0.31389 (0.17382–0.62325)); ($p = 0.0327$). Also in the tumor samples, the KRT6C protein had a higher level in the p16(+) group than in p16(−) (0.19478 (0.18870–0.26409) vs. 0.10560 (0.06749–0.17158)); ($p = 0.0451$). Results are presented in Figure 5CD. Tables with characteristics of the p16 and HPV groups are presented in Supplement S1. For HPV groups there was no association with other parameters observed (Table S9). For the p16 groups there was an association with the smoking status ($p = 0.014$), group p16(−) included more patients who were smokers (Table S10).

2.7. Correlation of KRT6 Proteins

KRT6C protein from the margin samples was significantly mildly correlating with all three KRT6 proteins in the tumor samples. With KRT6A, a negative correlation was observed, presented in Table 1 in blue (−0.34; $p = 0.0156$), and with KRT6B (0.39; $p = 0.01193$) and KRT6C (0.35; $p = 0.01458$) positive correlations were observed as presented in orange in Table 1. KRT6C showed also a stronger correlation (0.53; $p = 0.000195$) with KRT6B in the margin samples; a positive correlation was presented in Table 1 in orange.

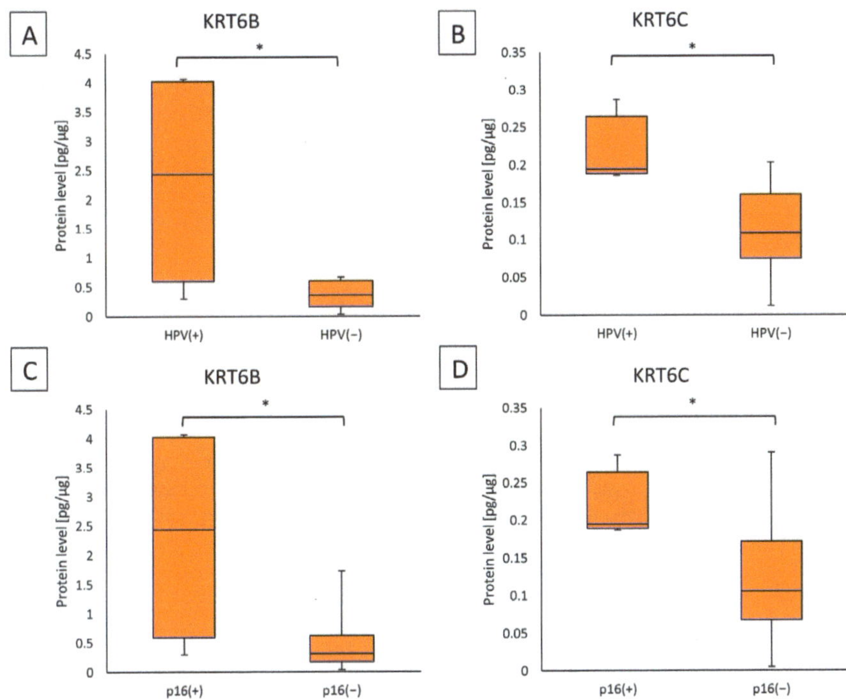

Figure 5. (**A**) The KRT6B protein level in the tumor samples according to the HPV status; (**B**) the KRT6C protein level in the tumor samples according to the HPV status; (**C**) the KRT6B protein level in the tumor samples according to p16 status; (**D**) the KRT6C protein level in the tumor samples according to p16 status. Statistical analysis was performed with the Mann–Whitney U test and differences with $p \leq 0.05$ are considered statistically significant. The orange blocks indicate the tumor. * Indicate significant differences.

Table 1. p-value and R^2 of Spearman correlation for KRT6 proteins.

		Tumor			Margin		
p-Value		KRT6A	KRT6B	KRT6C	KRT6A	KRT6B	KRT6C
tumor	KRT6A	1.00	0.45	0.23	0.48	0.16	**0.02**
	KRT6B	0.45	1.00	0.15	0.37	0.36	**0.01**
	KRT6C	0.23	0.15	1.00	0.94	0.37	**0.01**
margin	KRT6A	0.48	0.37	0.94	1.00	0.48	0.86
	KRT6B	0.16	0.36	0.37	0.48	1.00	**0.00**
	KRT6C	**0.02**	**0.01**	**0.01**	0.86	**0.00**	1.00
Spearman's Rank Coefficient		Tumor			Margin		
		KRT6A	KRT6B	KRT6C	KRT6A	KRT6B	KRT6C
tumor	KRT6A	1.00	0.12	−0.18	−0.11	−0.21	−0.34
	KRT6B	0.12	1.00	0.23	0.15	0.16	0.39
	KRT6C	−0.18	0.23	1.00	0.01	−0.14	0.35
margin	KRT6A	−0.11	0.15	0.01	1.00	−0.12	0.03
	KRT6B	−0.21	0.16	−0.14	−0.12	1.00	0.53
	KRT6C	−0.34	0.39	0.35	0.03	0.53	1.00

The bold p-value indicates significance. Orange color indicates a positive correlation, and blue indicates a negative correlation.

3. Discussion

The exact role of KRT6A, KRT6B, and KRT6C in HNSCC is still unclear. Based on our knowledge (databases PubMed and Medline), this has been the first study to analyze the concentration of these proteins by ELISA in the tumor and margin samples, obtained from patients with HNSCC.

In our study, the median KRT6A protein level was significantly higher in the HNSCC tumor sample than in the margin. In HNSCC and OSCC in other types of KRT (KRT17, KRT19), similarly to ours, increased protein levels were observed in the tumor compared to the healthy sample or surgical margin [21–23]. Some studies found upregulated KRT6A levels in other types of the cancer sample, compared to matched normal samples [13,24–26]. The high KRT6A concentration was reported in the analysis of non-small-cell lung cancer (NSCLC) and lung adenocarcinoma (LADC) and is associated with lymph node metastasis and advanced T stage cancer [13,24,25]. In addition, the authors reported that KRT6A overexpression can affect the upregulation of G6PD (glucose-6-phosphate dehydrogenase), resulting in activation of the metabolic pathway promoting invasion and the growth of cancer cells [25]. Additionally, the increased KRT6A expression was observed in the nasopharyngeal carcinoma cell line, and it was noted that KRT6A silencing was associated with inhibition of cell invasion and metastasis formation via the β-catenin pathway [18]. Also, Chen et al. suggested that KRT6A played a prominent role in promoting proliferation and adhesion, with simultaneous inhibiting tumor cell apoptosis in bladder cancer [27]. In contrast, a study on the immune microenvironment in pancreatic ductal adenocarcinoma showed that KRT6A could modulate the function of tumor-associated macrophages (TAMs), an important part of the leukocyte infiltrate, through other proteins and molecular pathways [28]. In addition, evidence suggests that keratin proteins may be involved in migration, adhesion, cell proliferation, and regulation of keratinocyte inflammation, and play an important physiological role in cell repair [29]. Therefore, based on our observation, we suggested that KRT6A could play an important role in HNSCC, however, in order to confirm this hypothesis, it would be important to conduct studies on a larger group of patients than ours.

On the other hand, in our study, lower levels of KRT6B and KRT6C proteins were observed in HNSCC samples, compared to the resected surgical margin. In the case of other types of KRT (KRT13, KRT14, KRT24) in HNSCC, OSCC, and neoplastic oral mucosa, decreased protein levels were observed in the tumor sample compared to the margin/healthy tissue, which is consistent with our results [30–34]. Similar to our results, the in silico breast cancer studies found reduced KRT6B and KRT6C levels in the tumor tissues, compared to the cancer-free samples [20,35,36]. On the contrary, increased KRT6B expression was demonstrated in colorectal cancer, compared to healthy colorectal mucosal tissue, based on the results of bioinformatics analyses [16]. Higher levels of KRT6C in saliva were observed in OSCC compared to the control group [37]. A recent study confirmed that OSCC tumor margin cells had unique transcriptomic profiles and ligand–receptor interactions [38,39]. In addition, it is suspected that KRT6B may be involved in the process of immune response, including M2 polarization of macrophages [40]. The role of KRT6B and KRT6C proteins in cancerogenesis is unclear. KRT protein levels may vary according to different types of cancer, which could be associated with specific molecular characteristics of the tumor as well as the demographic, clinical, and pathological parameters [34]. Moreover, the cited studies may have used different protein detection methods, varying in sensitivity and specificity, which resulted in different KRT level results obtained [20,34–39]. The increased levels of KRT6B and KRT6C proteins in the tumor margin observed in our study might therefore indicate that these proteins may affect the immune response and the immune microenvironment.

In our research, KRT6B and KRT6C protein concentrations in the margin tissue samples in the group of patients with the nodal status N2 or N3 were significantly higher, as compared to samples in the group of patients with the nodal status N0. In a study evaluating the salivary proteins as potential biomarkers for the early diagnosis of OSCC, it was reported that KRT6C was significantly elevated at the tumor stage T3/T4, compared to T1/T2, whereas in our study we did not note any differences [37]. Importantly, Liu et al. and Song et al. detected elevated KRT6B levels in the bladder tumors, which correlated positively with the metastatic status and the stage of the disease [40,41]. Other authors reported the elevated KRT6C levels in lung adenocarcinoma cell lines to be associated with cell proliferation, migration, and invasion [42]. Our study demonstrated the elevated levels of these proteins, which may be related to their functions in the epithelial-mesenchymal transition—an important process for epithelial cells to achieve invasiveness, whereas it would be important to repeat the research on a larger number of samples.

Our study reported that smokers showed significantly increased concentrations of KRT6A in the tumor samples and significantly decreased concentrations of KRT6B in the margin tissue. Importantly, it has been shown that exposure to cigarette smoke is able to inhibit DNA repair, immunosuppression, induction of oxidative stress, and induction changes in the proteome of oral keranocytes, resulting in both, upregulation and downregulation of selected proteins [43,44]. The authors suggest that the consequences of smoking may therefore be oral cancerous lesions [45]. The studies based on proteomic analysis of exhaled breath condensate in the group of patients with lung cancer, in the healthy volunteers and additionally on primary human gingival epithelial cells, showed that KRT6A and KRT6B proteins were elevated in 58.1% of smokers in both groups [46,47]. The study based on bioinformatic analyses in NSCLC showed that the increased KRT6A expression was associated with current and past smoking habits and KRT6As in NSCLC function as oncogenes and may be useful as potential prognostic diagnostic biomarkers of NSCLC in smokers [48]. In another study, the authors suspect that keratins are proteins that respond to oxidative stress, and the impaired expression of KRT may be a response to disruption of the oral mucosal barrier exposed to tobacco smoke [49]. Based on our results and observations, the effects of cigarette smoke may take place through modulation of the expression of cytokeratin type II protein in response to epithelial damage. It is possible that additional genetic and epigenetic mechanisms are present in smokers that may be involved in the altered expression of KRT family genes and proteins.

Moreover, we reported the increased concentration of KRT6B protein in the tumor and the margin samples of regular alcohol drinkers compared to occasional drinkers. A higher KRT6C protein level was noted in the margin samples collected from regular drinkers, in comparison to the abstinent patients. Subsequent publications confirm alcohol drinking as a risk factor in malignancies of the oral cavity, pharynx, and larynx [50–54]. As one potential factor, non-pathogenic strains colonizing the oral cavity have been shown to be responsible for the processes that convert ethanol to acetaldehyde, which is able to modify the processes of methylation, synthesis, and DNA repair, and could interact with proteins, which ultimately implies cell damage and proliferation. In addition, the authors also reported the ability of acetaldehyde to activate the oncogenic transcription factors in oral keratinocytes [51–54]. Therefore, we suggested that the upregulation of KRT6B and KRT6C protein levels in the tumor and margin samples of regular alcohol drinkers, compared to occasional drinkers, could be related to increased keratinization, in response to the tissue damage caused by alcohol and its metabolites.

In this study, we observed that tumor samples with proliferation index Ki-67 > 20 showed significantly higher KRT6C concentration than those with Ki-67 $\leq$ 20. The Ki-67 protein is one of the most commonly used markers due to its correlation with tumor proliferation. Various studies have shown higher Ki-67 expression in OSCC tissues than in normal tissues and increased with the progression of dysplasia in oral mucosa tissues [55–57]. Keratins are the main structural proteins of the epithelial cells, so we suspect that the increased level of KRT6C may be related to the role of these proteins in proliferation and repair processes in response to disorders in the epithelial cells.

Our study showed that the median protein concentrations of KRT6B and KRT6C were higher in HPV(+) patients than in the group of HPV(−) in tumor tissues. Moreover, we found a similar correlation in the case of p16 status, where we obtained higher levels of KRT6B and KRT6C proteins in p16-positive tumor samples. In the study by Woods et al., similar to ours, increased levels of KRT7 and KRT19 were observed in HPV-positive OPSCC compared to HPV-negative cases, which may indicate the involvement of keratins in the etiopathogenesis of HPV-related OPSCC [58]. Studies have shown that HPV is associated with infection of the basal layer of the epithelium, and then to complete the life cycle of the virus; it uses the pathways of epithelial proliferation and differentiation into keratinocytes [59,60]. The study by Zhang et al. showed that HPV(+) HNSCC tumors are characterized by increased expression of genes related to keratinocyte differentiation processes [60]. In an in vitro studies, dysregulated interferon signaling, DNA replication, and DNA damage response pathways were observed in HPV(+) keranocyte cells [61,62]. In addition, in cervical cancer, it was reported that the presence of HPV16, and thus the effect of E6/E7 oncoproteins, was associated with increased expression of keratinization genes [63]. As it is known, HNSCC is a heterogeneous disease entity due to HPV infection status, so we suggest that there are differences in oncogenic pathways, probably in addition to E6 and E7 oncoproteins, related to HPV genes associated with various processes that can modulate the expression of a large number of genes and proteins, including KRT6B and KRT6C, which in response can result in increased keratinization.

4. Materials and Methods

4.1. Study Population

The study group consisted of 54 patients (66.67% male and 33.33% female) diagnosed with HNSCC. Specimens of the tumor and of the corresponding margins were collected after surgical resection at the Department of Otolaryngology and Maxillary Surgery, St. Vincent De Paul Hospital, Gdynia, Poland. The collected samples were examined histologically and classified as primary HNSCC. Marginal samples were taken from the surgical margin at least 10 mm from the tumor border and were histologically confirmed as being free of cancer. Classification and staging of tumor specimens were performed according to the 8th edition of the AJCC Cancer Staging Manual [64]. All the resected specimens were secured and then transported in dry ice to the Laboratory of Medical and Molecular Biology, Medical University of Silesia in Katowice, Poland, where they were stored at $-80\,^{\circ}$C until further analysis. The HNSCC comprised 30 cases of oral squamous cell carcinoma (OSCC), 2 cases of oropharyngeal squamous cell carcinoma (OPSCC), 17 cases of laryngeal squamous cell carcinoma (LSCC), 2 cases of hypopharyngeal squamous cell carcinoma (HPSCC), 2 cases of nasal cavity squamous cell carcinoma (NCSCC), and 1 case of skin squamous cell carcinoma (SSCC). Primary tumor regions used in this study were localized in the larynx (17 samples, 31.48%), the floor of the mouth (14, 25.93%), the tongue (10, 18.52%), the jaw (5, 9.26%), the oral cavity (2, 3.70%), the retromolar trigone (4, 7.40%), and one each in the cheek (1, 1.85%), and the soft palate (1, 1.85%). The main inclusion criteria for the HNSCC group included a diagnosis of a primary tumor and no preoperative radio/chemotherapy. All patients participating in the study gave written informed consent. Approval for the study was obtained from the Bioethical Committee, Regional Medical Chamber in Gdansk (no. KB-42/21). Table 2 shows the characteristics of the study group.

Table 2. Parameters characterizing the HNSCC patients.

Parameter	N (%)
Sex	
female	18 (33.33)
male	36 (66.67)
Average age (range)	64 (53–72)
Smoking	
yes	45 (83.33)
no	9 (16.67)

Table 2. *Cont.*

Parameter	N (%)
Drinking	
yes occasionally	36 (66.67)
yes regularly	15 (27.78)
no	3 (5.56)
Clinical T-classification (T)	
T1	3 (5.56)
T2	9 (16.67)
T3	19 (35.19)
T4	20 (37.04)
NA *	3 (5.56)
Nodal status (N)	
N0	31 (57.41)
N1	6 (11.11)
N2	12 (22.22)
N3	2 (3.70)
NA *	3 (5.56)
Metastasis (M)	
M0	52 (96.30)
M1	0 (0.00)
NA *	2 (3.70)
Histological grading (G)	
G1	15 (27.78)
G2	32 (59.26)
G3	6 (11.11)
G4	1 (1.85)

* NA—not assessed.

Smokers were using tobacco for 32.61 ± 11.83 pack-years on average. Drinking status was determined by patient surveys (abstinent, occasionally) and information about alcohol abuse or alcoholism disease (regularly).

4.2. Homogenization and Total Protein Concentration

First, 10% tissue homogenates were obtained in PRO 200 mechanical homogenizer (PRO Scientific Inc., Oxford, CT, USA) at the rate of 10,000 rpm, in the presence of an appropriate volume of cooled PBS buffer (Eurx, Gdansk, Poland). The commercial AccuOrange™ Protein Quantitation Kit (Biotium, Fremont, CA, USA) was used to quantify the total protein. The determinations were carried out in the previously prepared tissue homogenates, in duplicate, according to the manufacturer's instructions, without dilutions. The detection range of the assay was 0.1–15 µg/mL of protein. Fluorescence was measured at the excitation wavelength of 480 nm and the emission wavelength of 598 nm (SYNERGY H1 microplate reader; BIOTEK, Winooski, VT, USA using the Gen5 2.06 software). Total protein determinations were performed to express the KRT6A, KRT6B, and KRT6C concentrations in units per µg of total protein.

The homogenates were dissected and then frozen at $-80\ °C$ until further analysis.

4.3. Enzyme-Linked Immunosorbent Assay (ELISA) Kits for Proteins Concentration

The protein levels of KRT6A, KRT6B, and KRT6C were assayed in homogenates by the ELISA method, according to the standard instruction (ELISA Kits Cloud-Clone Corp., Katy, TX, USA); (Assay ID: SED234Hu for KRT6A, SEA486Hu for KRT6B and SED758Hu for KRT6C). To determine the concentrations of the tested samples, a calibration curve was prepared using the standards included in the kit. Absorbance readings were recorded at 450 nm wavelength and calibrated according to the standard curve in ng/mL (SYNERGY H1 microplate reader; BIOTEK, Winooski, VT, USA). The tests were characterized by the following sensitivities: 6.1 pg/mL for KRT6A, 0.060 ng/mL for KRT6B, and 13.3 pg/mL for KRT6C. The intra-assay variation was below 10% and the inter-assays were below 12% for all the determined proteins. All standards and samples were run in duplicate. The results obtained were calculated corresponding to the total protein concentration and presented in pg/μg.

4.4. DNA Isolation and HPV Detection

The tissue samples were homogenized with Lysing Matrix A (MP Biomedicals, Irvine, CA, USA). DNA was isolated with a GeneMATRIX Tissue DNA Purification Kit (Eurx, Gdansk, Poland) according to the protocol. The quality and quantity of isolated DNA were assessed on a spectrophotometer (NanoPhotometer Pearl, Implen, Munich, Germany) and DNA was kept at $-20\ ^\circ$C until HPV analysis.

HPV status was assessed with GeneFlow^{TM} HPV Array Test Kit (DiagCor Bioscience Ltd., Kowloon Bay, Hongkong) with FT^{PRO} Flow-through System (DiagCor Bioscience Ltd., Kowloon Bay, Hongkong) and FT^{PRO} Auto System (DiagCor Bioscience Ltd., Kowloon Bay, Hongkong) according to instruction. First, isolated DNA was used in PCR reaction on Mastercycler Personal Thermal Cycler (Eppendorf, Hamburg, Germany). Next PCR products were denatured and hybridized, and after enzyme conjugation and color development, the results were scanned with FT^{PRO} Auto System (DiagCor Bioscience Ltd., Kowloon Bay, Hongkong). Positive and negative controls provided by the manufacturer were performed in all runs.

4.5. Evaluation of p16 and Ki-67 by Immunohistochemical Staining

Immunohistochemical analysis for p16 was performed by commercial kit according to the manufacturer's instructions (CINtec p16 Histology, Roche MTM Laboratories, Mannheim, Germany). The staining made use of BenchMark ULTRA automated system (Roche Diagnostics, Basel, Switzerland). The immunohistochemical analysis for Ki-67 was evaluated using the CONFIRM anti-Ki-67 (30-9) Rabbit Monoclonal Primary Antibody (Ventana Medical Systems, Inc., Tucson, AZ, USA) using BenchMark ULTRA in automatic mode (Roche Diagnostics, Basel, Switzerland). The expression of proliferation index Ki-67 was categorized into 2 groups: Ki-67 > 20 and Ki-67 $\leq$ 20.

p16 and Ki-67 expression status were evaluated following the instructions provided in the kit package.

4.6. Statistical Analysis

The results from ELISA and the patient surveys were tested with Shapiro–Wilk to determine normality for groups with less than or equal to 5 cases non-normal distribution was assumed. Differences between the groups and the tissue type were tested with the Mann–Whitney U test or Kruskal–Wallis with Dunn–Sidak post hoc. Data in Supplementary S1 were processed with the chi^2 test with Bonferroni correction for multiple comparisons should more than one comparison be carried out. Differences in age were tested with the Students *t*-test or One-Way ANOVA with Tukey HSD post hoc or Mann–Whitney U test or Kruskal–Wallis with Dunn–Sidak in case of small groups ($\leq$5 cases). Correlations were performed with the Spearman rank coefficient method. Significant results were indicated with $p \leq 0.05$ or $p \leq 0.05/$(number of comparisons). All calculations were performed with Statistica 13.1 (TIBCO Software Inc., Palo Alto, CA, USA) or with Excel 2019 (Microsoft, Redmond, WA, USA) software. Significant data are presented as box plots with the median in the middle and the 1st and 3rd quartile as a box with minimum and maximum values as whiskers. Data mentioned in the text are presented as median with the 1st and 3rd quartile as follows, median (quartile 1st–quartile 3rd).

5. Conclusions

Our results showed that concentrations of KRT6s were different in the tumor and the margins samples and varied in relation to clinical and demographic parameters. We add information to the current knowledge about the role of KRT6s isoforms in HNSCC. Moreover, our study identified the changes in concentration of selected KRT proteins in both, tumor and surgical margin samples from HNSCC patients, and can provide more accurate information on the function in normal and cancer cells and regulation in response to various factors. Observation of the levels of KRT6 isoforms in the tumor and the margin shows that the isoforms play different roles and functions in diverse cell types which also depends on the HPV/p16 status, the proliferative index Ki-67, and the tobacco and alcohol habits. There is also an association between the levels of these proteins and clinical and pathological variables, such as tumor clinical stage and nodal status, which could be due to the presence and development of the tumor and its microenvironment.

The main limitation of the study was the small size of the samples. Therefore, to confirm the associations more accurately, future studies should be conducted on larger cohorts. In addition, tests with cell lines and animal models are required, which could provide valuable information for a better understanding of the role in tumorgenesis of KRT6s proteins. Furthermore, considering the analyzed cytokeratins type II proteins are needed to understand fully their impact on cancer prognosis and progression. Our future studies will then focus on the analysis of the disease-free and the overall survival in HNSCC patients.

Supplementary Materials: The following supporting information can be downloaded at: https://www.mdpi.com/article/10.3390/ijms25137356/s1.

Author Contributions: Conceptualization, D.N. and J.K.S.; methodology, D.N. and J.K.S.; formal analysis, D.N.; investigation, D.N. and A.Ś.; resources, D.N. and Z.Z.; writing—original draft preparation D.N., A.Ś., D.H. and K.W.; writing—review and editing, Z.Z. and J.K.S.; visualization, D.N. and D.H.; supervision, J.K.S.; project administration, D.N. All authors have read and agreed to the published version of the manuscripts.

Funding: This research received no external funding.

Institutional Review Board Statement: The study was conducted according to the guidelines of the Declaration of Helsinki, and approved by the Bioethical Committee, Regional Medical Chamber in Gdańsk (no. KB-42/21).

Informed Consent Statement: Informed consent was obtained from all subjects involved in the study.

Data Availability Statement: The data used to support the findings of this research are available upon request.

Acknowledgments: We thank all our patients for their voluntary participation in this study. We wish to thank Radosław Lenckowski, Department of Pathology, Szpitale Pomorskie Sp. z o.o., Gdynia for the immunohistochemical analyses performed.

Conflicts of Interest: The authors declare no conflicts of interest. The funders had no role in the design of the study; in the collection, analyses, or interpretation of data; in the writing of the manuscript; or in the decision to publish the results.

References

1. Mody, M.D.; Rocco, J.W.; Yom, S.S.; Haddad, R.I.; Saba, N.F. Head and neck cancer. *Lancet* **2021**, *398*, 2289–2299. [CrossRef]
2. Gawełko, J.; Cierpiał-Wolan, M.; Kawecki, A.; Wilk, K.; Pięciak-Kotlarz, D.B.; Sikorski, D. Comparative analysis of the incidence of head and neck cancer in south-eastern Poland and in Poland in the years 1990–2012. *Contemp. Oncol.* **2017**, *21*, 77–82. [CrossRef] [PubMed]
3. Butkiewicz, D.; Gdowicz-Kłosok, A.; Krześniak, M.; Rutkowski, T.; Krzywon, A.; Cortez, A.J.; Domińczyk, I.; Składowski, K. Association of Genetic Variants in ANGPT/TEK and VEGF/VEGFR with Progression and Survival in Head and Neck Squamous Cell Carcinoma Treated with Radiotherapy or Radiochemotherapy. *Cancers* **2020**, *12*, 1506. [CrossRef] [PubMed]
4. Cohen, N.; Fedewa, S.; Chen, A.Y. Epidemiology and demographics of the head and neck cancer population. *Oral Maxillofac. Surg. Clin. N. Am.* **2018**, *30*, 381–395. [CrossRef] [PubMed]
5. Yan, F.; Knochelmann, H.M.; Morgan, P.F.; Kaczmar, J.M.; Neskey, D.M.; Graboyes, E.M.; Nguyen, S.A.; Ogretmen, B.; Sharma, A.K.; Day, T.A. The evolution of care of cancers of the head and neck region: State of the science in 2020. *Cancers* **2020**, *12*, 1543. [CrossRef] [PubMed]
6. Li, Q.; Tie, Y.; Alu, A.; Ma, X.; Shi, H. Targeted therapy for head and neck cancer: Signaling pathways and clinical studies. *Signal Transduct. Target. Ther.* **2023**, *8*, 31. [CrossRef]

7.   Rothenberger, N.J.; Stabile, L.P. Hepatocyte growth factor/c-met signaling in head and neck cancer and implications for treatment. *Cancers* **2017**, *9*, 39. [CrossRef]
8.   Johnson, D.E.; Burtness, B.; Leemans, C.R.; Lui, V.W.Y.; Bauman, J.E.; Grandis, J.R. Head and neck squamous cell carcinoma. *Nat. Rev. Dis. Primers* **2020**, *6*, 92. [CrossRef] [PubMed]
9.   Van den Bossche, V.; Zaryouh, H.; Vara-Messler, M.; Vignau, J.; Machiels, J.P.; Wouters, A.; Schmitz, S.; Corbet, C. Microenvironment-driven intratumoral heterogeneity in head and neck cancers: Clinical challenges and opportunities for precision medicine. *Drug Resist. Updates* **2022**, *60*, 100806. [CrossRef]
10.   Moll, R.; Divo, M.; Langbein, L. The human keratins: Biology and pathology. *Histochem. Cell Biol.* **2008**, *129*, 705–733. [CrossRef]
11.   Werner, S.; Keller, L.; Pantel, K. Epithelial keratins: Biology and implications as diagnostic markers for liquid biopsies. *Mol. Asp. Med.* **2020**, *72*, 100817. [CrossRef] [PubMed]
12.   Sharma, S.; Alsharif, A.; Fallatah, B.M. Chung Intermediate filaments as effectors of cancer development and metastasis: A focus on keratins, vimentin, and nestin. *Cells* **2019**, *8*, 497. [CrossRef] [PubMed]
13.   Xiao, J.; Lu, X.; Chen, X.; Zou, Y.; Liu, A.; Li, W.; He, B.; He, S.; Chen, Q. Eight potential biomarkers for distinguishing between lung adenocarcinoma and squamous cell carcinoma. *Oncotarget* **2017**, *8*, 71759–71771. [CrossRef] [PubMed]
14.   Lin, J.B.; Feng, Z.; Qiu, M.L.; Luo, R.G.; Li, X.; Liu, B. KRT 15 as a prognostic biomarker is highly expressed in esophageal carcinoma. *Future Oncol.* **2020**, *16*, 1903–1909. [CrossRef] [PubMed]
15.   Han, W.; Hu, C.; Fan, Z.J.; Shen, G.L. Transcript levels of keratin 1/5/6/14/15/16/17 as potential prognostic indicators in melanoma patients. *Sci. Rep.* **2021**, *11*, 1023. [CrossRef] [PubMed]
16.   Qin, Y.; Chen, L.; Chen, L. Identification and verification of key cancer genes associated with prognosis of colorectal cancer based on bioinformatics analysis. *Zhong Nan Da Xue Xue Bao Yi Xue Ban* **2021**, *46*, 1063–1070. [PubMed]
17.   Miñoza, J.M.A.; Rico, J.A.; Zamora, P.R.F.; Bacolod, M.; Laubenbacher, R.; Dumancas, G.G.; de Castro, R. Biomarker Discovery for Meta-Classification of Melanoma Metastatic Progression Using Transfer Learning. *Genes* **2022**, *13*, 2303. [CrossRef] [PubMed]
18.   Chen, C.; Shan, H. Keratin 6A gene silencing suppresses cell invasion and metastasis of nasopharyngeal carcinoma via the β-catenin cascade. *Mol. Med. Rep.* **2019**, *19*, 3477–3484. [CrossRef] [PubMed]
19.   Xu, Q.; Yu, Z.; Mei, Q.; Shi, K.; Shen, J.; Gao, G.; Liu, S.; Li, M. Keratin 6A (KRT6A) promotes radioresistance, invasion, and metastasis in lung cancer via p53 signaling pathway. *Aging* **2024**, *16*, 7060–7072. [CrossRef]
20.   Modi, A.; Purohit, P.; Gadwal, A.; Ukey, S.; Roy, D.; Fernandes, S.; Banerjee, M. In-Silico Analysis of Differentially Expressed Genes and Their Regulating microRNA Involved in Lymph Node Metastasis in Invasive Breast Carcinoma. *Cancer Investig.* **2022**, *40*, 55–72. [CrossRef]
21.   Zhong, L.P.; Chen, W.T.; Zhang, C.P.; Zhang, Z.Y. Increased CK19 expression correlated with pathologic differentiation grade and prognosis in oral squamous cell carcinoma patients. *Oral Surg. Oral Med. Oral Pathol. Oral Radiol. Endodontol.* **2007**, *104*, 377–384. [CrossRef] [PubMed]
22.   Wei, K.J.; Zhang, L.; Yang, X.; Zhong, L.P.; Zhou, X.J.; Pan, H.Y.; Li, J.; Chen, W.T.; Zhang, Z.Y. Overexpression of cytokeratin 17 protein in oral squamous cell carcinoma in vitro and in vivo. *Oral Dis.* **2009**, *15*, 111–117. [CrossRef] [PubMed]
23.   Coelho, B.A.; Peterle, G.T.; Santos, M.; Agostini, L.P.; Maia, L.L.; Stur, E.; Silva, C.V.; Mendes, S.O.; Almança, C.C.; Freitas, F.V. et al. Keratins 17 and 19 expression as prognostic markers in oral squamous cell carcinoma. *Genet. Mol. Res.* **2015**, *14*, 15123–15132. [CrossRef] [PubMed]
24.   Yang, B.; Zhang, W.; Zhang, M.; Wang, X.; Peng, S.; Zhang, R. KRT6A Promotes EMT and Cancer Stem Cell Transformation in Lung Adenocarcinoma. *Technol. Cancer Res. Treat.* **2020**, *19*, 1533033820921248. [CrossRef]
25.   Che, D.; Wang, M.; Sun, J.; Li, B.; Xu, T.; Lu, Y.; Pan, H.; Lu, Z.; Gu, X. KRT6A Promotes Lung Cancer Cell Growth and Invasion through MYC-Regulated Pentose Phosphate Pathway. *Front. Cell Dev. Biol.* **2021**, *9*, 694071. [CrossRef] [PubMed]
26.   Wang, H.; Liu, J.; Li, J.; Zang, D.; Wang, X.; Chen, Y.; Gu, T.; Su, W.; Song, N. Identification of gene modules and hub genes in colon adenocarcinoma associated with pathological stage based on WGCNA analysis. *Cancer Genet.* **2020**, *242*, 1–7. [CrossRef]
27.   Chen, Y.; Ji, S.; Ying, J.; Sun, Y.; Liu, J.; Yin, G. KRT6A expedites bladder cancer progression, regulated by miR-31-5p. *Cell Cycle* **2022**, *21*, 1479–1490. [CrossRef] [PubMed]
28.   Zhang, J.; Sun, H.; Liu, S.; Huang, W.; Gu, J.; Zhao, Z.; Qin, H.; Luo, L.; Yang, J.; Fang, Y.; et al. Alteration of tumor-associated macrophage subtypes mediated by KRT6A in pancreatic ductal adenocarcinoma. *Aging* **2020**, *12*, 23217–23232. [CrossRef]
29.   Zhang, X.; Yin, M.; Zhang, L.J. Keratin 6, 16 and 17-Critical Barrier Alarmin Molecules in Skin Wounds and Psoriasis. *Cells* **2019**, *8*, 807. [CrossRef]
30.   Farrar, M.; Sandison, A.; Peston, D.; Gailani, M. Immunocytochemical analysis of AE1/AE3, CK 14, Ki-67 and p53 expression in benign, premalignant and malignant oral tissue to establish putative markers for progression of oral carcinoma. *Br. J. Biomed. Sci.* **2004**, *61*, 117–124. [CrossRef]
31.   Lv, X.Z.; Chen, W.T.; Zhang, C.P. Expression and significance of cytokeratin 13 gene in oral squamous cell carcinoma. *Shanghai Kou Qiang Yi Xue* **2005**, *14*, 378–381.
32.   Farrukh, S.; Syed, S.; Pervez, S. Differential Expression of Cytokeratin 13 in Non-Neoplastic, Dysplastic and Neoplastic Oral Mucosa in a High Risk Pakistani Population. *Asian Pac. J. Cancer Prev.* **2015**, *16*, 5489–5492. [CrossRef]
33.   Xu, E.S.; Yang, M.H.; Liu, C.Y.; Liu, K.W.; Yang, T.T.; Chou, T.Y.; Hwang, T.Z.; Hsu, C.T. Decreasing cytokeratin 17 expression in head and neck cancer predicts nodal metastasis and poor prognosis: The first evidence. *Clin. Otolaryngol.* **2018**, *43*, 1010–1018. [CrossRef]

34. Gül, D.; Habtemichael, N.; Dietrich, D.; Dietrich, J.; Gößwein, D.; Khamis, A.; Deuss, E.; Künzel, J.; Schneider, G.; Strieth, S.; et al. Identification of cytokeratin24 as a tumor suppressor for the management of head and neck cancer. *Biol. Chem.* **2021**, *403*, 869–890. [CrossRef]

35. Shen, D.; He, J.; Chang, H.R. In silico identification of breast cancer genes by combined multiple high throughput analyses. *Int. J. Mol. Med.* **2005**, *15*, 205–212. [CrossRef] [PubMed]

36. Khan, M.S.; Hanif, W.; Alsakhen, N.; Jabbar, B.; Shamkh, I.M.; Alsaiari, A.A.; Almehmadi, M.; Alghamdi, S.; Shakoori, A.; Al Farraj, D.A.; et al. Isoform switching leads to downregulation of cytokine producing genes in estrogen receptor positive breast cancer. *Front. Genet.* **2023**, *14*, 1230998. [CrossRef] [PubMed]

37. Jain, A.; Kotimoole, C.N.; Ghoshal, S.; Bakshi, J.; Chatterjee, A.; Prasad, T.S.K.; Pal, A. Identification of potential salivary biomarker panels for oral squamous cell carcinoma. *Sci. Rep.* **2021**, *11*, 3365. [CrossRef] [PubMed]

38. Barkley, D.; Moncada, R.; Pour, M.; Liberman, D.A.; Dryg, I.; Werba, G.; Wang, W.; Baron, M.; Rao, A.; Xia, B.; et al. Cancer cell states recur across tumor types and form specific interactions with the tumor microenvironment. *Nat. Genet.* **2022**, *54*, 1192–1201. [CrossRef] [PubMed]

39. Arora, R.; Cao, C.; Kumar, M.; Sinha, S.; Chanda, A.; McNeil, R.; Samuel, D.; Arora, R.K.; Matthews, T.W.; Chandarana, S.; et al. Spatial transcriptomics reveals distinct and conserved tumor core and edge architectures that predict survival and targeted therapy response. *Nat. Commun.* **2023**, *14*, 5029. [CrossRef]

40. Song, Q.; Yu, H.; Cheng, Y.; Han, J.; Li, K.; Zhuang, J.; Lv, Q.; Yang, X.; Yang, H. Bladder cancer-derived exosomal KRT6B promotes invasion and metastasis by inducing EMT and regulating the immune microenvironment. *J. Transl. Med.* **2022**, *20*, 308. [CrossRef]

41. Liu, Y.R.; Yin, P.N.; Silvers, C.R.; Lee, Y.F. Enhanced metastatic potential in the MB49 urothelial carcinoma model. *Sci. Rep.* **2019**, *9*, 7425. [CrossRef] [PubMed]

42. Hu, H.B.; Yang, X.P.; Zhou, P.X.; Yang, X.A.; Yin, B. High expression of keratin 6C is associated with poor prognosis and accelerates cancer proliferation and migration by modulating epithelial-mesenchymal transition in lung adenocarcinoma. *Genes Genom.* **2020**, *42*, 179–188. [CrossRef] [PubMed]

43. Weng, M.W.; Lee, H.W.; Park, S.H.; Hu, Y.; Wang, H.T.; Chen, L.C.; Rom, W.N.; Huang, W.C.; Lepor, H.; Wu, X.R.; et al. Aldehydes are the predominant forces inducing DNA damage and inhibiting DNA repair in tobacco smoke carcinogenesis. *Proc. Natl. Acad. Sci. USA* **2018**, *115*, 6152–6161. [CrossRef]

44. Khowal, S.; Wajid, S. Role of Smoking-Mediated molecular events in the genesis of oral cancers. *Toxicol. Mech. Methods* **2019**, *29*, 665–685. [CrossRef]

45. Rajagopalan, P.; Nanjappa, V.; Patel, K.; Jain, A.P.; Mangalaparthi, K.K.; Patil, A.H.; Nair, B.; Mathur, P.P.; Keshava Prasad, T.S.; Califano, J.A.; et al. Role of protein kinase N2 (PKN2) in cigarette smoke-mediated oncogenic transformation of oral cells. *J. Cell Commun. Signal.* **2018**, *12*, 709–721. [CrossRef] [PubMed]

46. López-Sánchez, L.M.; Jurado-Gámez, B.; Feu-Collado, N.; Valverde, A.; Cañas, A.; Fernández-Rueda, J.L.; Aranda, E.; Rodríguez-Ariza, A. Exhaled breath condensate biomarkers for the early diagnosis of lung cancer using proteomics. *Am. J. Physiol. Lung Cell. Mol. Physiol.* **2017**, *313*, 664–676. [CrossRef]

47. Alharbi, I.A.; Rouabhia, M. Repeated exposure to whole cigarette smoke promotes primary human gingival epithelial cell growth and modulates keratin expression. *J. Periodontal. Res.* **2016**, *51*, 630–638. [CrossRef]

48. Zhou, J.; Jiang, G.; Xu, E.; Zhou, J.; Liu, L.; Yang, Q. Identification of SRXN1 and KRT6A as Key Genes in Smoking-Related Non-Small-Cell Lung Cancer through Bioinformatics and Functional Analyses. *Front. Oncol.* **2022**, *11*, 810301. [CrossRef]

49. Donetti, E.; Gualerzi, A.; Bedoni, M.; Volpari, T.; Sciarabba, M.; Tartaglia, G.; Sforza, C. Desmoglein 3 and keratin 10 expressions are reduced by chronic exposure to cigarette smoke in human keratinised oral mucosa explants. *Arch. Oral Biol.* **2010**, *55*, 815–823. [CrossRef]

50. Godlewski, D.; Wojtyś, P.; Bury, P. Alcohol as a cancer risk factor. *Contemp. Oncol. Współczesna Onkol.* **2000**, *4*, 13–15.

51. Warnakulasuriya, S.; Parkkila, S.; Nagao, T.; Preedy, V.R.; Pasanen, M.; Koivisto, H.; Niemelä, O. Demonstration of ethanol-induced protein adducts in oral leukoplakia (pre-cancer) and cancer. *J. Oral Pathol. Med.* **2008**, *37*, 157–165. [CrossRef] [PubMed]

52. Reidy, J.; McHugh, E.; Stassen, L.F. A review of the relationship between alcohol and oral cancer. *Surgeon* **2011**, *9*, 278–283. [CrossRef] [PubMed]

53. Waszkiewicz, N.; Zalewska, A.; Szulc, A.; Kepka, A.; Konarzewska, B.; Zalewska-Szajda, B.; Chojnowska, S.; Waszkiel, D.; Zwierz, K. Wpływ alkoholu na jame ustna, ślinianki oraz śline [The influence of alcohol on the oral cavity, salivary glands and saliva]. *Pol. Merkur. Lek.* **2011**, *30*, 69–74.

54. Hettmann, A.; Demcsák, A.; Decsi, G.; Bach, Á.; Pálinkó, D.; Rovó, L.; Nagy, K.; Takács, M.; Minarovits, J. Infectious Agents Associated with Head and Neck Carcinomas. *Adv. Exp. Med. Biol.* **2016**, *897*, 63–80. [PubMed]

55. Gissi, D.B.; Gabusi, A.; Tarsitano, A.; Badiali, G.; Marchetti, C.; Morandi, L.; Foschini, M.P.; Montebugnoli, L. Ki67 Overexpression in mucosa distant from oral carcinoma: A poor prognostic factor in patients with long-term follow-up. *J. Craniomaxillofac. Surg.* **2016**, *44*, 1430–1435. [CrossRef] [PubMed]

56. Takkem, A.; Barakat, C.; Zakaraia, S.; Zaid, K.; Najmeh, J.; Ayoub, M.; Seirawan, M.Y. Ki-67 Prognostic Value in Different Histological Grades of Oral Epithelial Dysplasia and Oral Squamous Cell Carcinoma. *Asian Pac. J. Cancer Prev.* **2018**, *19*, 3279–3286. [CrossRef] [PubMed]

57. Jing, Y.; Zhou, Q.; Zhu, H.; Zhang, Y.; Song, Y.; Zhang, X.; Huang, X.; Yang, Y.; Ni, Y.; Hu, Q. Ki 67 is an independent prognostic marker for the recurrence and relapse of oral squamous cell carcinoma. *Oncol. Lett.* **2019**, *17*, 974–980. [CrossRef] [PubMed]

58. Woods, R.S.R.; Callanan, D.; Jawad, H.; Molony, P.; Werner, R.; Heffron, C.; Feeley, L.; Sheahan, P. Cytokeratin 7 and 19 expression in oropharyngeal and oral squamous cell carcinoma. *Eur. Arch. Oto-Rhino-Laryngol.* **2022**, *279*, 1435–1443. [CrossRef] [PubMed]

59. Moody, C.A.; Laimins, L.A. Human papillomavirus oncoproteins: Pathways to transformation. *Nat. Rev. Cancer* **2010**, *10*, 550–560. [CrossRef]

60. Zhang, Y.; Koneva, L.A.; Virani, S.; Arthur, A.E.; Virani, A.; Hall, P.B.; Warden, C.D.; Carey, T.E.; Chepeha, D.B.; Prince, M.E.; et al. Subtypes of HPV-Positive Head and Neck Cancers Are Associated with HPV Characteristics, Copy Number Alterations, PIK3CA Mutation, and Pathway Signatures. *Clin. Cancer Res.* **2016**, *22*, 4735–4745. [CrossRef]

61. Dust, K.; Carpenter, M.; Chen, J.C.; Grant, C.; McCorrister, S.; Westmacott, G.R.; Severini, A. Human Papillomavirus 16 E6 and E7 Oncoproteins Alter the Abundance of Proteins Associated with DNA Damage Response, Immune Signaling and Epidermal Differentiation. *Viruses* **2022**, *14*, 1764. [CrossRef] [PubMed]

62. Qin, T.; Li, S.; Henry, L.E.; Chou, E.; Cavalcante, R.G.; Garb, B.F.; D'Silva, N.J.; Rozek, L.S.; Sartor, M.A. Whole-genome CpG resolution DNA Methylation Profiling of HNSCC Reveals Distinct Mechanisms of Carcinogenesis for Fine-scale HPV+ Cancer Subtypes. *Cancer Res. Commun.* **2023**, *3*, 1701–1715. [CrossRef] [PubMed]

63. Lu, X.; Jiang, L.; Zhang, L.; Zhu, Y.; Hu, W.; Wang, J.; Ruan, X.; Xu, Z.; Meng, X.; Gao, J.; et al. Immune Signature-Based Subtypes of Cervical Squamous Cell Carcinoma Tightly Associated with Human Papillomavirus Type 16 Expression, Molecular Features and Clinical Outcome. *Neoplasia* **2019**, *21*, 591–601. [CrossRef] [PubMed]

64. Amin, M.B.; Greene, F.L.; Edge, S.B.; Compton, C.C.; Gershenwald, J.E.; Brookland, R.K.; Meyer, L.; Gress, D.M.; Byrd, D.R.; Winchester, D.P. The Eighth Edition AJCC Cancer Staging Manual: Continuing to build a bridge from a population-based to a more "personalized" approach to cancer staging. *CA Cancer J. Clin.* **2017**, *67*, 93–99. [CrossRef] [PubMed]

International Journal of
Molecular Sciences

Article

A Transcriptomic Analysis of Laryngeal Dysplasia

Fausto Maffini [1,*,†], Daniela Lepanto [1], Francesco Chu [2,†], Marta Tagliabue [2,3], Davide Vacirca [1], Rita De Berardinis [2], Sara Gandini [4], Silvano Vignati [4], Alberto Ranghiero [1], Sergio Taormina [1], Alessandra Rappa [1], Maria Cossu Rocca [5], Daniela Alterio [6], Susanna Chiocca [7], Massimo Barberis [1], Lorenzo Preda [8,9], Fabio Pagni [10], Nicola Fusco [1,11] and Mohssen Ansarin [2]

[1] Department of Surgical Pathology, European Institute of Oncology IRCCS, 20141 Milan, Italy
[2] Division of Otolaryngology Head and Neck Surgery, European Institute of Oncology IRCCS, 20141 Milan, Italy
[3] Department of Biomedical Sciences, University of Sassari, 07100 Sassari, Italy
[4] Molecular and Pharmaco-Epidemiology Unit, Department of Experimental Oncology, European Institute of Oncology IRCCS, 20141 Milan, Italy
[5] Medical Oncology Division of Urogenital and Head and Neck Tumors, European Institute of Oncology IRCCS, 20141 Milan, Italy
[6] Department of Radiotherapy, European Institute of Oncology IRCCS, 20141 Milan, Italy
[7] Department of Experimental Oncology, European Institute of Oncology IRCCS, 20139 Milan, Italy
[8] Diagnostic Imaging Unit, National Center of Oncological Hadron-Therapy (CNAO), 27100 Pavia, Italy; lorenzo.preda@unipv.it
[9] State University School of Medicine, University of Pavia, 27100 Pavia, Italy
[10] Department of Medicine and Surgery, Pathology, IRCCS Fondazione San Gerardo dei Tintori, University of Milano-Bicocca, 20126 Milan, Italy
[11] State University School of Medicine, University of Milan, 20122 Milan, Italy
* Correspondence: fausto.maffini@ieo.it
† These authors contributed equally to this work.

Abstract: This article describes how the transcriptional alterations of the innate immune system divide dysplasias into aggressive forms that, despite the treatment, relapse quickly and more easily, and others where the progression is slow and more treatable. It elaborates on how the immune system can change the extracellular matrix, favoring neoplastic progression, and how infections can enhance disease progression by increasing epithelial damage due to the loss of surface immunoglobulin and amplifying the inflammatory response. We investigated whether these dysregulated genes were linked to disease progression, delay, or recovery. These transcriptional alterations were observed using the RNA-based next-generation sequencing (NGS) panel Oncomine Immune Response Research Assay (OIRRA) to measure the expression of genes associated with lymphocyte regulation, cytokine signaling, lymphocyte markers, and checkpoint pathways. During the analysis, it became apparent that certain alterations divide dysplasia into two categories: progressive or not. In the future, these biological alterations are the first step to provide new treatment modalities with different classes of drugs currently in use in a systemic or local approach, including classical chemotherapy drugs such as cisplatin and fluorouracile, older drugs like fenretinide, and new checkpoint inhibitor drugs such as nivolumab and pembrolizumab, as well as newer options like T cell therapy (CAR-T). Following these observed alterations, it is possible to differentiate which dysplasias progress or not or relapse quickly. This information could, in the future, be the basis for determining a close follow-up, minimizing surgical interventions, planning a correct and personalized treatment protocol for each patient and, after specific clinical trials, tailoring new drug treatments.

Keywords: laryngeal dysplasia; cancer biology; cancer treatment

1. Introduction

Laryngeal dysplasia is a premalignant lesion that significantly impacts patients' social life and quality of life. Despite advancements in surgical techniques and minimally invasive surgical treatments, the dysplasia in many cases progresses to laryngeal squamous cell carcinoma (LSCC), with a great number of sequels being related to swallowing and speech disorders due to the disease and repeated treatments.

Therefore, identifying biological alterations in the laryngeal mucosa provides the basis for hypothesizing different therapeutic possibilities compared with surgical intervention, aiming to preserve laryngeal function—an ambitious but desirable objective.

LSCC is a stepwise process, progressing from laryngeal premalignant dysplasia (LDy) to invasive carcinoma [1,2]. The diagnosis of LDy is based on cytoarchitectural changes in the squamous epithelium laryngeal wall. However, the histopathologic grading alone is not sufficient to define the risk of malignant progression towards invasive cancer. It also does not allow for standardized management, which can range from simple observation to biopsy under endoscopic control, radical excision with a cold blade, or transoral laser microsurgery with a CO_2 laser [3,4].

Recently, the risk of malignant progression from LDy to LSCC has been found to be related to a series of immunogenetic mutations and extracellular matrix alterations [5].

We recently reported on the different immunogenetic landscape of LDy progressing to invasive cancer over time (progressing dysplasia, PDy) and LDy not evolving towards LSCC (non-progressing dysplasia, NPDy). We found that the difference observed in the two sub-groups, in terms of risk of malignant progression, was significantly related to transcriptional alterations, thus resulting in an up-regulation or over-expression of some genes (over-expression assumed mRNA transcript expression value of >2-fold). Some of these genes were positively related to an augmented risk of LSCC, while others were more expressed in NPDy, thus having a protective role [5].

Furthermore, tumor-infiltrating lymphocytes and altered expressions of specific genes associated with the tumor–host innate immune system have been shown to be independent risk factors for the risk of malignant progression towards LSCC [5,6].

Moreover, 20% of hyperplastic laryngeal lesions with loss of heterozygosis (LOH) involving p16ink4 and p53 might be considered potentially dysplastic. Hyperplasia does not show the same morphological, cytological, and architectural atypia as low-grade dysplasia and could be underestimated [3,7,8].

The decision making in LDy is well defined and unchanged overtime. The therapeutic treatments considered are surgery, radiotherapy (RT), and in some cases, planned a close follow-up [7]. Therefore, if it was possible to identify PDy and NPDy early, we could use different therapeutic and control planning based on the risk of malignant transformation and propose adequate treatment choices [3].

The histopathological report, including grading and the type of lesion (verrucose, mixed or basaloid), is not enough to recognize an NPDy or a PDy [6,9]. It is necessary to identify prognostic factors that can help find the best personalized therapeutic approach, achieving disease control and avoiding or delaying unnecessary surgical or radiotherapy treatments. Understanding the dysplasia transcriptomic alterations may help define the biological behavior of these premalignancies, assuming possible medical therapies improving patients' prognosis [3,5,10].

In this work, we aim at providing new insights into LDy transcriptomic alterations for a better understanding of the cancerogenesis-leading mechanisms.

2. Results

There are significant differences between the NPDy and PDy. Previous reports highlighted the role of the adaptative–innate immune system and TILs in LDy and advanced LSCC. Specifically, the immune system reaction and extracellular matrix factors (ECMs) greatly influenced patients' prognosis in advanced laryngeal cancer and the risk of malignant progression from LDy to invasive cancer [5,10].

Genes belonging to different families with different functions were differentially expressed in NPDy or PDy (Figures 1–3).

S-PLS discriminant genes: genes that descriminate between PD and NPD

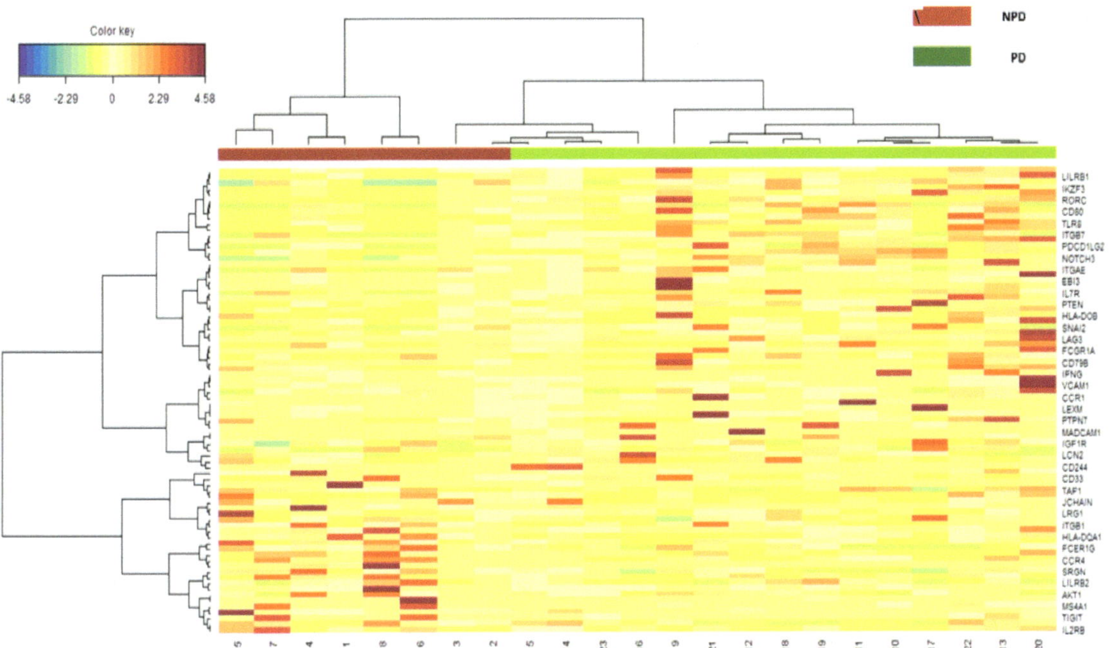

Figure 1. The hierarchical heatmap of the S-PLS discriminant analysis showing the difference in the expression of genes that better discriminate between the PDy (green) and NPDy (red). The color intensity also shows the value of expression. The authors are grateful to Cancers for kindly allowing the use of this table.

Gene expression by PD status

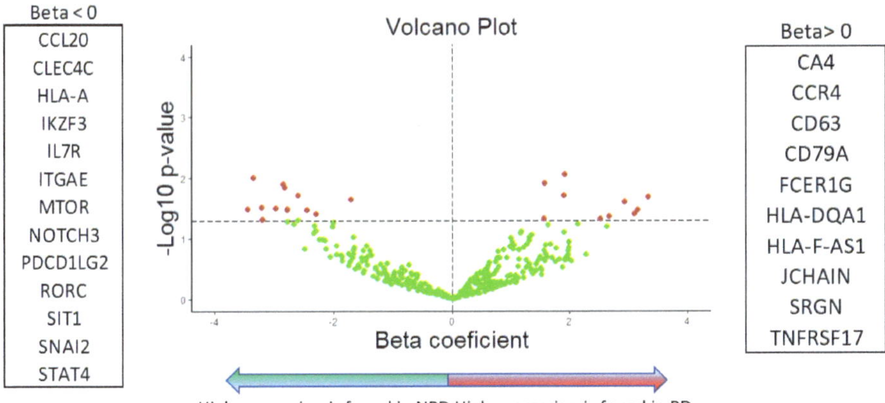

Figure 2. Volcano plot of gene expression by PD (ANOVA models), the red dots indicate gene expression values that show statistical significance, characterised by (*p*-value on the y-axis), versus the magnitude of change (fold change on the x-axis). They are located above the line of significance (horizontal dashed line), allowing us to quickly identify the most biologically significant genes.

Association of gene expression with PD status

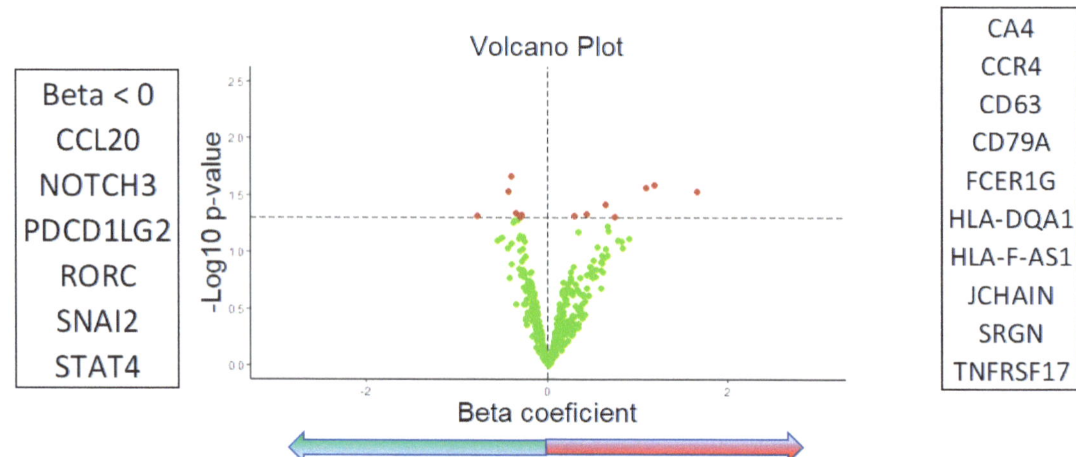

Figure 3. Volcano plot of gene expression association with PDy (Logistic models), also in this image, the red dots indicate gene expression values that show statistical significance, characterised by (*p*-value on the y-axis) versus the magnitude of change (fold change on the x-axis). They are located above the line of significance (horizontal dashed line), allowing us to identify the most biologically significant genes.

Many of them were involved in the chemokine signaling, tumor markers, antigen processing, B cell marker, helper T cells, neutrophil, checkpoint inhibitors, and lymphocyte infiltrate. In cases of PDy, there was a more than 2-fold over-expression of genes compared with the NPDy (Table 1).

Table 1. This table shows the genes involved in NPDy and PDy that results in a significant statistical value (*p* < 0.05), and for all genes, we can observe the function of genes associated with dysplasia; the nomenclature NCBI accession is: https://www.ncbi.nlm.nih.gov/nuccore/NM_005060.4 (accessed on 3 September 2024). (UR, up-regulated; OE, over-expression >2-fold).

NCBI-Accession CODE	Gene Function	PDy	NPDy	Gene
NM_004591	Chemokine signaling		UR	*CCL20*
NM_005508	Chemokine signaling	OE		*CCR4*
NM_000435	Tumor marker		UR	*NOTCH3*
NM_003068	Tumor marker steamness		UR	*SNAI2*
NM_002122	Antigen processing	OE		*HLA-DQA1*
NR_026972	Antigen-processing	OE		*HLA-F-AS1*
NM_001192	B cell marker	OE		*TNFRSF17*
NM_144646	B cell marker	OE		*JCHAIN*
NM_001783	B cell-receptor signaling	OE		*CD79A*
NM_025239	Checkpoint pathway		UR	*PDCD1LG2*
NM_003151	Helper T cell		UR	*STAT4*

Table 1. *Cont.*

NCBI-Accession CODE	Gene Function	PDy	NPDy	Gene
NM_005060	Helper T cell		UR	*RORC*
NM_000717	Neutrophil	OE		*CA4*
NM_001780	Lymphocyte infiltrate	OE		*CD63*
NM_002727	Lymphocyte infiltrate	OE		*SRGN*
NM_004106	Lymphocyte infiltrate	OE		*FCER1G*

2.1. Dysregulated Genes Involving the Chemokine Signaling

The motif chemokine ligand 20 (*CCL20*) regulating the chemokine signaling was up-regulated in NPDy, while G protein-coupled receptor family 4 (*CCR4*) was over-expressed in PDy.

CCL20 is a secretory protein involved in inflammation, modulating the immune response and showing chemotactic activity among the lymphocytes and myeloid series. In our cases, this protein was observed to be up-regulated in NPDy. No reports indicated an up-regulation of *CCL20* in the dysplastic lesion, but it has been reported to be up-regulated in lung adenocarcinoma as a poor prognostic marker [11]. In lung adenocarcinoma, *CCL20* promotes tumor progression by increasing the epithelial–mesenchymal transition (EMT), facilitating invasion, and metastasizing. This effect is linked to TGF-β and its receptor *CCR6*. Tao Fan et al. (2022) suggested that the activity of *CCL20* appears to be linked to the tumor's non-response to immunotherapy when linking to TGF-β, blocking the immune checkpoint receptor, and promoting the metastatic diffusion [11,12]. Qian Song et al. (2020) suggested another pathway in which *CCL20* was up-regulated by *RUNX3*, recruiting the CD8$^+$ T cells. [12]. The CC motif chemokine receptor 6 (*CCR6*) was observed to be increased in high TILs in PDy (recruiting the lymphocyte and dendritic cells) and modified the extracellular matrix (ECM). *CCR6* binds with its ligand *CCL20*, promoting cancer progression. The *CCL20–CCR6* axis was involved in neoangiogenesis due to ischemic damage [13,14]. In NPDy, the axis could be inactive, blocking or slowing down the progression of dysplasia. It is noteworthy that *CCL20* was reported to be up-regulated in patients who smoke cigarettes [5,13].

Additionally, the G protein-coupled receptor family 4 (*CCR4*) is a chemokine signaling-related receptor. This receptor for CC motif chemokine is responsible for homeostasis, and it regulates many types of leukocytes, the immune system, and angiogenesis; it was found to be more expressed in PDy. *CCR4* was reported to be over-expressed in many hematological and solid tumors and also in head and neck tumors. Zihang Ling et al. (2022) showed that *CCR4* was over-expressed in tumors with poor prognosis in head and neck squamous cell carcinoma (HNSCC). Their work demonstrates that the axis with its ligand CC motif chemokine ligand 2 (*CCL2*) promotes motility and progression with increased metastatic ability. The authors, by inhibiting this characteristic with siRNA (silencing RNA), reported a lowered ability to exhibit motility, progression, and metastatic potential in HNSCC [15]. The over-expression in PDy could increase the motility of dysplastic cells, thereby enhancing tumor progression.

2.2. Major Histocompatibility Complex Dysregulated Genes

The histocompatibility complex *HLA-DQA1* and *HLA-F-AS1* were found to be over-expressed in cases of PDy. The *HLA-DQA1* complex belongs to a major histocompatibility complex (HLA), class II, DQ alpha 1, mRNA. This membrane receptor protein plays a role in presenting antigens to the antigen-presenting cells (APCs) of the immune system for CD4 T cell recognition. There are limited data highlighting the role of *HLA-DQA1* in carcinoma progression. Sheng-Chien Tsai et al. (2011) described the protective role of *HLA-DQA1* in carcinoma development, but they did not find an up-regulation between control and tumor cases [16]. *HLA-DQA1* was reported to be over-expressed in sarcoma, and this over-expression was associated with long-term survival [4]. However, an over-expression of this gene in PDy was observed in the cases under consideration, suggesting that *HLA-DQA1* may attempt to enhance APC functions and present antigens to CD4 T cell but fails, possibly due to a blockage in its pathways.

HLA-F-AS1 is another major histocompatibility system (HLA-F) antisense RNA 1 transcript variant 1, a long non-coding RNA (lncRNA). This was over-expressed in the cases of PDy under study, suggesting a role in dysplasia progression with a mechanism similar to that observed for *HLA-DQA1* but involving different pathways. This observation aligns with the findings of Huang Y. et al. (2020), where the *HLA-F-AS1* was up-regulated in colorectal adenocarcinoma, promoting carcinoma progression via miR-330-3p. The *HLA-F-AS1* lncRNA FGD5-AS1-miR-330-3p axis was reported to be altered. This axis enhances the resistance of neoplastic cells to 5-fluorouracile drugs (5-FU) via epithelial growth factor receptor (EGFR) in colorectal cancer [17,18].

2.3. Dysregulated Genes of B Cell Markers

The B cell marker *TNFRSF17*, *JCHAIN*, and CD79A were observed to be over-expressed and associated with PDy in the cases under consideration.

The tumor necrosis factor receptor superfamily 17 mRNA (*TNFRSF17*), belonging to a superfamily of TNF-receptor, was expressed in mature B lymphocytes. Its significance lies in enhancing the immune response through the nuclear factor kappa-light-chain-enhancer of activated B cell (NF-kB) activation. This transcription factor activates MAPK/JNK and reprograms the B cell [19]. *TNFRSF17* can use a different pathway in the transduction of the signal, employing a tumor necrosis factor receptor-associated factor (TRAF) member family, thereby increasing cell survival and proliferation.

Another gene that was over-expressed in PDy was a *JCHAIN* (Homo sapiens joining chain of multimeric IgA and IgM, mRNA) [20]. This gene regulates the function of polymeric IgA (pIgA) and pentameric IgM. IgA and IgM are secretory immunoglobulins (SIgs) secreted by mucosal plasma cells associated with lymphoid tissue in the oral cavity (MALT). IgA and IgM were created by plasma cells in the rough endoplasmic reticulum and were synthesized in the Golgi apparatus. When synthesized, the Golgi apparatus formed a dimer of IgA with *JCHAIN*, creating a polymeric IgA (pIgA) and linking the IgM in a pentameric structure. After polymerizing with *JCHAIN*, the pIgA and a pentameric IgM were delivered and secreted into the extracellular matrix. This dimeric pIgA and pentameric IgM strongly binded with their receptor (pIgR) through *JCHAIN* on the epithelial cell. The epithelial cell then transported the pIgA and a pentameric IgM from the basal layer to surface layer for secretion on the luminal surface [21]. Why *JCHAIN* was up-regulated in PDy is unknown. We suppose that the secretory activity of epithelial cells was lost, resulting in the accumulation of pIgA and pentameric IgM, possibly due to increased TILs, as reported in our previous work [5]. We speculate that in PDy, the loss of SIg does not allow for the agglutination and cytolyzing of bacteria, leading to damage on the natural barrier on the surface of the larynx and increasing the possibility of infection. It is known that SIg is important in preventing the infection by agglutinating and cytolyzing the bacteria. The deficiency in the immune system to infection increases the production of IgA and IgM. In dysplastic lesions, blockage and slowing down of the delivery activity of immunoglobulin transport from the basal side to luminal surface favor bacteria proliferation and inflammation [22].

The human immunoglobulin-associated alpha transcript variant 1 gene (CD79A), encoding for a Sig receptor-α on the B cells, is necessary for the antigenic B cell receptor function. According to our data, CD79A is over-expressed in PDy. We speculate that this could be due to the decreased SIg secretion in response to epithelial damage, followed by increased infection.

Our previous work showed that CD79A was up-regulated and associated with better prognosis in invasive laryngeal carcinoma, and this was related to the stage of the disease; CD79A was more up-regulated in pT1-3 then in pT4. This up-regulation was also demonstrated by Yan Chen et al. (2022) in their work [10,21].

These genes (*JCHAIN* and CD79A*)* were reported to be lower expressed altogether in colorectal carcinoma and in others non-squamous cell carcinoma by Pan J. et al. (2021), suggesting a potential indicator for immunotherapy. The role of the B cell regulatory pattern can improve the immune response against cancer [22,23]. The high expression of *JCHAIN* and CD79A could decrease the immune response in PDy lesions and may be ineffective in preventing dysplasia progression.

2.4. Dysregulation of Lymphocyte Infiltrate Genes

In PDy, we observed an over-expression of CD63, *SRGN*, and *FCER1G*, which are markers of lymphocyte infiltration.

Fc ε-receptor Ig (*FCER1G*) encodes a protein similar to the Fc fragment of γ-Ig and has a pivotal role in infection by activating the phagocytosis in myeloid cells and inducing an antineoplastic activity. The immune system also induces the up-regulation of *FCER1G* in PDy [24,25]. *FCER1G* is a high-affinity IgE receptor and was found to be over-expressed in PDy in our cases. It regulates many aspects of the immune response by binding to many FcR α-chain receptors among CD8$^+$T cells, CD4$^+$T cells, B cells, macrophages, and dendritic cells (DCs). This boosts the inflammatory response by recognizing neoplastic antigens and providing an innate-like immune response with high cytotoxic activity [26,27]. *FCER1G* was involved in the activation via the neutrophil of collagen, mediated by integrins. Although *FCER1G* dysregulation has been reported in multiple myelomas with a good prognosis [28], in some reports, it's up-regulation was associated with poor overall survival and disease-free survival in head and neck carcinoma [29,30]. The exact mechanism of *FCER1G* in dysplastic lesions of the head and neck is unknown, but chronic inflammation is suggested as an etiopathogenetic factor, as proposed by Mantovani et al. (2008) [31].

CD63 is a surface protein belonging to the transmembrane 4 superfamily, regulating motility, growth, and cell development. Down-regulation of CD63 was associated with metastasis and neoplastic progression in melanoma and other tumors [32,33], as well as increased motility in tumor cells and enhanced matrix-degrading ability. This surface receptor is crucial in ECM control, and its down-regulation boosts ECM degradation linked with interring beta1 [34,35]. Takino T. et al. (2003) demonstrated that CD63 has the ability to link membrane-type 1 matrix metalloproteinase (MT1-MMP) and influence the lysosomal proteolysis [35]. In the early phase of melanoma progression, CD63 was up-regulated; however, during disease progression, there is a down-regulation in CD63, increasing a metastatic potential [34]. In PDy, the up-regulation of CD63 aims to inhibit the development of the metastatic potential and mobility of dysplastic cells while simultaneously inhibiting the proteolysis of ECM against cancer progression.

The serglycin transcript variant 1 mRNA (*SRGN*) is a protein stored in secretory granules of many hematopoietic cells, neutralizing proteolytic enzymes. It was over-expressed in PDy in our cases. In the literature, *SRGN* is an important poor prognostic marker in lung adenocarcinoma negative for TTF1 expression (thyroid transcription factor 1), while in nasopharyngeal carcinoma, it was reported to be highly expressed in metastatic cases [36]. *SRGN* has been associated with monocytes that secrete *SRGN*. The over-expression of *SRGN* in PDy suggests an action against ECM degradation by neutralizing proteolytic enzymes.

2.5. Dysregulated Neutrophil Genes

Carbonic anhydrase 4 (CAIV or CA4) is a mRNA transcriptor codifying for a zinc metalloenzyme. It has catalytic properties that participate in several biological activities, such as collagen and bone resorption, acid–base balance, respiratory function, gastric acid regulation, cerebrospinal fluid regulation, and others. This enzyme was reported to be up-regulated in chronic obstructive pulmonary disease (COPD). Nava et al. (2021), using the immunohistochemical method, found that increased expressions of CA4 in chondrocytes and collagen in patients with COPD were followed by fibrosis and cartilaginous calcification of the bronchial wall. This is because of a higher pH promoted by CA4 and a long-standing lung inflammation [37]. As of our knowledge cutoff date, there is no available literature that directly compares CA4 up-regulation with prognosis and progression in HNSCC and dysplasia. The only work demonstrating a predictive value was found in colorectal carcinoma (CRC), where CA4 and CA1 were predictive factors for prognosis [38]. In our cases of PDy, we speculate that inflammation promotes the production of CA4, leading to collagen degeneration and desmoplasia and, in advanced cases, allowing for the dysplastic progression to infiltrate the carcinoma by altering the ECM. However, further research is needed to understand the specific role of ECM in dysplasia.

2.6. Dysregulated Helper T Cell Genes

The signal traducer and activator of transcription 4 mRNA (*STAT4*) was found to be up-regulated in NPDy. This gene is linked to a surface receptor mediated by interleukin-12 (IL-12), which activates its transcription. IL-12 plays a key role in regulating T-helper cells. According to W.E. Thierfelder et al. (1996), IL-12, through INF-γ, contributes to the defense against bacterial and parasitic infection. Disruption of IL-12 and INF-γ deprives mitogenesis and enhances the cytolytic property of natural killer cells (NKs) due to the loss of thyrosinase phosphorylation activation in T lymphocytes [39–42]. We hypothesize that the up-regulation of *STAT4* activates CD4⁺T helper cells and increases NK cytolytic function, preventing neoplastic growth.

Retinoic acid-related orphan receptor C mRNA (*RORC*) is a DNA-binding protein receptor located in the nuclear membrane belonging to a subfamily of nuclear hormone receptors (NR1). The function of this gene is still unknown, but in vitro studies indicate its activity in lymphoid and thymus development, particularly in the subtype of CD4⁺ T cells known as T-helper 17 (Th17) [41,42].

While only one study reported a better prognosis for *RORC* up-regulation in ovarian surface carcinoma, indicating a role in regulating and increasing pro-apoptotic activity and suppressing the neoplastic growth, a down-regulation of *RORC* is associated with poor prognosis in bladder carcinoma. An over-expression of *RORC* has been linked to the down-regulation of PD-L1 mRNA [43–45]. The up-regulation of *RORC* in the cases under study enables the Th17 cell activity, delaying dysplastic progression and potentially down-regulating PD-L1 [45]. Our cases did not exhibit an up- or down-regulation of PD-L1 mRNA. Further studies are necessary to understand the mechanisms of *RORC* action in laryngeal dysplasia.

2.7. Dysregulated Genes Linked to Checkpoint Inhibitors

Programmed cell death 1 ligand 2 mRNA (*PDCD1LG2*) was found to be up-regulated in NPDy. Checkpoint inhibitors have emerged as crucial drugs in the treatment of HNSCC. PD-L2 works synergistically with PD-L1 in the control of the disease. PD-L2 was negatively correlated with CD4⁺ and CD8⁺ T cells through DCs, which belong to APCs and its ligand PD-L2 in T cells [46,47]. Research by Yearley J.H. et al. (2017) demonstrated that PD-L2's prevalence, independently of PD-L1, predicts the clinical response to pembrolizumab in neoplastic, stromal, and immune cells, leading to long progression-free survival (PFS). The association between PD-L1 and PD-L2 has been shown to improve the response to pembrolizumab by up to 27.5% [48]. The only difference between this study and Yearly's was the method employed to evaluate PD-L2. We evaluated the up-regulation of mRNA, while Yearley evaluated the expression of PD-L2 with an immunohistochemical assay [48].

Moratin J. et al. (2019) showed an over-expression of PD-L2, evaluated through immunohistochemistry, in patients with HNSCC, which is predictive of a poor prognosis [49]. Another mechanism linked to *PDCD1LG2* involves its fusion with CD274 (PD-L1). Bossi P. et al. (2017) reported that the presence of fusion genes CD274-*PDCD1LG2*, found in 50% of their cases, was associated with a shorter progression-free survival [50].

This unexpected result opens the door to interesting speculations, suggesting that immunotherapeutic drugs such as nivolumab and/or pembrolizumab could be used in H&N dysplasia therapy, particularly when the surgery fails to control the disease. This concept holds promise and should be investigated further to explore its potential in clinical applications.

2.8. Dysregulated Tumor Genes

The notch receptor 3 mRNA (*NOTCH3*) gene belongs to the family of transmembrane receptors that includes *NOTCH* gene 1, 2, and 4 and three transmembrane receptors, that are important in the embryonic development of mammals. The alteration of NOTCH3 was observed in patients with inherited cerebral autosomal-dominant arteriopathy disease with subcortical infarct and leukoencephalopathy syndrome (CADASIL), characterized by cerebral developments dysfunction [51]. In NPDy cases, this gene was up-regulated. Zhang Y.Q. et al. (2021) reported that *NOTCH3* up-regulation inhibits the AKT–mTOR pathway and regulates the PTEN expression in breast carcinoma [52]. This *NOTCH3*-mediated pathway inhibits the proliferation and malignancy in breast cancer [50], suggesting a role for *NOTCH3* in delaying cancer progression in dysplastic lesions of the larynx. Additionally, the up-regulation of *NOTCH3* was reported in SCC, and it has been associated with resistance to the 5-FU drug and reduced sensitivity to cisplatin in nasopharyngeal carcinoma [52–55].

Regarding *SNAIL* family transcriptional repressor 2 mRNA (*SNAI1* and *SNAI2*), a member of the C2H2 zinc finger transcription factors, *SNAI2* was found to be up-regulated in our cases of NPDy. In a previous study, it was reported that *SNAI1* was up-regulated in PDy in high TIL cases [5]. This gene is involved in epithelial–mesenchymal transition (EMT), a process in which the differentiating cell acquires stem cell characteristics and the ability to differentiate into various cell lineages [47,56]. This transition results in a hybrid cell with a mesenchymal and epithelial phenotype.

Slug (*SNAI2*) belongs to the same family; however, it has a distinct role in regulating cells with stem-like features by inhibiting them. It's up-regulation and/or over-expression could inhibit EMT due to *SNAI1* [56]. This may occur because SNAI2 exerts its regulatory influence over the epithelial cell adhesion molecule (EPCAM), a molecule associated with stem cells during embryogenesis [56,57].

3. Discussion

This paper aimed to answers and insights on the following: (1) the observed differences in dysplastic lesions related to alterations of transcriptional genes belonging to the innate immune system; (2) how these alterations could be used for medical therapy; (3) the role of inflammation, due to local bacterial infection, in increasing neoplastic progression; (4) The differences between PDy and NPDy related to the host's innate immune system which could be subjects for future medical therapy when standard treatments, such as surgery or RT, fail to control the disease.

These results suggest that every dysplasia can be considered a unique disease with different biological alterations involving the dysplasia itself, the host immune system, the ECM, and bacterial infection. These observations in the progression of dysplasia were reported by Chai Peng Gan et al. (2022), who noted that dysplastic lesions show different evolutions due to different transcriptional analyses [58].

It is well established that oral cavity and larynx tumors arise after precancerous damage. These types of alterations involving the upper areal digestive system were termed "Field Cancerization" and have been demonstrated since 1953 by Slaughter et al. [59] (Figure 4A–F). The genesis of dysplasia is also well established and attributable to the LOH of 9p21 involving $p16^{ink4a}$ and 17p13 involving p53 [59,60], both of which are proteins involved in cell cycle control and pro-apoptotic factors. From a histopathological point of view, this genetic LOH causes overgrowth, leading to a change in the epithelial morphology. The staining properties of cells become darker due to an increase in nuclear DNA. This alteration has also been observed in hyperplastic reactive lesions [1,2,9].

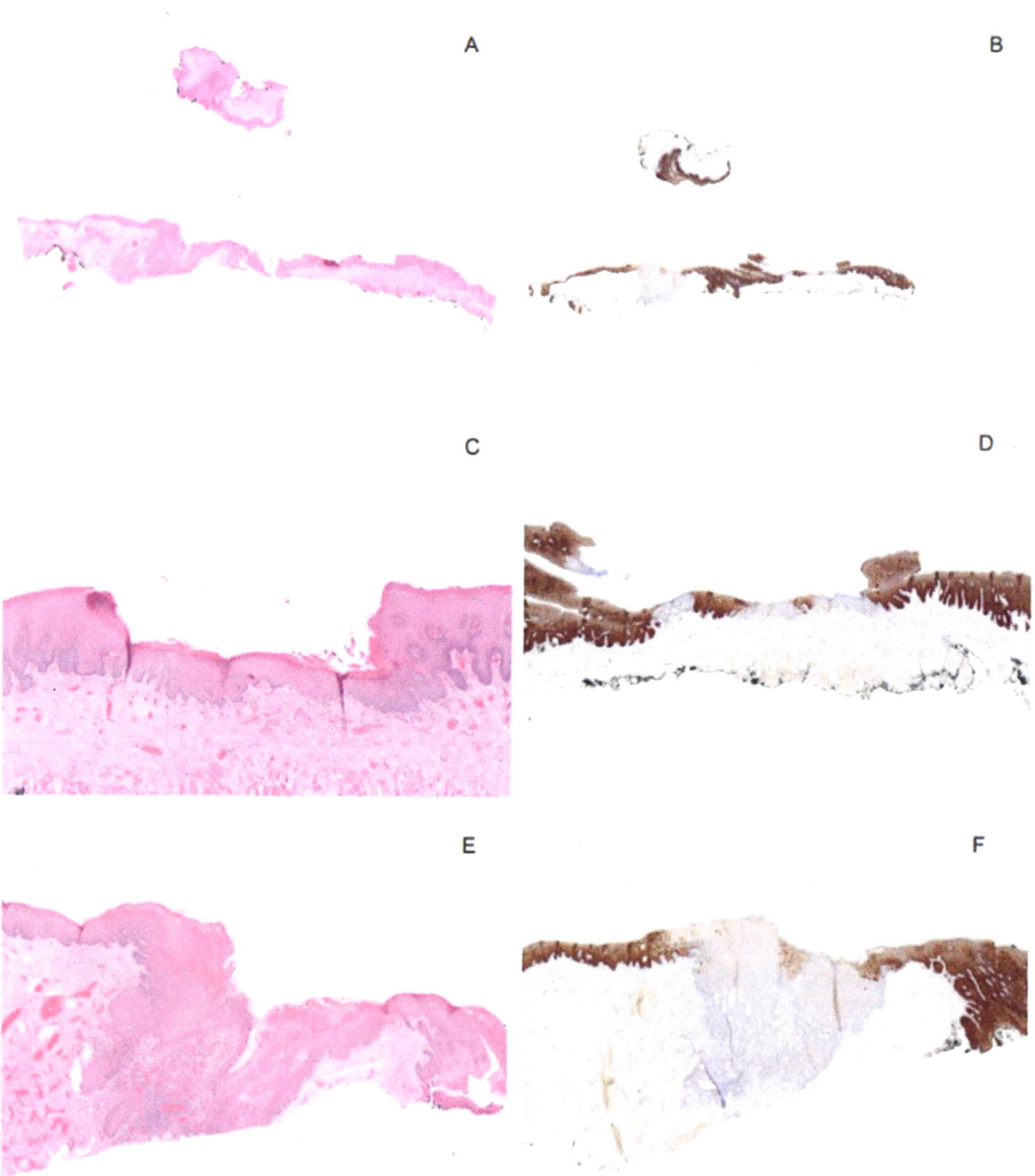

Figure 4. Field cancerization, (**A–F**): This example image shows the "field cancerization" in a mucosal epithelium of the oral cavity; the H&E stain in the left of the image shows how the pathologist observes the dysplasia (**A,C,E**). The overall image showed (in (**A**)) has four points of dysplasia; one of these easier to evaluate morphologically (in (**E**)), while the others are difficult to differentiate from a hyperplastic reactive epithelium that show the same alteration (hyperkeratosis with hypergranulosis and elongation of the rete ridge) of a low-grade dysplasia (in (**C**)). The right images show how staining with keratin 13 (according to the factory guideline Supplementary S3) helps the pathologist to differentiate a hyperplastic-reactive-normal epithelium (brown staining) from dysplasia (unstained epithelium) (in (**B,D,F**)). Of interest are the foci of dysplasia, where the normal epithelium is spaced;

this is a situation of a large number of cases where the dysplastic epithelium involves the entire epithelial surface as a "leopard patch". This distribution of dysplasia creates a difficult therapeutic choice and leads to difficult recovery and control of the disease. While the image (in (**E**)) was observed by the physicians as a leukoplakia "white spot", the others (in (**C**)) could disappear at the physician's evaluation. The same morphological alteration can affect the laryngeal epithelium with the same histological and immunohistochemical alteration reported here in the oral cavity. The magnification of all images is 100X (Objective 5× and Eyepiece ×20).

Defining the degree of dysplasia could be subject to many differences in interpretation by the pathologist and may be prone to errors, especially in the case of low-grade dysplasia and hyperplastic lesions (Figure 4A–F). However, it remains as the only information about tumor progression and the type of tumor that might develop [6]. Actually, the type of dysplasia (I.E basaloid- vs. simplex-type) shows different progression timing and histopathological correlation [6]. This results in different dysplastic lesions, each with different prognoses and treatment results [3,6].

Following the microsatellite instability, Jang S. et al. (2001) showed that the dysplasia may involve different clonal cells [61–63], reinforcing the idea of a multiple lesion process in neoplastic premalignant lesions that are "ab-initio" different [63]. In HNSCC, the dysplastic lesions progress to invasive carcinoma in 17.6% of cases at 4 years and 36,4% at 8 years [64]. This has a significant impact on therapeutic choices. In fact, we know that "Field Cancerization" can reduce the recovery chance due to many dangerous points in the oral cavity and laryngeal wall [60,63]. Over time, this decreases surgical therapy with curative intentions, which still remains as the best treatment. The diagnosis of low-grade dysplasia (SIN 1) is difficult and sometimes merges with a hyperplastic lesion. An amount of 20% of cases of hyperplastic lesions are true early dysplasia and need to be treated with curative intentions. At the end of the last century, the idea arose that these patients could be treated with medical therapy [3,8,64]. Costa A. et al. (1994) published that fenretinide N-(4-hydroxypheniyl) retinamide (4-HPR), an analog of vitamin A, was able to control and prevent the recurrences and new localization of oral dysplasia [65–67].

In 2005, F. Chiesa et al. conducted a clinical trial on the adjuvant chemotherapy of dysplastic lesions of the oral cavity using fenretinide. After one year of therapy, the authors found that fenretinide was effective in controlling recurrence but was associated with side effects [65]. Discriminating between NPDy and PDy and identifying cases that quickly progress to carcinoma versus those with delayed or no progression to carcinoma is now imperative. This work underscores the importance of transcriptomic analysis in dysplasia, suggesting that up-regulation of certain genes can help differentiate the latency of progression. This information is crucial for future medical therapy, especially in cases where surgical therapy fails to control the disease.

In our previous work, we demonstrated that TILs and gene alterations linked to TILs could assist physicians in selecting the most effective treatment, stratifying patients with high-risk evolving dysplasia based solely on TIL characteristics [5,10]. In this study, independent of TILs, we sought to assess transcriptional differences among patients with NDPy and PDy [5,10].

Among the 395 genes evaluated, certain genes were up-regulated or over-expressed, dividing the population in PDy and NPDy We focused on genes that showed statistical significance, derived from various classes such as chemokine signaling, B cell markers, lymphocyte infiltration, helper T cells, antigen processing, checkpoint inhibitors, neu-trophils, and tumor markers (Table 2). The up-regulation of *CCL20* in NPDy has been associated with smoking habits [68]. Additionally, this gene imparts resistance to certain drugs such as 5-FU, platinum-based drugs, and some immunotherapeutic drugs used in HNSCC treatment [1,5,12,68,69]. In invasive carcinoma, *CCL20* has been reported to be up-regulated in association with fenretinide therapy (2.03-fold) and antifolate treatment, suggesting a synergistic action between *CCL20* and fenretinide and an inverse correlation with thymidylate synthetase (TS) [69–71]. In cases of methotrexate therapy (MTX), there is an up-regulation of *CCL20* together to TS reduction, increasing the pro-inflammatory response and secondary platinum sensitivity due to TS reduction [71].

Table 2. Patient characteristics by PD status.

p-Value [1]	PD n = 15 (100%)	NPD n = 31 (100%)	Characteristics
0.2			Sex
	1 (6.7%)	8 (26%)	Female
	14 (93%)	23 (74%)	Male
0.2	66 (56, 70)	58 (54, 64)	Age, median (IQR)
0.8			Smoking
	3 (20%)	4 (13%)	Never
	6 (40%)	13 (42%)	Current
	6 (40%)	14 (45%)	Former
0.6			Cancer site
	13 (87%)	27 (87%)	Glottis
	0 (0%)	2 (6.5%)	Supraglottis
	2 (13%)	2 (6.5%)	Glottis + Supraglottis
0.2			Side
	11 (73%)	28 (90%)	Monolateral
	4 (27%)	3 (9.7%)	Bilateral
0.029			Histological grade
	3 (20%)	11 (35%)	SIN 1
	1 (6.7%)	10 (32%)	SIN 2
	11 (73%)	10 (32%)	SIN 3
0.4			Multifocal
	11 (73%)	26 (84%)	No
	4 (27%)	5 (16%)	Yes

[1] *p*-values: Fisher's exact or chi-square test or Wilcoxon rank sum test for the continuous variable: we are grateful to Cancers for their kindly concession to use this picture.

We are aware that a high level of TS is responsible for drug treatment failure in laryngeal carcinoma, as reported in 2015 by Cossu Rocca et al. [72]. In cases of NPDy, an up-regulation of *CCL20* was observed, suggesting a potential therapeutic role for fenretinide, as reported in clinical trials by Chiesa F. et al. (2005), and the folate inhibitor drugs that improve treatment results [64,71,73]. These data help to identify dysregulated genes that could guide the best chance of therapeutic proposals as alternatives to surgery [64,73]. Of particular interest is that this gene, involving the immune system, alters neoplastic features and provides the ability to resist certain antineoplastic drugs.

While all literature works consistently demonstrated that *CCL20* up-regulation in invasive carcinoma, leading to a poor prognosis, the reasons behind its occurrence in dysplasia, particularly in association with NPDy, remain unknown. Notably, *RORC* is up-regulated in NPDy, whereas it has been reported to be down-regulated in carcinomas [71]. The interplay among ECM, tumors, and the inflammatory system plays a crucial role in the progression from dysplasia to carcinoma.

Studies by Lekva et al. (2013) have demonstrated that low levels of *RORC* in pituitary adenomas were associated with progression, tumor growth, and unfavorable response to treatment with somatostatin analog drugs [71]. Other studies showed a correlation between a low expression of *RORC* and a negative correlation with PD-L1 in bladder carcinoma, suggesting the suppression of the PD-L1/ITGB6 pathway [45,46,74]. The level of *RORC* has been identified as a predictor of prognosis and therapy response in bladder carcinoma. Additionally, a high intake of dietary retinoids has been linked to a 15% reduction in the incidence of bladder carcinoma [46,74]. Therefore, the association between *RORC* and fenretinide could be considered as a potential therapeutic strategy [45,46,74]. Given this, the possible role of fenretinide in the therapy of laryngeal dysplasia should be reappraised and highlighted through new transcriptional evaluations.

CCR4 over-expression in PDy is correlated with platinum resistance and resistance to immunotherapeutic drugs such as mogamulizumab. The *CCR4* receptor is expressed by neoplastic cells and CCR4$^+$Treg cells. CCR4$^+$Treg cells modify the environment, allowing tumor cells to escape from the immune host response. Suppression of *CCR4* due to mogamulizumab, a monoclonal antibody IgG1-k, reduces the tumor burden and enhances antitumoral activity by decreasing CCR4$^+$Treg cells and increasing the immune response [75,76]. In cases of PDy, we speculate that controlling the disease with mogamulizumab could be possible. Another option in these cases could be the potential association of mogamulizumab with another immunotherapeutic drug, such as nivolumab, as suggested by D.S. Hong et al. (2022) in their phase I/II study, although without an observed improvement in efficacy [77,78].

Of particular interest is the new T cell therapy for solid tumors, similar to the Car-T in hematological malignancies, which could be used in small dysplastic lesions of the larynx, especially when *CCL20* and *CCR4* are the most important gene alterations observed in dysplasia. [79,80]. In laboratory experiments, isolated cells expressing *CCL20* and *CCR4*, stimulated by IL-5 and IL-7, demonstrated an increased ability of CD8$^+$ T cells to penetrate dysplastic cells, thereby improving therapeutic efficiency in fighting dysplasia. This therapy could be associated with mogamulizumab, an anti-CCR4 antibody, to amplify therapeutic efficiency. This immunotherapeutic drug has shown efficacy in T cell lymphomas [81].

PD-L2, encoded by the *PDCD1LG2* gene and belonging to checkpoint inhibitors that regulate the immune host response, has been found in APCs such as DCs, macrophages, helper, and cytotoxic T cells. It's up-regulation in NPDy suggested a potential blockage of the host immune response, promoting neoplastic growth and tumor sensitivity to anti PD-L1 and PD-L2 drugs. We observed that this gene is up-regulated in cases of NPDy, suggesting an early phase of precocious blocking of the immune system response against neoplastic cell transformation. In this early phase, using immune drugs such as nivolumab could be a rational method to control dysplasia progression or delay or even recover from the disease with low side effects. A potential therapeutic approach involves the association of PD-L1 and PD-L2. Yearley et al. (2017) suggest that therapy is more effective when both PD-L1 and PD-L2 are positive. The prognosis and response to pembrolizumab were found to be greater when PD-L1 and PD-L2 were positive compared with PD-L1 positivity alone [48,78,82,83]. We observed that certain genes, which modify the host immune response, alter the tissue microenvironment (TME) acquiring staminal cell status by modifying the EMT. Our study indicated that the protective role of SIgs such as pIgA and pentameric IgM is lost when *JCHAIN* is up-regulated because there is a loss of SIg delivery through epithelial cells. Dysplastic cells are unable to transport pIgA and the pentameric IgM to the surface, compromising mucosal protection against infection. Infection in these cases could favor neoplastic progression due to mucosal inflammation, recruitment of inflammatory cells, and subsequent cellular damage [23,78,83,84]. A recent study by Junling Ren et al. (2023) showed that Porphyromonas Gengivalis infection promotes neoplastic growth by up-regulating the PD-L1 of DCs and CD8$^+$ T cells, suggesting a potential medical therapy with anti-PD-L1. *JCHAIN*, reported to be up-regulated in patients responding to anti-PD-L1 therapy [85], raises speculation about the potential improvement of treatments with antibacterial topical or systemic therapy.

We underlined how various alterations linked to dysplasias can exist in the same patient. The standard treatment should be reappraised; while histopathological grading alone suggests possible treatments, mRNA transcriptional analysis provides interesting information about the stratus of the dysplasia and the potential cancer evolution. Distinguishing between NPDy and PDy in the future could lead to a change in therapy. The evaluation of targetable genes may reduce the need for re-treatment surgery and improve patients' prognoses by selecting cases that require more aggressive therapy based on their transcriptional risk of malignant transformation.

We are aware of the potential for adverse events, the cost, and the lack of studies on the use of immune-focused drugs in the treatment of laryngeal dysplasia.

To date, there have been few studies on the use of drugs with a local focus, such as phototherapy [86]. In order to reduce costs and possible side effects, a local approach could only be chosen by planning specific prospective and multi-center clinical trials.

We acknowledge that this is a pilot study. While the evaluation involves only the mRNA transcripts and does not assess complete pathways and host–tumor interactions, it underscores a tight correlation with the immune system and tumor growth. This correlation may be due to poorly understood pathways involved in pre-neoplastic progression.

An additional point that differentiates our work from others is the selection of patients that all have laryngeal dysplasia. This is crucial because dysplastic lesions in the larynx differ from those in the oral cavity and do not exist for oropharyngeal carcinoma. We are aware that our study involved a small number of cases, implying low statistical power. Larger studies, including not only observational but also clinical trials that consider gene expression, should be planned for the future. Therefore, understanding how the immune system works and which genes are involved in the progression or delay of growth in LDy is pivotal for a better comprehension of the mechanisms controlling pre-neoplastic diseases in the head and neck.

4. Materials and Methods

We conducted a retrospective review of all patients treated with transoral laser micro-surgery (TLM) between January 2005 and December 2020 in the Department of Otolaryn-gology and Head and Neck Surgery at the European Institute of Oncology (EIO), a tertiary comprehensive cancer center. All patients were referred to transoral laser microsurgery with radical intent, following the internal and current international NCCN standard of care [87].

Based on our previous doubled-matched case–control study, we considered 15 cases of LDy that evolved towards invasive LSCC during follow-up (PDy) and 31 cases of LDy that did not degenerate to LSCC (NPDy) for comparison [3].

The collected data included: age, sex, past medical history, pre-operative smoking and alcohol habits, site of the laryngeal lesion, surgical procedure adopted, histopathological findings, severity of squamous intraepithelial neoplasia (SIN), and progression to LSCC during follow-up.

Patients were excluded from the current study in cases of the following: previous LSCC, previous laryngeal surgery, immune diseases, or concurrent medical treatments interfering with immune system function (i.e., scleroderma, rheumatoid arthritis, steroid immunosuppressive treatments, etc.).

To comprehensively address the immune system factors and prognostic immuno-genetic alterations for malignant progression, we assessed the presence of stromal tumor-infiltrating lymphocytes (TILs) in post-surgery specimens. Moreover, we used the RNA-based next-generation sequencing (NGS) panel Oncomine Immune Response Research Assay (OIRRA) (ThermoFisher, Waltham, MA, USA) to measure the expression of genes associated with lymphocyte regulation, cytokine signaling, lymphocyte markers, and check-point pathways in all cases of PDy. We also included two matched pair NPDy cases for each PDy case.

All patients signed an informed consent form for data use for scientific purposes, and the study was conducted in accordance with the Declaration of Helsinki and the guidelines provided by our ethical committee (ID Hospital trial: 2519).

4.1. Histopathological Analysis

We collected 46 cases of LDy from our surgical pathology department archive fol-lowing the guidelines of our ethical committee. Standard staining with Hematoxylin and Eosin (H&E) was performed on formalin fixed paraffin-embedded blocks (FFPE) from the patients. The pathologists (FM, DL) evaluated the LDy grading following the system recommended by the International Agency Research on Cancer of the World Health Organization (IARC-WHO IARC 4th edition 2017) [88].

This system comprises three tailored grading levels: SIN 1 for low-grade dysplasia, SIN 2 for medium-grade dysplasia, and SIN 3 for high-grade dysplasia and carcinoma "in-situ". This classification assesses the level of cell atypia within the epithelium. SIN 1 is characterized by the presence of dysplastic cells in the basal layer of the squamous epithelium, SIN 2 involves dysplastic alterations in the lower two-thirds of the epithelium while maintaining an upper level of maturation, and SIN 3 or "carcinoma in-situ" features dysplastic cells in all epithelial layers without any breaches in the basal lamina. According to this classification, SIN3 and "carcinoma in-situ" have the same high risk of malignant transformation, indicating a lesion that requires prompt treatment [88,89] (Table 2).

From each H&E stain, we studied TILs by selecting an area inclusive of dysplasia with underlying inflammatory cells (lymphocyte, macrophage, plasma cells avoiding the necrosis). Two pathologists (F.M. and D.L.) independently selected these areas in a double-blinded manner, following the methods outlined in our previous work [5].

For genomic analysis, we used the "NCBI-accession code" as the nucleotide analysis data base from GenBank®; we used the NIH genetic sequence database part of the International Nucleotide Sequence Database Collaboration, housed at the National Library Medicine (NLM), Bethesda, MD, USA [90].

4.2. Genetic Analysis

Of the 46 LDy specimens, we successfully extracted RNA transcript from 24 specimens, comprising 9 PDys and 15 NPDys. The RNA-based next-generation sequencing panel OIRRA (TermoFisher, Waltham, MA, USA) was used to assess the expression of 395 genes related to immune system activation. These genes included those associated with lymphocyte regulation, cytokine signaling, lymphocyte markers, checkpoint pathways, and tumor characterization. For the extraction, five unstained slides at 7 μm thicknesses were obtained from representative FFPE blocks. Manual microdissection was performed before nucleic acid isolation to selectively isolate dysplastic tissue. The RNA extraction was automatically performed with the Promega Maxwell instrument (Promega, Madison, WI, USA) using the Promega Maxwell RSC RNA FFPE kit. Sample purification was obtained using paramagnetic particles, providing a mobile solid phase for optimized sample capture, washing, and purification. Each sample was de-paraffinized in 300 μL of mineral oil for 2 min at 80 °C, followed by lysis with a master mix containing 224 μL of lysis buffer, 25 μL of proteinase K, and 1 μL of blue dye. After centrifugation at $10,000\times g$ rpm for 30 s, samples were heated on a 56 °C heat block for 15 min and then transferred on an 80° heat block for one hour. Subsequently, each sample was treated with DNase cocktail containing 26 μL of MnCl2 (0.09 M), 14 μL of DNase Buffer, and 10 μL of reconstituted DNase I to catalyze the hydrolysis of DNA. Samples were incubated for 15 min at room temperature, then centrifuged at maximum speed for 5 min. Finally, the blue phase was transferred to well #1 of a Maxwell® FFPE Cartridge. Each RNA sample was quantified using the QuantiFluor RNA system on the Quantus fluorometer. A quantitative RT-PCR was conducted in triplicate to determine the quality of the RNA samples using serial dilutions of the RNA standard (HL60 total RNA) for calibration (i.e., 50, 12.5, 3.13, 0.78, 0.2, and 0.05 ng/μL). Subsequently, each sample was treated with a reaction mix containing 8.75 μL of 4X TaqMan® Fast Virus 1- Step Master Mix, 1.75 μL of 20X TaqMan® Gene Expression Assay, GUSB, and 21 μL of nuclease-free water. An amount of 31.5 μL of the reaction mix and 3.5 μL of each sample, as well as standard and negative control (NTC), were added to a 96-well plate. An amount of 10 μL of each well was added to adjacent wells to obtain a triplicate. The plate was then sealed with a new MicroAmp™ Clear Adhesive Film (Life Technologies Quality Assurance, Carlsbad, CA, USA) and placed on 7900HT Fast Real-Time PCR System. Samples with a threshold value ≥ 0.2 were considered suitable for library preparation.

The RNA library preparation used an amplicon-based technology, enabling the sequencing of specific regions of interest (ROIs) only. Gene expression analysis was performed using the OIRRA NGS assay (Thermo Fisher Scientific, Waltham, MA, USA), targeting 395 immune-related genes. The RNA fragments underwent reverse transcription to cDNA using the SuperScript™ IV VILO™ (SSIV VILO) Master Mix (Thermo Fisher Scientific, Waltham, MA, USA). For each sample, 3 μL of 5X VILO™ Reaction Mix and 12 μL of total RNA (25 ng) 50 were dispensed onto Ion Code plates and loaded onto the thermal cycle. The final product was loaded onto the Ion Chef instrument (Thermo Fisher Scientific, Waltham, MA, USA) for automatized library amplification with the Ion AmpliSeq DL8 kit (ThermoFisher Scientific, Waltham, MA, USA) with specified thermal conditions (25 cycles for the amplification step, 4 min each). Libraries were then automatically loaded onto the Ion 530™ Chip and sequenced on Ion S5™ System (ThermoFisher Scientific, Waltham, MA, USA), following the manufacturer instructions. Primary analysis of the data was performed through the S5 torrent server. These data included chip well details, such as the percentage of ISP loading and enrichment, percentage of monoclonal and polyclonal DNA fragments, total number of reads, total bases, percentage of usable sequence meeting requirements for polyclonal fragments, low-quality and adapter dimers, as well as final library. Additionally, targeted RNA-sequencing data were analyzed using the Torrent Suite Immune Response RNA plugin, which produced gene transcript quantification from sequence read data.

For gene expression analysis, samples with mapped reads >1,000,000 and valid reads >800,000 were deemed adequate for further analysis. The data were processed using the Affymetrix Transcriptome Analysis Console software (TAC) v4.0. The gene expression sequencing data were transformed into logs and normalized for reads per million.

4.3. Statistical Analysis

Demographic and clinical characteristics were described using descriptive statistics based on PDy status, the main outcome measure. The only histopathological characteristic associated with the progression of dysplasia is the grading $p = 0.029$ (Table 2).

4.4. Gene Expression Analysis

RNA-sequencing data were obtained for the evaluation of gene expression levels. The gene expression read per million (RPM) data were centered log-ratio transformed after a non-parametric multivariate imputation of zeros for compositional data. After excluding one gene with low variability, data for 399 genes were available for this analysis. First, gene expression levels were categorized based on the presence or absence of expression. Second, gene expression levels were classified as "high" or "low" based on whether the expression level was higher or lower/equal to the sample median gene expression, respectively. Lastly, gene expression was considered on a continuous scale.

A heatmap was generated by performing a sparse partial least square differential analysis (sPLS-DA) with 7-fold cross-validation and 100 repeats. The most discriminative genes were selected based on the first and second component loading vectors using MixOmics Package 6.18.1. In particular, sPLS-DA was assessed using 7-fold cross-validation with 100 repeats.

Differences in gene expressions between PDy and NPDy were initially analyzed using univariate ANOVA tests, while univariate logistic regression models were also employed to assess the association of gene expression with PDy. Volcano plots were generated to visualize the results, presenting the beta coefficients obtained from ANOVA and logistic models against significance indicated by a p-value < 0.05.

In order to investigate the association with time to PDy, univariate Cox proportional hazards models were adopted, considering the sample median gene expression values as the cutoff. For genes found to be significantly associated with PDy, boxplots with Wilcoxon tests and survival curves with log-rank tests were presented (Supplementary Figures S1 and S2).

Supplementary Materials: The following supporting information can be downloaded at: https://www.mdpi.com/article/10.3390/ijms25179685/s1.

Author Contributions: Conceptualization, F.M., M.T. and F.C.; methodology, F.M., D.L., F.C., M.T., S.G., S.C., A.R. (Alberto Ranghiero), D.V., A.R. (Alessandra Rappa), R.D.B., S.V., M.B., L.P., S.T., D.A. and M.C.R.; software, S.G., S.V., A.R. (Alberto Ranghiero), A.R. (Alessandra Rappa) and D.V.; validation, F.M., F.C., M.T., S.G. and M.A.; formal analysis, F.M., F.C., M.T., D.L., S.G., S.V., D.V., S.T., A.R. (Alberto Ranghiero) and A.R. (Alessandra Rappa); investigation, D.V., A.R. (Alberto Ranghiero), A.R. (Alessandra Rappa), S.G. and S.V.; data curation, F.M., D.L., F.C., M.T., S.C., S.G. and S.V.; writing—original draft preparation, F.M. and F.C.; writing—review and editing, F.M., D.L., M.T., S.G., N.F., F.P. and M.A.; supervision, M.B., N.F., L.P., F.P., D.A., M.C.R. and M.A. All authors have read and agreed to the published version of the manuscript.

Funding: This research received no external funding.

Institutional Review Board Statement: This study was conducted in accordance with the Declaration of Helsinki and approved by the Institutional Ethics Committee of European Institute of Oncology (protocol code ID Hospital trial: 2519 and date of approval, 14 September 2020).

Informed Consent Statement: Informed consent was obtained from all subjects involved in the study.

Data Availability Statement: The data presented in this study are available upon request from the corresponding author.

Acknowledgments: The authors would like to thank Frialdi for her English editing.

Conflicts of Interest: The authors declare no existing competing interests or financial relationships.

References

1. Califano, J.; Van Der Riet, P.; Westra, W.; Nawroz, H.; Clayman, G.; Piantadosi, S.; Corio, R.; Lee, D.; Greenberg, B.; Koch, W.; et al. Genetic progression model for head and neck cancer: Implications for field cancerization. *Cancer Res.* **1996**, *56*, 2488–2492. [CrossRef]
2. Califano, J.; Westra, W.H.; Meininger, G.; Corio, R.; Koch, W.M.; Sidransky, D. Genetic progression and clonal relationship of recurrent premalignant head and neck lesions. *Clin. Cancer Res. Off. J. Am. Assoc. Cancer Res.* **2000**, *6*, 347–352.
3. Chu, F.; De Santi, S.; Tagliabue, M.; De Benedetto, L.; Zorzi, S.; Pietrobon, G.; Herman, I.; Maffini, F.; Chiocca, S.; Corso, F.; et al. Laryngeal dysplasia: Oncological outcomes in a large cohort of patients treated in a tertiary comprehensive cancer centre. *Am. J. Otolaryngol.* **2021**, *42*, 102861. [CrossRef]
4. Bae, J.Y.; Choi, K.U.; Kim, A.; Lee, S.J.; Kim, K.; Kim, J.Y.; Lee, I.S.; Chung, S.H.; Kim, J.I. Evaluation of immune-biomarker expression in high-grade soft-tissue sarcoma: HLA-DQA1 expression as a prognostic marker. *Exp. Ther. Med.* **2020**, *20*, 107. [CrossRef] [PubMed]
5. Chu, F.; Maffini, F.; Lepanto, D.; Vacirca, D.; Taormina, S.V.; De Berardinis, R.; Gandini, S.; Vignati, S.; Ranghiero, A.; Rappa, A.; et al. The Genetic and Immunologic Landscape Underlying the Risk of Malignant Progression in Laryngeal Dysplasia. *Cancers* **2023**, *15*, 1117. [CrossRef] [PubMed]
6. Arsenic, R.; Kurrer, M.O. Differentiated dysplasia is a frequent precursor or associated lesion in invasive squamous cell carcinoma of the oral cavity and pharynx. *Virchows Arch.* **2013**, *462*, 609–617. [CrossRef]
7. Van Hulst, A.M.; Kroon, W.; van der Linden, E.S.; Nagtzaam, L.; Ottenhof, S.R.; Wegner, I.; Gunning, A.C.; Grolman, W.; Braunius, W. Grade of dysplasia and malignant transformation in adults with premalignant laryngeal lesions. *Head. Neck.* **2016**, *38* (Suppl. S1), E2284–E2290. [CrossRef]
8. Lee, D.H.; Yoon, T.M.; Lee, J.K.; Lim, S.C. Predictive factors of recurrence and malignant transformation in vocal cord leukoplakia. *Eur. Arch. Otorhinolaryngol.* **2015**, *272*, 1719–1724. [CrossRef]
9. Thompson, L.D.R. Laryngeal Dysplasia, Squamous Cell Carcinoma, and Variants. *Surg. Pathol. Clin.* **2017**, *10*, 15–33. [CrossRef]
10. Tagliabue, M.; Maffini, F.; Fumagalli, C.; Gandini, S.; Lepanto, D.; Corso, F.; Cacciola, S.; Ranghiero, A.; Rappa, A.; Vacirca, D.; et al. A role for the immune system in advanced laryngeal cancer. *Sci. Rep.* **2020**, *10*, 18327, Erratum in *Sci. Rep.* **2021**, *11*, 9760. [CrossRef]
11. Fan, T.; Li, S.; Xiao, C.; Tian, H.; Zheng, Y.; Liu, Y.; Li, C.; He, J. CCL20 promotes lung adenocarcinoma progression by driving epithelial-mesenchymal transition. *Int. J. Biol. Sci.* **2022**, *18*, 4275–4288. [CrossRef]
12. Song, Q.; Shang, J.; Zhang, C.; Chen, J.; Zhang, L.; Wu, X. Transcription factor RUNX3 promotes CD8+ T cell recruitment by CCL3 and CCL20 in lung adenocarcinoma immune microenvironment. *J. Cell Biochem.* **2020**, *121*, 3208–3220. [CrossRef] [PubMed]
13. Kadomoto, S.; Izumi, K.; Mizokami, A. The CCL20-CCR6 Axis in Cancer Progression. *Int. J. Mol. Sci.* **2020**, *21*, 5186. [CrossRef] [PubMed]
14. Lörchner, H.; Cañes Esteve, L.; Góes, M.E.; Harzenetter, R.; Brachmann, N.; Gajawada, P.; Günther, S.; Doll, N.; Pöling, J.; Braun, T. Neutrophils for Revascularization Require Activation of CCR6 and CCL20 by TNFα. *Circ. Res.* **2023**, *133*, 592–610. [CrossRef]
15. Ling, Z.; Li, W.; Hu, J.; Li, Y.; Deng, M.; Zhang, S.; Ren, X.; Wu, T.; Xia, J.; Cheng, B.; et al. Targeting CCL2-CCR4 axis suppress cell migration of head and neck squamous cell carcinoma. *Cell Death Dis.* **2022**, *13*, 158. [CrossRef]

16. Tsai, S.C.; Sheen, M.C.; Chen, B.H. Association between HLA-DQA1, HLA-DQB1 and oral cancer. *Kaohsiung J. Med. Sci.* **2011**, *27*, 441–445. [CrossRef]

17. Huang, Y.; Sun, H.; Ma, X.; Zeng, Y.; Pan, Y.; Yu, D.; Liu, Z.; Xiang, Y. HLA-F-AS1/miR-330-3p/PFN1 axis promotes colorectal cancer progression. *Life Sci.* **2020**, *254*, 117180. [CrossRef]

18. Gao, S.J.; Ren, S.N.; Liu, Y.T.; Yan, H.W.; Chen, X.B. Targeting EGFR sensitizes 5-Fu-resistant colon cancer cells through modification of the lncRNA-FGD5-AS1-miR-330-3p-Hexokinase 2 axis. *Mol. Ther. Oncolytics.* **2021**, *23*, 14–25. [CrossRef]

19. Lim, K.H.; Yang, Y.; Staudt, L.M. Pathogenetic importance and therapeutic implications of NF-κB in lymphoid malignancies. *Immunol. Rev.* **2012**, *246*, 359–378. [CrossRef]

20. Johansen, F.E.; Braathen, R.; Brandtzaeg, P. Role of J chain in secretory immunoglobulin formation. *Scand. J. Immunol.* **2000**, *52*, 240–248. [CrossRef]

21. Chen, Y.; Jiang, N.; Chen, M.; Sui, B.; Liu, X. Identification of tumor antigens and immune subtypes in head and neck squamous cell carcinoma for mRNA vaccine development. *Front. Cell Dev. Biol.* **2022**, *10*, 1064754. [CrossRef]

22. Ting, H.S.L.; Chen, Z.; Chan, J.Y.K. Systematic review on oral microbial dysbiosis and its clinical associations with head and neck squamous cell carcinoma. *Head Neck* **2023**, *45*, 2120–2135. [CrossRef] [PubMed]

23. Pan, J.; Weng, Z.; Xue, C.; Lin, B.; Lin, M. The Bioinformatics-Based Analysis Identifies 7 Immune-Related Genes as Prognostic Biomarkers for Colon Cancer. *Front. Oncol.* **2021**, *11*, 726701. [CrossRef] [PubMed]

24. Mancardi, D.A.; Albanesi, M.; Jönsson, F.; Iannascoli, B.; Van Rooijen, N.; Kang, X.; England, P.; Daëron, M.; Bruhns, P. The high-affinity human IgG receptor FcγRI (CD64) promotes IgG-mediated inflammation, anaphylaxis, and antitumor immunotherapy. *Blood* **2013**, *121*, 1563–1573. [CrossRef]

25. Mladenov, R.; Hristodorov, D.; Cremer, C.; Gresch, G.; Grieger, E.; Schenke, L.; Klose, D.; Amoury, M.; Woitok, M.; Jost, E.; et al. CD64-directed microtubule associated protein tau kills leukemic blasts ex vivo. *Oncotarget* **2016**, *7*, 67166–67174. [CrossRef]

26. Chou, C.; Zhang, X.; Krishna, C.; Nixon, B.G.; Dadi, S.; Capistrano, K.J.; Kansler, E.R.; Steele, M.; Han, J.; Shyu, A.; et al. Programme of self-reactive innate-like T cell-mediated cancer immunity. *Nature* **2022**, *605*, 139–145. [CrossRef]

27. Dadi, S.; Chhangawala, S.; Whitlock, B.M.; Franklin, R.A.; Luo, C.T.; Oh, S.A.; Toure, A.; Pritykin, Y.; Huse, M.; Leslie, C.S.; et al. Cancer Immunosurveillance by Tissue-Resident Innate Lymphoid Cells and Innate-like T Cells. *Cell* **2016**, *164*, 365–377. [CrossRef]

28. Fu, L.; Cheng, Z.; Dong, F.; Quan, L.; Cui, L.; Liu, Y.; Zeng, T.; Huang, W.; Chen, J.; Pang, Y.; et al. Enhanced expression of FCER1G predicts positive prognosis in multiple myeloma. *J. Cancer* **2020**, *11*, 1182–1194. [CrossRef]

29. Andreu, P.; Johansson, M.; Affara, N.I.; Pucci, F.; Tan, T.; Junankar, S.; Korets, L.; Lam, J.; Tawfik, D.; DeNardo, D.G.; et al. FcRgamma activation regulates inflammation-associated squamous carcinogenesis. *Cancer Cell* **2010**, *17*, 121–134. [CrossRef]

30. Zhang, X.; Cai, J.; Song, F.; Yang, Z. Prognostic and immunological role of FCER1G in pan-cancer. *Pathol. Res. Pract.* **2022**, *240*, 154174. [CrossRef]

31. Mantovani, A.; Allavena, P.; Sica, A.; Balkwill, F. Cancer-related inflammation. *Nature* **2008**, *454*, 436–444. [CrossRef] [PubMed]

32. Radford, K.J.; Thorne, R.F.; Hersey, P. Regulation of tumor cell motility and migration by CD63 in a human melanoma cell line. *J. Immunol.* **1997**, *158*, 3353–3358. [CrossRef]

33. Chen, Z.; Mustafa, T.; Trojanowicz, B.; Brauckhoff, M.; Gimm, O.; Schmutzler, C.; Köhrle, J.; Holzhausen, H.; Kehlen, A.; Klonisch, T.; et al. CD82, and CD63 in thyroid cancer. *Int. J. Mol. Med.* **2004**, *14*, 517–544. [CrossRef] [PubMed]

34. Jang, H.I.; Lee, H. A decrease in the expression of CD63 tetraspanin protein elevates invasive potential of human melanoma cells. *Exp. Mol. Med.* **2003**, *35*, 317–323. [CrossRef]

35. Takino, T.; Miyamori, H.; Kawaguchi, N.; Uekita, T.; Seiki, M.; Sato, H. Tetraspanin CD63 promotes targeting and lysosomal proteolysis of membrane-type 1 matrix metalloproteinase. *Biochem. Biophys. Res. Commun.* **2003**, *304*, 160–166. [CrossRef]

36. Wang, Y.L.; Ren, D.; Lu, J.L.; Jiang, H.; Wei, J.Z.; Lan, J.; Liu, F.; Qu, S.H. STAT3 regulates SRGN and promotes metastasis of nasopharyngeal carcinoma through the FoxO1-miR-148a-5p-CREB1 axis. *Lab. Investig.* **2022**, *102*, 919–934. [CrossRef] [PubMed]

37. Nava, V.E.; Khosla, R.; Shin, S.; Mordini, F.E.; Bandyopadhyay, B.C. Enhanced carbonic anhydrase expression with calcification and fibrosis in bronchial cartilage during COPD. *Acta Histochem.* **2022**, *124*, 151834. [CrossRef]

38. Liu, Z.; Bai, Y.; Xie, F.; Miao, F.; Du, F. Comprehensive Analysis for Identifying Diagnostic and Prognostic Biomarkers in Colon Adenocarcinoma. *DNA Cell Biol.* **2020**, *39*, 599–614. [CrossRef]

39. Thierfelder, W.E.; van Deursen, J.M.; Yamamoto, K.; Tripp, R.A.; Sarawar, S.R.; Carson, R.T.; Sangster, M.Y.; Vignali, D.A.; Doherty, P.C.; Grosveld, G.C.; et al. Requirement for Stat4 in interleukin-12-mediated responses of natural killer and T cells. *Nature* **1996**, *382*, 171–174. [CrossRef]

40. Bacon, C.M.; Petricoin, E.F.; Ortaldo, J.R.; Rees, R.C.; Larner, A.C.; Johnston, J.A.; O'Shea, J.J. Interleukin 12 Induces Tyrosine Phosphorylation and Activation of STAT4 in Human Lymphocytes. *Proc. Natl. Acad. Sci. USA* **1995**, *92*, 7307–7311. [CrossRef]

41. Yamamoto, K.; Quelle, F.W.; Thierfelder, W.E.; Kreider, B.L.; Gilbert, D.J.; Jenkins, N.A.; Copeland, N.G.; Silvennoinen, O.; Ihle, J.N. Stat4, a novel gamma interferon activation site-binding protein expressed in early myeloid differentiation. *Mol. Cell Biol.* **1994**, *14*, 4342–4349. [PubMed]

42. Wiche Salinas, T.R.; Zhang, Y.; Sarnello, D.; Zhyvoloup, A.; Marchand, L.R.; Fert, A.; Planas, D.; Lodha, M.; Chatterjee, D.; Karwacz, K.; et al. Th17 cell master transcription factor RORC2 regulates HIV-1 gene expression and viral outgrowth. *Proc. Natl. Acad. Sci. USA* **2021**, *118*, e2105927118. [CrossRef]

43. Villey, I.; de Chasseval, R.; de Villartay, J.-P. RORγT, a thymus-specific isoform of the orphan nuclear receptor RORγ / TOR, is up-regulated by signaling through the pre-T cell receptor and binds to the TEA promoter. *Eur. J. Immunol.* **1999**, *29*, 4072–4080. [CrossRef]

44. Marchenko, S.; Piwonski, I.; Hoffmann, I.; Sinn, B.V.; Kunze, C.A.; Monjé, N.; Pohl, J.; Kulbe, H.; Schmitt, W.D.; Darb-Esfahani, S.; et al. Prognostic value of regulatory T cells and T helper 17 cells in high grade serous ovarian carcinoma. *J. Cancer Res. Clin. Oncol.* **2023**, *149*, 2523–2536. [CrossRef] [PubMed]

45. Tratnjek, L.; Jeruc, J.; Romih, R.; Zupančič, D. Vitamin A and Retinoids in Bladder Cancer Chemoprevention and Treatment: A Narrative Review of Current Evidence, Challenges and Future Prospects. *Int. J. Mol. Sci.* **2021**, *22*, 3510. [CrossRef]

46. Cao, D.; Qi, Z.; Pang, Y.; Li, H.; Xie, H.; Wu, J.; Huang, Y.; Zhu, Y.; Shen, Y.; Zhu, Y.; et al. Retinoic Acid-Related Orphan Receptor C Regulates Proliferation, Glycolysis, and Chemoresistance via the PD-L1/ITGB6/STAT3 Signaling Axis in Bladder Cancer. *Cancer Res.* **2019**, *79*, 2604–2618. [CrossRef]

47. Battula, V.L.; Evans, K.W.; Hollier, B.G.; Shi, Y.; Marini, F.C.; Ayyanan, A.; Wang, R.Y.; Brisken, C.; Guerra, R.; Andreeff, M.; et al. Epithelial-mesenchymal transition-derived cells exhibit multilineage differentiation potential similar to mesenchymal stem cells. *Stem Cells* **2010**, *28*, 1435–1445. [CrossRef]

48. Yearley, J.H.; Gibson, C.; Yu, N.; Moon, C.; Murphy, E.; Juco, J.; Lunceford, J.; Cheng, J.; Chow, L.Q.M.; Seiwert, T.Y.; et al. PD-L2 Expression in Human Tumors: Relevance to Anti-PD-1 Therapy in Cancer. *Clin. Cancer Res.* **2017**, *23*, 3158–3167. [CrossRef]

49. Moratin, J.; Metzger, K.; Safaltin, A.; Herpel, E.; Hoffmann, J.; Freier, K.; Hess, J.; Horn, D. Upregulation of PD-L1 and PD-L2 in neck node metastases of head and neck squamous cell carcinoma. *Head Neck* **2019**, *41*, 2484–2491. [CrossRef]

50. Bossi, P.; Siano, M.; Bergamini, C.; Cossu Rocca, M.; Sponghini, A.P.; Giannoccaro, M.; Tonella, L.; Paoli, A.; Marchesi, E.; Perrone, F.; et al. Are Fusion Transcripts in Relapsed/Metastatic Head and Neck Cancer Patients Predictive of Response to Anti-EGFR Therapies? *Dis. Markers* **2017**, *2017*, 6870614. [CrossRef]

51. Hosseini-Alghaderi, S.; Baron, M. Notch3 in Development, Health and Disease. *Biomolecules* **2020**, *10*, 485. [CrossRef] [PubMed]

52. Zhang, Y.Q.; Liang, Y.K.; Wu, Y.; Chen, M.; Chen, W.L.; Li, R.H.; Zeng, Y.Z.; Huang, W.H.; Wu, J.D.; Zeng, D.; et al. Notch3 inhibits cell proliferation and tumorigenesis and predicts better prognosis in breast cancer through transactivating PTEN. *Cell Death Dis.* **2021**, *12*, 502. [CrossRef]

53. Aburjania, Z.; Jang, S.; Whitt, J.; Jaskula-Stzul, R.; Chen, H.; Rose, J.B. The Role of Notch3 in Cancer. *Oncologist* **2018**, *23*, 900–911. [CrossRef]

54. Man, C.H.; Wei-Man Lun, S.; Wai-Ying Hui, J.; To, K.F.; Choy, K.W.; Wing-Hung Chan, A.; Chow, C.; Tin-Yun Chung, G.; Tsao, S.W.; Tak-Chun Yip, T.; et al. Inhibition of NOTCH3 signalling significantly enhances sensitivity to cisplatin in EBV-associated nasopharyngeal carcinoma. *J. Pathol.* **2012**, *226*, 471–481. [CrossRef] [PubMed]

55. Zhang, T.H.; Liu, H.C.; Zhu, L.J.; Chu, M.; Liang, Y.J.; Liang, L.Z.; Liao, G.Q. Activation of Notch signaling in human tongue carcinoma. *J. Oral. Pathol. Med.* **2011**, *40*, 37–45. [CrossRef]

56. Wang, H.; Chirshev, E.; Hojo, N.; Suzuki, T.; Bertucci, A.; Pierce, M.; Perry, C.; Wang, R.; Zink, J.; Glackin, C.A.; et al. The Epithelial-Mesenchymal Transcription Factor SNAI1 Represses Transcription of the Tumor Suppressor miRNA let-7 in Cancer. *Cancers* **2021**, *13*, 1469. [CrossRef]

57. Liu, X.; Zhang, N.; Chen, Q.; Feng, Q.; Zhang, Y.; Wang, Z.; Yue, X.; Li, H.; Cui, N. SNAI2 Attenuated the Stem-like Phenotype by Reducing the Expansion of EPCAMhigh Cells in Cervical Cancer Cells. *Int. J. Mol. Sci.* **2023**, *24*, 1062. [CrossRef]

58. Gan, C.P.; Lee, B.K.B.; Lau, S.H.; Kallarakkal, T.G.; Zaini, Z.M.; Lye, B.K.W.; Zain, R.B.; Sathasivam, H.P.; Yeong, J.P.S.; Savelyeva, N.; et al. Transcriptional analysis highlights three distinct immune profiles of high-risk oral epithelial dysplasia. *Front. Immunol.* **2022**, *13*, 954567. [CrossRef]

59. Slaughter, D.P.; Southwick, H.W.; Smejkal, W. Field cancerization in oral stratified squamous epithelium; clinical implications of multicentric origin. *Cancer* **1953**, *6*, 963–968. [CrossRef]

60. Tang, Y.; Weiss, S.J. Snail/Slug-YAP/TAZ complexes cooperatively regulate mesenchymal stem cell function and bone formation. *Cell Cycle* **2017**, *16*, 399–405. [CrossRef]

61. Tabor, M.P.; Braakhuis, B.J.; van der Wal, J.E.; van Diest, P.J.; Leemans, C.R.; Brakenhoff, R.H.; Kummer, J.A. Comparative molecular and histological grading of epithelial dysplasia of the oral cavity and the oropharynx. *J. Pathol.* **2003**, *199*, 354–360. [CrossRef]

62. González, M.V.; Pello, M.F.; López-Larrea, C.; Suárez, C.; Menéndez, M.J.; Coto, E. Loss of heterozygosity and mutation analysis of the p16 (9p21) and p53 (17p13) genes in squamous cell carcinoma of the head and neck. *Clin. Cancer Res.* **1995**, *1*, 1043–1049. [PubMed]

63. Jang, S.J.; Chiba, I.; Hirai, A.; Hong, W.K.; Mao, L. Multiple oral squamous epithelial lesions: Are they genetically related? *Oncogene* **2001**, *20*, 2235–2242. [CrossRef] [PubMed]

64. O'Shaughnessy, J.A.; Kelloff, G.J.; Gordon, G.B.; Dannenberg, A.J.; Hong, W.K.; Fabian, C.J.; Sigman, C.C.; Bertagnolli, M.M.; Stratton, S.P.; Lam, S.; et al. Treatment and prevention of intraepithelial neoplasia: An important target for accelerated new agent development. *Clin. Cancer Res.* **2002**, *8*, 314–346. [PubMed]

65. Chiesa, F.; Tradati, N.; Grigolato, R.; Boracchi, P.; Biganzoli, E.; Crose, N.; Cavadini, E.; Formelli, F.; Costa, L.; Giardini, R.; et al. Randomized trial of fenretinide (4-HPR) to prevent recurrences, new localizations and carcinomas in patients operated on for oral leukoplakia: Long-term results. *Int. J. Cancer* **2005**, *115*, 625–629. [CrossRef]

66. Costa, A.; Formelli, F.; Chiesa, F.; Decensi, A.; De Palo, G.; Veronesi, U. Prospects of chemoprevention of human cancers with the synthetic retinoid fenretinide. *Cancer Res.* **1994**, *54* (Suppl. S7), 2032s–2037s.

67. Tradati, N.; Chiesa, F.; Rossi, N.; Grigolato, R.; Formelli, F.; Costa, A.; de Palo, G. Successful topical treatment of oral lichen planus and leukoplakias with fenretinide (4-HPR). *Cancer Lett.* **1994**, *76*, 109–111. [CrossRef]
68. Wang, G.Z.; Cheng, X.; Li, X.C.; Liu, Y.Q.; Wang, X.Q.; Shi, X.; Wang, Z.Y.; Guo, Y.Q.; Wen, Z.S.; Huang, Y.C.; et al. Tobacco smoke induces production of chemokine CCL20 to promote lung cancer. *Cancer Lett.* **2015**, *363*, 60–70. [CrossRef]
69. Ferrari, N.; Pfeffer, U.; Dell'Eva, R.; Ambrosini, C.; Noonan, D.M.; Albini, A. The transforming growth factor-beta family members bone morphogenetic protein-2 and macrophage inhibitory cytokine-1 as mediators of the antiangiogenic activity of N-(4-hydroxyphenyl)-retinamide. *Clin. Cancer Res.* **2005**, *11*, 4610–4619. [CrossRef]
70. Mlynska, A.; Salciuniene, G.; Zilionyte, K.; Garberyte, S.; Strioga, M.; Intaite, B.; Barakauskiene, A.; Lazzari, G.; Dobrovolskiene, N.; Krasko, J.A.; et al. Chemokine profiling in serum from patients with ovarian cancer reveals candidate biomarkers for recurrence and immune infiltration. *Oncol. Rep.* **2019**, *41*, 1238–1252. [CrossRef]
71. Lekva, T.; Berg, J.P.; Heck, A.; Lyngvi Fougner, S.; Olstad, O.K.; Ringstad, G.; Bollerslev, J.; Ueland, T. Attenuated RORC expression in the presence of EMT progression in somatotroph adenomas following treatment with somatostatin analogs is associated with poor clinical recovery. *PLoS ONE* **2013**, *8*, e66927. [CrossRef] [PubMed]
72. Cossu Rocca, M.; Maffini, F.; Chiocca, S.; Massaro, M.; Santoro, L.; Cattaneo, A.; Verri, E.; Chiesa, F.; Preda, L.; Nole, F.; et al. Induction chemotherapy followed by transoral laser microsurgery: A mutimodal approach to improve outcomes for locally advanced laryngeal cancer patients? *J. Clin. Oncol.* **2015**, *33* (Suppl. S15), e17039. [CrossRef]
73. Municio, C.; Soler Palacios, B.; Estrada-Capetillo, L.; Benguria, A.; Dopazo, A.; García-Lorenzo, E.; Fernández-Arroyo, S.; Joven, J.; Miranda-Carús, M.E.; González-Álvaro, I.; et al. Methotrexate selectively targets human proinflammatory macrophages through a thymidylate synthase/p53 axis. *Ann. Rheum. Dis.* **2016**, *75*, 2157–2165. [CrossRef]
74. Wu, S.; Liu, Y.; Michalek, J.E.; Mesa, R.A.; Parma, D.L.; Rodriguez, R.; Mansour, A.M.; Svatek, R.; Tucker, T.C.; Ramirez, A.G. Carotenoid Intake and Circulating Carotenoids Are Inversely Associated with the Risk of Bladder Cancer: A Dose-Response Meta-analysis. *Adv Nutr.* **2020**, *11*, 630–643. [CrossRef]
75. Ishida, T.; Ueda, R. CCR4 as a novel molecular target for immunotherapy of cancer. *Cancer Sci.* **2006**, *97*, 1139–1146, Erratum in *Cancer Sci.* **2007**, *98*, 1137. [CrossRef]
76. Zhang, T.; Sun, J.; Li, J.; Zhao, Y.; Zhang, T.; Yang, R.; Ma, X. Safety and efficacy profile of mogamulizumab (Poteligeo) in the treatment of cancers: An update evidence from 14 studies. *BMC Cancer* **2021**, *21*, 618. [CrossRef]
77. Mollica Poeta, V.; Massara, M.; Capucetti, A.; Bonecchi, R. Chemokines and Chemokine Receptors: New Targets for Cancer Immunotherapy. *Front. Immunol.* **2019**, *10*, 379. [CrossRef]
78. Hong, D.S.; Rixe, O.; Chiu, V.K.; Forde, P.M.; Dragovich, T.; Lou, Y.; Nayak-Kapoor, A.; Leidner, R.; Atkins, J.N.; Collaku, A.; et al. Mogamulizumab in Combination with Nivolumab in a Phase I/II Study of Patients with Locally Advanced or Metastatic Solid Tumors. *Clin. Cancer Res.* **2022**, *28*, 479–488. [CrossRef]
79. Zhong, C.; Chen, J. CAR-T cell engineering with CCR6 exhibits superior anti-solid tumor efficacy. *Sci. Bull.* **2021**, *66*, 755–756. [CrossRef]
80. Jin, L.; Cao, L.; Zhu, Y.; Cao, J.; Li, X.; Zhou, J.; Liu, B.; Zhao, T. Enhance anti-lung tumor efficacy of chimeric antigen receptor-T cells by ectopic expression of C-C motif chemokine receptor 6. *Sci. Bull.* **2021**, *66*, 803–812. [CrossRef]
81. Ueda, R. Clinical Application of Anti-CCR4 Monoclonal Antibody. *Oncology* **2015**, *89* (Suppl. S1), 16–21. [CrossRef]
82. Zhang, Y.; Chung, Y.; Bishop, C.; Daugherty, B.; Chute, H.; Holst, P.; Kurahara, C.; Lott, F.; Sun, N.; Welcher, A.A.; et al. Regulation of T cell activation and tolerance by PDL2. *Proc. Natl. Acad. Sci. USA* **2006**, *103*, 11695–11700. [CrossRef] [PubMed]
83. Pardoll, D.M. The blockade of immune checkpoints in cancer immunotherapy. *Nat. Rev. Cancer* **2012**, *12*, 252–264. [CrossRef]
84. Zhang, L.; Liu, Y.; Zheng, H.J.; Zhang, C.P. The Oral Microbiota May Have Influence on Oral Cancer. *Front. Cell Infect. Microbiol.* **2020**, *9*, 476. [CrossRef] [PubMed]
85. Ren, J.; Han, X.; Lohner, H.; Hoyle, R.G.; Li, J.; Liang, S.; Wang, H.P. gingivalis Infection Upregulates PD-L1 Expression on Dendritic Cells, Suppresses CD8+ T-cell Responses, and Aggravates Oral Cancer. *Cancer Immunol. Res.* **2023**, *11*, 290–305. [CrossRef] [PubMed]
86. Kübler, A.; Haase, T.; Rheinwald, M.; Barth, T.; Mühling, J. Treatment of oral leukoplakia by topical application of 5-aminolevulinic acid. *Int. J. Oral Maxillofac. Surg.* **1998**, *27*, 466–469. [CrossRef]
87. National Comprehensive Cancer Network. Available online: https://www.nccn.org/guidelines/recently-published-guidelines (accessed on 3 September 2024).
88. Gale, N.; Hille, J.; Jordan, R.C.; Nadal, A.; William, M.D. *WHO Classification of Head and Neck Tumor*, 4th ed.; Tumor of the Hypopharynx, Larynx and Parapharyngeal Space; International Agency Reserch on Cancer (IARC): Lyon, France, 2017; pp. 91–93, Chapter 3.
89. Zhang, H.K.; Liu, H.G. Is severe dysplasia the same lesion as carcinoma in situ? 10-Year follow-up of laryngeal precancerous lesions. *Acta Otolaryngol.* **2012**, *132*, 325–328. [CrossRef]
90. GenBank ®—The NIH Genetic Sequence Database Part of the International Nucleotide Sequence Database Collaboration Housed at the National Library Medicine (NLM), Bethesda, MD, USA. Available online: https://www.ncbi.nlm.nih.gov/nuccore (accessed on 3 September 2024).

Article

Inflammatory Biomarkers and Oral Health Disorders as Predictors of Head and Neck Cancer: A Retrospective Longitudinal Study

Amr Sayed Ghanem [1], Kitti Sipos [2], Ágnes Tóth [3] and Attila Csaba Nagy [1,*]

[1] Department of Health Informatics, Faculty of Health Sciences, University of Debrecen, 4028 Debrecen, Hungary; aghanem@etk.unideb.hu

[2] Department of Operative Dentistry and Endodontics, Faculty of Dentistry, University of Debrecen, 4032 Debrecen, Hungary; sipos.kitti@dental.unideb.hu

[3] Department of Integrative Health Sciences, Faculty of Health Sciences, University of Debrecen, 4028 Debrecen, Hungary; toth.agnes@etk.unideb.hu

* Correspondence: nagy.attila@etk.unideb.hu

Abstract: Head and neck cancers (HNCs) are often diagnosed late, leading to poor prognosis. Chronic inflammation, particularly periodontitis, has been linked to carcinogenesis, but systemic inflammatory markers remain underexplored. This study was the first to examine whether elevated C-reactive protein (CRP) can serve as a cost-effective adjunct in HNC risk assessment, alongside oral health indicators. A retrospective cohort study analysed 23,742 hospital records (4833 patients, 2015–2022) from the University Hospital of Debrecen. HNC cases were identified using ICD-10 codes, with CRP and periodontitis as key predictors. Kaplan–Meier survival analysis, log-rank tests, and Weibull regression were used to assess risk, with model performance evaluated via AIC/BIC and ROC curves. Periodontitis was significantly associated with HNC (HR 5.99 [1.96–18.30]), while elevated CRP (>15 mg/L) independently increased risk (HR 4.16 [1.45–12.00]). Females had a significantly lower risk than males (HR 0.06 [0.01–0.50]). CRP may serve as a cost-effective, easily accessible biomarker for early HNC detection when combined with oral health screening. Integrating systemic inflammation markers into HNC risk assessment models could potentially improve early diagnosis in high-risk populations.

Keywords: C-reactive protein; CRP; eGFR; estimated glomerular filtration rate; head and neck cancers; oral cavity cancer; oral squamous cell carcinoma; malignant

1. Introduction

Head and neck cancer (HNC), which includes malignancies of the lip and oral cavity, is among the most prevalent cancers worldwide, with lip and oral cavity cancers ranking 13th [1] globally. It is more common in men and in the elderly population [1]. Approximately 90% of HNCs are squamous cell carcinomas, while the remaining 10% include sarcomas, melanomas, lymphomas, salivary gland tumours, and odontogenic tumours [2]. Sores or lumps on the lips or inside the mouth are common and simple indicators of neoplasms [3]. Despite the fact that the disease is easily recognisable, many people seek medical help only at advanced stages, which plays a crucial role in the high mortality rate of the disease, which is nearly 50% [3]. In order to enhance prognosis and reduce the number of fatal cases, early diagnosis is essential [4]. The most common methods for diagnosis are visual inspection and palpation, which may be accompanied by biopsy and histological analysis (e.g., toluidine staining). Additionally, imaging techniques may also be involved in the detection and precise staging of the primary tumour(s) or regional lymph nodes [5]. In cases of identified disorders, the conventional therapy is surgical removal, potentially supported by radiation and/or chemotherapy or other postoperative adjuvants [6,7].

There is a large amount of evidence that periodontitis is associated with an increased risk of developing certain non-communicable chronic diseases [8–10], such as several types of cancers, including HNCs [11]. Periodontitis is the inflammation of the periodontium [12], which is typically a chronic condition and followed by clinical attachment and alveolar bone loss. It affects tooth location, leading to drifting, and tooth instability, resulting in increased mobility [13]. Periodontitis may include periodontal pocketing, recession, or enlargement of the gingiva and bleeding, and it is a major cause of tooth loss in society [12,13]. It is estimated that approximately 19% of the adult population or over 1 billion people is affected by severe periodontal disease [1]. Due to the multifactorial nature of the disease [9] a number of factors have been identified as associated with the higher likelihood of periodontal disease, which involves the dysbiosis of the oral microbiota as a major causative factor [11].

The composition of the oral microbiome can be altered by different lifestyle habits such as cigarette smoking, alcohol consumption, diet, and poor oral health, and it can be influenced by several medical conditions as well, e.g., diabetes mellitus or chronic kidney disease [14–17], resulting in the overpopulation of certain pathogenic microorganisms [18] such as the Gram-negative and anaerobic bacterial species, Porphyromonas gingivalis, Aggregatibacter actinomycetecomitans, and Fusobacterium nucleatum [9], among which Porphyromonas gingivalis is considered to be the primary causative bacterial strain [19]. Persistent infections have the ability to induce tumorigenesis; however, it is important to note that bacteria work synergistically, and one bacterium is insufficient to determine the entire process in an organism [20]. The transcription factor nuclear factor kappa B (NF-κB) is a central regulator of inflammation-driven carcinogenesis [21]. Pathogenic bacteria release endotoxins such as lipopolysaccharide (LPS), which activates toll-like receptor 4 (TLR4), initiating NF-κB-mediated signalling cascades that promote cytokine release [14]. This pathway drives cellular transformation, proliferation, apoptosis evasion, invasion, angiogenesis, and metastasis—hallmarks of cancer progression [21]. A positive feedback loop between NF-κB and pro-inflammatory mediators, particularly tumour necrosis factor-alpha (TNF-α), interleukin-6 (IL-6), and IL-1β, further amplifies chronic inflammation, perpetuating a tumour-promoting microenvironment [22].

Inflammatory factors can result in periodontal tissue damage [9,11,23] and induce the production of C-reactive protein (CRP) by hepatocytes after systemic dispersion [24,25]. IL-6 is the major cytokine inducing the expression of CRP during the acute phase response [26] by transcription factor signal transducer and activator of transcription 3 (STAT3); additionally, IL-1β acts as a synergistic enhancer in the process [27,28]. CRP secreted from hepatocytes is a pentameric molecule and circulates in the systemic vasculature [29]. After binding to lipid rafts of cells engaged in inflammatory reactions, it dissociates to a highly active monomeric form and interacts with many cell types at the sites of inflammation and the components of the extracellular matrix (ECM) [29]. Pentameric CRP is a measurable quantity, suitable for diagnostic testing [29]. Epidemiological studies have reported that serum CRP levels are elevated in patients suffering from chronic periodontitis. CRP is currently regarded as one of the most relevant biomarkers of systemic inflammation [25].

Emerging evidence suggests that ferroptosis, an iron-dependent form of regulated cell death, plays a pivotal role in oral squamous cell carcinoma (OSCC) pathogenesis and progression. The dysregulation of ferroptosis pathways contributes to tumour survival, immune evasion, and therapy resistance, underscoring its potential as both a biomarker and therapeutic target [30]. Notably, pro-inflammatory cytokines, particularly interleukin-

6 (IL-6), promote ferroptosis resistance via JAK2/STAT3-mediated upregulation of xCT (SLC7A11), an antiporter critical for glutathione homeostasis, reinforcing the role of chronic inflammation in OSCC carcinogenesis [31].

Ferroptosis-related genes (FRGs) and their associated non-coding RNAs have been identified as prognostic biomarkers in OSCC, with glutathione peroxidase 4 (GPX4) and adipocyte enhancer-binding protein 1 (AEBP1) serving as key mediators of ferroptosis suppression [32]. In addition, FTH1 (ferritin heavy chain 1), a critical iron storage protein, has been implicated in OSCC proliferation and epithelial–mesenchymal transition (EMT), with its inhibition inducing ferroptotic cell death, highlighting its therapeutic potential [33].

Given that HNC can remain asymptomatic for long periods, its early detection remains a major clinical challenge [3], contributing to delayed diagnosis and poor prognosis [4]. While chronic inflammation, particularly periodontitis [11], has been implicated in oral carcinogenesis, limited research has evaluated systemic inflammatory markers as potential adjuncts in HNC risk assessment. Despite evidence linking periodontitis to malignant transformation, no widely accessible biomarker has been integrated into routine screening for HNC risk stratification. This study aimed to bridge this gap by investigating whether CRP, a readily available inflammatory biomarker, could serve as a cost-effective tool for early detection, alongside oral health indicators.

Since longitudinal investigations are essential to establish causalities, our study aimed to assess the associations of CRP and periodontitis with HNC while identifying other potential predictors of the disease, including the estimated glomerular filtration rate (eGFR), since patients with chronic kidney disease might be more susceptible to oral health issues due to oral microbial changes. Furthermore, apart from periodontitis, additional oral conditions were investigated, such as gingival disorders, dental development disorders, eruption issues, hard tissue, pulp, periapical tissue diseases, dental caries that can act as plaque retention factors and promote the growth of anaerobic bacteria, and gingivitis, which is a mild form of periodontal disease and, if untreated, might also lead to periodontitis and, ultimately, malignancy [30].

2. Results

2.1. Baseline Characteristics

The baseline characteristics of the 4833 participants are summarised in Table 1. The mean age was 50.43 years (SD = 15.40), with a median of 51 years (IQR = 40–62). The cohort consisted of 2193 males (45.40%) and 2637 females (54.60%). Periodontitis was present in 1514 participants (31.33%). Among those with available data, elevated CRP levels (>15 mg/L) were observed in 102 participants (25.69%), while 295 (74.31%) had normal levels ($\leq$15 mg/L). Kidney disease (eGFR < 60 mL/min/1.73 m^2) was reported in 78 participants (19.75%) out of those with available eGFR data. Disorders of tooth development and eruption (K00) were present in 66 participants (1.37%), Embedded and impacted teeth (K01) were noted in 40 participants (0.83%). Dental caries (K02) affected 135 participants (2.79%), other diseases of hard tissues of teeth (K03) was reported in 770 participants (15.93%), and diseases of pulp and periapical tissues (K04) were documented in 147 participants (3.04%). Other disorders of gingiva and edentulous alveolar ridge (K06) were observed in 77 participants (1.59%), and other specified disorders of teeth and supporting structures (K08) were present in 167 participants (3.46%).

Table 1. Baseline characteristics of participants (N = 4833).

Variable	Category	N (%)
Age, years	Mean (SD)	50.43 (15.40)
	Median (IQR)	51 (40–62)
Sex	Male	2193 (45.40)
	Female	2637 (54.60)
Periodontitis	No	3319 (68.67)
	Yes	1514 (31.33)
C-reactive protein	Normal ($\leq$15)	295 (74.31)
	High (>15)	102 (25.69)
eGFR	Normal ($\geq$60)	317 (80.25)
	Kidney Disease (<60)	78 (19.75)
Disorders of tooth development and eruption (K00)	No	4767 (98.63)
	Yes	66 (1.37)
Embedded and impacted teeth (K01)	No	4793 (99.17)
	Yes	40 (0.83)
Dental caries (K02)	No	4698 (97.21)
	Yes	135 (2.79)
Other diseases of hard tissues of teeth (K03)	No	4063 (84.07)
	Yes	770 (15.93)
Diseases of pulp and periapical tissues (K04)	No	4686 (96.96)
	Yes	147 (3.04)
Other disorders of gingiva and edentulous alveolar ridge (K06)	No	4756 (98.41)
	Yes	77 (1.59)
Other specified disorders of teeth and supporting structures (K08)	No	4666 (96.54)
	Yes	167 (3.46)

2.2. Survival Distributions and Log-Rank Test Results

The results of the log-rank test, comparing survival distributions across the different categories, are summarised in Table 2. A significant difference in survival probabilities was observed for periodontitis ($p < 0.001$), with 114 events in participants with periodontitis compared to an expected 38.15. Elevated C-reactive protein levels (>15 mg/L) were also significantly associated with survival outcomes ($p = 0.007$), with 20 observed events versus 8.1 expected. Gender showed a strong association ($p < 0.001$), with males experiencing 130 events compared to an expected 73.5, while females had 43 events compared to the 99.5 expected. Among dental conditions, significant associations were identified for disorders of tooth development and eruption (K00, $p = 0.042$), embedded and impacted teeth (K01, $p = 0.011$), dental caries (K02, $p < 0.001$), other diseases of hard tissues of teeth (K03, $p < 0.001$), and other specified disorders of teeth and supporting structures (K08, $p < 0.001$). Conversely, no significant differences were observed for diseases of pulp and periapical tissues (K04, $p = 0.74$), other disorders of gingiva and edentulous alveolar ridge (K06, $p = 0.971$), or eGFR ($p = 0.489$).

Table 2. Results of the log-rank test for survival distributions by key variables.

Variable	Category	Observed Events	Expected Events	*p*-Value
Periodontitis	No	59	134.85	**<0.001**
	Yes	114	38.15	
C-reactive protein	Normal ($\leq$15)	14	25.9	**0.007**
	High (>15)	20	8.1	
Gender	Male	130	73.5	**<0.001**
	Female	43	99.5	
Disorders of tooth development and eruption (K00)	No	164	168.33	**0.042**
	Yes	9	4.67	
Embedded and impacted teeth (K01)	No	169	171.79	**0.011**
	Yes	4	1.21	
Dental caries (K02)	No	149	164.94	**<0.001**
	Yes	24	8.06	
Other diseases of hard tissues of teeth (K03)	No	129	156.19	**<0.001**
	Yes	44	16.81	
Diseases of pulp and periapical tissues (K04)	No	168	167.22	0.74
	Yes	5	5.78	
Other disorders of gingiva and edentulous alveolar ridge (K06)	No	171	170.95	0.971
	Yes	2	2.05	
Other specified disorders of teeth and supporting structures (K08)	No	148	167.64	**<0.001**
	Yes	25	5.36	
eGFR	Normal ($\geq$60)	13	14.53	0.489
	Kidney Disease (<60)	9	7.47	

The table presents observed and expected events for the log-rank test, along with corresponding *p*-values. Statistically significant differences in survival distributions ($p < 0.05$) are highlighted in bold, suggesting varying survival probabilities across categories.

2.3. Cumulative Hazard Analysis

The cumulative hazard curves (Figure 1) illustrate the temporal progression of oral cancer risk across key variables identified as significant predictors in the log-rank tests. Figure 1A highlights the finding that participants with periodontitis exhibited higher cumulative hazards for HNC compared to those without periodontitis. Similarly, participants with elevated C-reactive protein levels (>15 mg/L) demonstrated higher cumulative hazards compared to those with normal CRP levels (Figure 1B). Gender-stratified cumulative hazard curves (Figure 1C) indicate that male participants had consistently higher cumulative hazards for HNC than females over the follow-up period. Lastly, Figure 1D depicts the cumulative hazards associated with dental developmental (DD) disorders, showing that participants with DD disorders had a markedly increased cumulative hazard compared to those without such disorders.

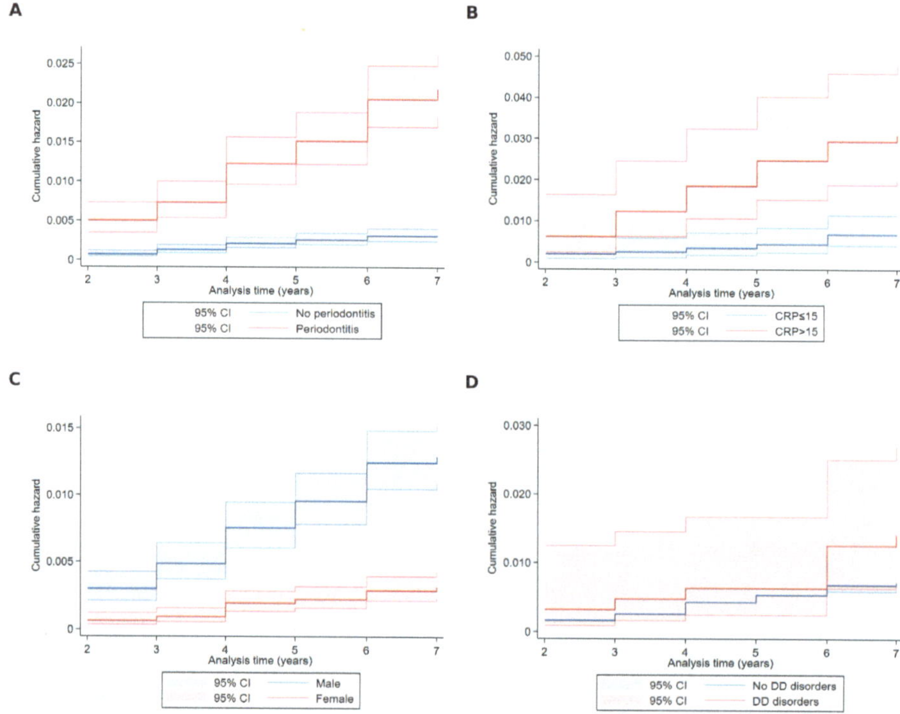

Figure 1. Cumulative hazard plots by periodontitis, CRP levels, gender, and dental developmental disorders. Note: Cumulative hazard functions illustrating the relationships between time to head and neck cancer diagnosis and key variables: (**A**) periodontitis (yes vs. no), (**B**) C-reactive protein levels (>15 mg/L vs. ≤15 mg/L), (**C**) gender (male vs. female), and (**D**) dental developmental (DD) disorders (present vs. absent). Hazard functions are stratified by categories with 95% confidence intervals (shaded areas), calculated using the Nelson–Aalen estimator. CRP, C-reactive protein; DD disorders, dental developmental disorders.

In Figure 2A, participants with embedded and impacted teeth exhibited a noticeably higher cumulative hazard over the follow-up period compared to those without this condition. Figure 2B demonstrates that participants with dental caries experienced an elevated cumulative hazard, particularly in later years of follow-up, compared to those without caries. Figure 2C illustrates the effect of disease of hard tissue (DHT) of teeth, where individuals with this condition showed a marked increase in cumulative hazard over time relative to those without DHT. Finally, Figure 2D highlights a significant cumulative hazard increase among participants with disorders of teeth and supporting structures (DTSSs) compared to those without these disorders, with a sharper rise observed during the later years of follow-up.

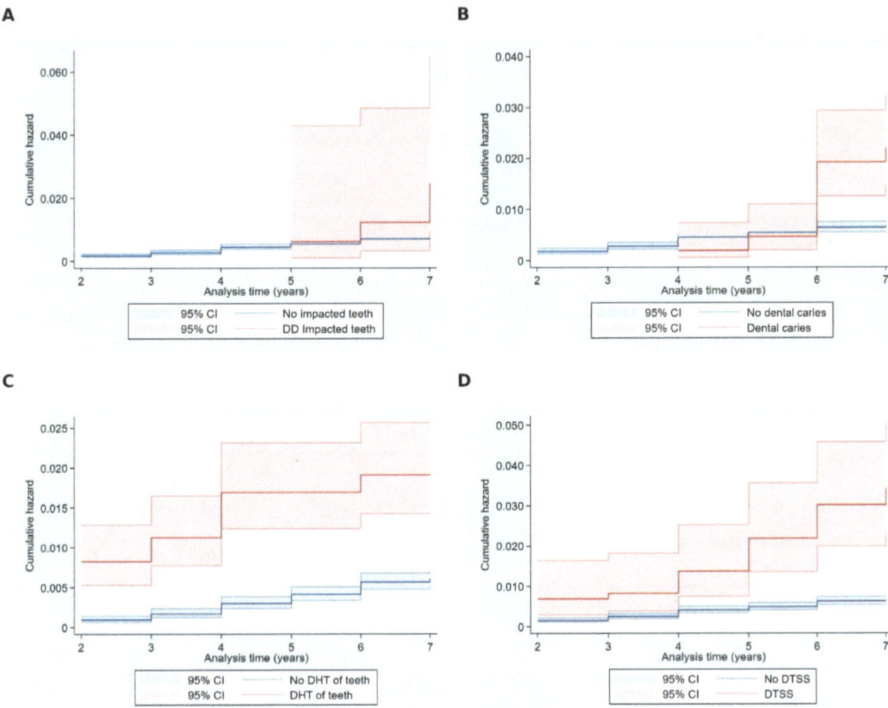

Figure 2. Cumulative hazard plots by embedded and impacted teeth, dental caries, disease of hard tissue of teeth, and disorders of teeth and supporting structures. Note: Cumulative hazard functions depicting the association between time to head and neck cancer diagnosis and key dental conditions: (**A**) embedded and impacted teeth (present vs. absent), (**B**) dental caries (present vs. absent), (**C**) disease of hard tissue (DHT) of teeth (present vs. absent), and (**D**) disorders of teeth and supporting structures (DTSS) (present vs. absent). Hazard functions are stratified by condition categories with 95% confidence intervals (shaded areas), calculated using the Nelson–Aalen estimator. DHT, disease of hard tissue; DTSS, disorders of teeth and supporting structures.

2.4. Weibull Regression Results

Participants with periodontitis demonstrated a markedly increased hazard, with an HR of 5.99 (95% CI: 1.96–18.30, p = 0.002). Similarly, elevated CRP levels (>15 mg/L) were associated with an increased hazard of HNC, with an HR of 4.16 (95% CI: 1.45–12.00, p = 0.008). Gender was also a significant predictor; females had a reduced hazard compared to males, with an HR of 0.06 (95% CI: 0.01–0.50, p = 0.009). Participants with embedded and impacted teeth (K01) had a markedly elevated hazard of HNC (HR = 12.52, 95% CI: 2.48–63.18, p = 0.002), the highest among all examined oral conditions. Although disorders of tooth development and eruption (K00) approached significance (HR: 3.97, 95% CI: 0.88–17.92, p = 0.073), it did not meet the threshold for statistical significance. Other oral health conditions, including dental caries (K02), other diseases of hard tissues of teeth (K03), diseases of pulp and periapical tissues (K04), and other specified disorders of teeth and supporting structures (K08), were not significantly associated with HNC hazard (all p > 0.05). Age, as a continuous variable, showed no significant association with HNC hazard (HR: 0.98, 95% CI: 0.95–1.02, p = 0.347). Kidney disease, defined by eGFR < 60 mL/min/1.73 m^2, was not significantly associated with the hazard of HNC (HR: 0.95, 95% CI: 0.35–2.61, p = 0.919) (Table 3).

Table 3. Results of Weibull regression analysis for predictors of oral cancer hazard.

Variable	Category	HR (95% CI)	*p*-Value
Periodontitis	Yes vs. No (ref)	**5.99 [1.96–18.30]**	**0.002**
C-reactive protein	High (>15) vs. Normal ($\leq$15, ref)	**4.16 [1.45–12.00]**	**0.008**
eGFR	Kidney Disease (<60) vs. Normal ($\geq$60, ref)	0.95 [0.35–2.61]	0.919
Age	Continuous (per year)	0.98 [0.95–1.02]	0.347
Gender	Female vs. Male (ref)	**0.06 [0.01–0.50]**	**0.009**
Disorders of tooth development and eruption (K00)	Yes vs. No (ref)	3.97 [0.88–17.92]	0.073
Embedded and impacted teeth (K01)	Yes vs. No (ref)	**12.52 [2.48–63.18]**	**0.002**
Dental caries (K02)	Yes vs. No (ref)	2.21 [0.56–8.75]	0.259
Other diseases of hard tissues of teeth (K03)	Yes vs. No (ref)	0.60 [0.13–2.86]	0.524
Diseases of pulp and periapical tissues (K04)	Yes vs. No (ref)	0.52 [0.06–4.61]	0.558
Other specified disorders of teeth and supporting structures (K08)	Yes vs. No (ref)	0.76 [0.18–3.17]	0.704

HR = hazard ratio; CI = confidence interval; ref = reference category. Statistically significant results ($p < 0.05$) are highlighted in bold.

In the Weibull regression analysis, the shape parameter (p) outlined in Table 4 was estimated to be 2.20 (95% CI: 1.39–3.49). This value indicates that the hazard of developing HNC increases over time, as $p > 1$ reflects a positive ageing effect. The increasing hazard aligns with the biological plausibility of cumulative risk exposure and age-related changes contributing to disease progression.

Table 4. Shape parameter (p) from Weibull regression and its interpretation.

Parameter	Estimate (95% CI)	Interpretation
p	2.20 [1.39–3.49]	Hazard increases over time (positive ageing effect)

The parameter estimate indicates the shape of the hazard function over time. A value greater than 1 suggests that the hazard increases with time, indicative of a positive ageing effect.

2.5. Model Validation and Discrimination Metrics

The discriminative performance of the Weibull regression model was evaluated using the area under the ROC curve and Harrell's concordance statistic. The ROC curve yielded an area under the curve (AUC) of 0.8646 (95% CI: 0.7933–0.9359), indicating excellent predictive accuracy (Figure 3). Harrell's c-statistic confirmed the model's robustness with a concordance coefficient of 0.8646 (95% CI: 0.7953–0.9339). These metrics demonstrate the model's strong ability to distinguish between participants who developed HNC and those who did not during the follow-up period.

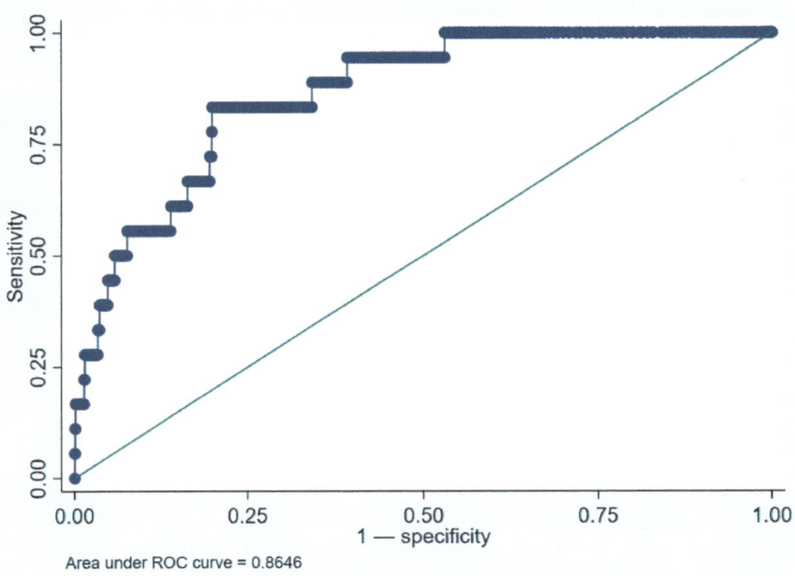

Figure 3. Receiver operating characteristic (ROC) curve of the Weibull regression model for predicting head and neck cancer. Note: The ROC curve shows an AUC of 0.8646, demonstrating excellent model discrimination.

Kaplan–Meier curves were constructed to illustrate the observed and predicted survival probabilities over time, stratified by significant covariates identified in the Weibull regression model. These curves were stratified by the presence or absence of embedded or impacted teeth (Figure 4A), periodontitis (Figure 4B), CRP levels (Figure 4C), and gender (Figure 4D). Panel A compares participants with and without embedded or impacted teeth, showing a steeper decline in survival probabilities among those with missing teeth, with Weibull model predictions closely aligning with observed data. Panel B examines participants with and without periodontitis, revealing a marked reduction in survival probability over time for those with periodontitis compared to their counterparts. Panel C presents survival probabilities based on CRP levels, where participants with high CRP levels (>15 mg/L) experienced noticeably lower survival probabilities, while those with normal CRP levels maintained higher probabilities throughout the follow-up period. Lastly, Panel D stratifies survival by sex, demonstrating that male participants exhibited lower survival probabilities than females, with the Weibull model predictions accurately reflecting observed trends in all cases.

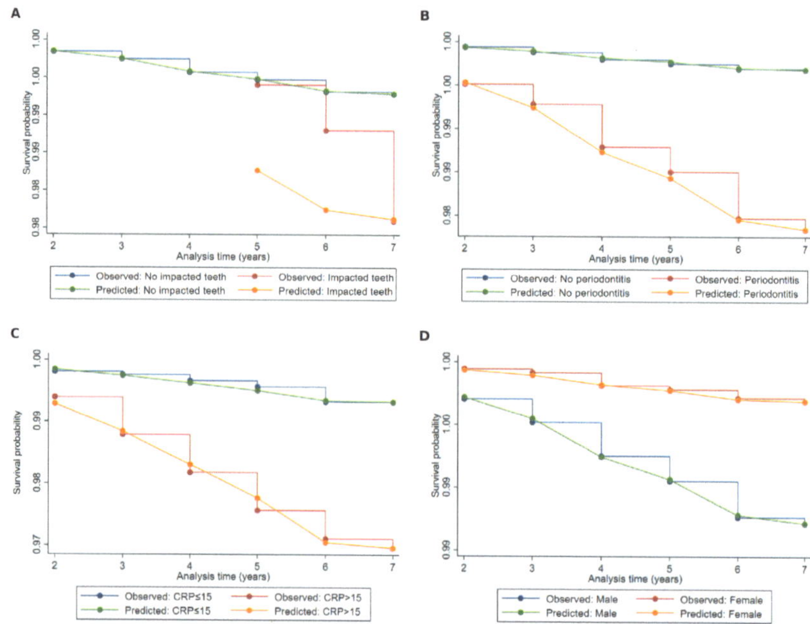

Figure 4. Kaplan–Meier survival curves stratified by significant predictors. Kaplan–Meier survival curves depicting observed and Weibull model-predicted probabilities over time. Stratification is based on significant covariates: (**A**) impacted teeth, (**B**) periodontitis, (**C**) C-reactive protein (CRP) levels, and (**D**) gender.

3. Discussion

The objective of the present study was to investigate the association between predictors of HNC, with a focus on identifying any predictive factors, such as CRP or eGFR, which have not been previously investigated in the context of such type of malignancy as a potential method for secondary prevention. This study is based on a large clinical database of cases and makes a significant contribution to the identification of risk factors by examining several concurrent factors. The study is of particular significance in the context of cancer research, as it focuses on a persistent and significant public health concern in Hungary. Specifically, HNC is the eighth most prevalent cancer type among both sexes in the country [34]. Whilst encouraging signs are evident in terms of a slight decline in the incidence and mortality rates of the aforementioned cancer types in the country, Hungary remains among the European countries with the highest rates of HNC including oral cavity cancers [34,35].

Risk factors for HNC represents a significant field of research [36,37]. In this context, Hungary exhibits comparatively weaker performance with regard to smoking and alcohol consumption in comparison to other European Union countries. Conversely, the country displays comparatively stronger rates of vaccination against the human papilloma virus (HPV) [38].

A review of the literature reveals a clear pattern indicating that HNC is more prevalent in males than in females [39,40], a trend that is also supported by Hungarian data [34]. The present study also demonstrated that the risk of developing HNC was lower for females than for males. This phenomenon may be attributed to disparities in tumour biology and hormonal influences, which render men more susceptible to the development of such malignancies. It has also been observed that women in Hungary exhibit superior oral hygiene [41] and men are more likely to engage in high-risk behaviours, such as smoking and excessive alcohol consumption. However, there is an increasing concern that the incidence of lung cancer among women is rising as a consequence of changes in smoking habits [34], which may also contribute to an increase in the incidence of HNC among women in the future. A previous study conducted at the Department of Oral Surgery of the University of Debrecen revealed that the majority of patients diagnosed with oral squamous cell carcinoma (OSCC) were smokers (65.5%) and reported alcohol consumption (75.5%) at the time of diagnosis. Only 12.6% of cases were found to have acceptable dental status [42].

The prevalence of dental caries in Hungary is indeed high [41], yet the present study found no association with oral cavity cancer. Tezal et al. found that patients with a high caries incidence, i.e., those with a predominantly cariogenic oral flora, displayed a reduced propensity for HNC development. Cariogenic bacteria have been shown to induce a Th1-mediated immune response, which has been demonstrated to result in a tumour suppressor effect [43].

In addition to dental caries, periodontal disease is the most prevalent bacterial infection of the oral cavity [44]. The present study also revealed a very high prevalence (31%) of periodontitis among the study participants. As demonstrated by numerous human and murine studies, the idea that Th1 cells and their cytokines characterize early/stable periodontal lesions is corroborated. By contrast, Th2 cells are implicated in the progression of the disease [45]. Periodontopathogenic flora has been observed to elicit a Th2/Th17 immune response, which has been associated with an elevated risk of tumour development [43]. The presence of Th2 cells and the cytokines they secrete has been linked to a poor prognosis in various malignancies, including HNC [46,47]. It has been hypothesized that Th2 responses accelerate the growth of tumours by inhibiting Th1-mediated anti-tumour action and boosting angiogenesis. However, findings suggest a more complex function for Th2 cells. Despite the identification of several key pathways, the precise mechanisms by which Th2 cells promote tumour growth remain to be elucidated [47]. Th17 cells, which originate from CD4+ T cells, have been shown to play a pivotal role in the progression and regulation of periodontal disease, with the cytokines they secrete (mainly IL-17 and IL-22) being a key factor in this process [48,49]. Th17 cells have also been shown to play a significant role in the promotion of inflammation in a variety of pathophysiological situations, including HNC [50]. These cells exhibit remarkable plasticity, allowing them to exhibit different phenotypes in the cancer microenvironment. The role of Th17 cells in cancer is multifaceted and dependent on the unique characteristics of the tumour. These cells can promote tumour progression through immunosuppressive activities and angiogenesis but also mediate anti-tumour immune responses through the use of immune cells in the tumour environment or by directly converting to the Th1 phenotype and producing interferon-γ (IFN-γ) [47]. The findings of this study demonstrated a substantial correlation between the occurrence of HNC and periodontitis, aligning with the findings of other researchers [51–55]. Therefore, dentists should consider that patients with periodontitis, especially those with coexisting lifestyle risk factors [41], have been linked to an increased risk of HNC.

The examination of factors associated with dental status revealed in this study that embedded and impacted teeth (K01) exhibited a notable oral health variable associated with an increased hazard. This represents a previously unreported association in the literature. This phenomenon can be attributed to the constant irritation and infection that occurs around the impacted teeth, which can result in chronic inflammation and persistent infections. A significant body of evidence exists, which indicates a clear association between chronic inflammation and cancer [56]. There is also a possibility that cystic lesions develop around the region of impacted wisdom teeth, which may subsequently transform into tumours [57]. This transformation carries with it a risk of malignancy [58]. A definitive causal relationship between impacted wisdom teeth and HNC has yet to be established. Consequently, these conditions can serve to increase the overall risk factors associated with cancer. Nevertheless, cysts and tumours have been observed in a small percentage of patients with impacted wisdom teeth. As a consequence, prophylactic removal of the wisdom tooth is not considered necessary [57]; however, subsequent follow-up is recommended.

The gold standard for the diagnosis of HNC is tissue biopsy and subsequent histological evaluation [4]. Nevertheless, there is the possibility of false-positive or false-negative biopsy results. HNC is often diagnosed at an advanced stage, which has a negative impact on patient survival rates [42]. There is mounting evidence to support the hypothesis that chronic inflammation is a contributing factor to the development of cancer [53]. Consequently, a significant number of studies have been conducted with the objective of identifying easily accessible early diagnostic biomarkers [4,59]. For instance, CRP has been shown to have elevated levels that correlate positively with tumour status [59,60]. The precise mechanisms underlying the association CRP levels and the survival of patients diagnosed with HNC remain to be elucidated; however, the following factors may be contributing factors [60]. It has been hypothesized that chronic inflammation may play a role in the development and progression of HNC [4]. Inflammation exerts a significant influence on the composition of the tumour microenvironment. In response to inflammation, CRP is synthesized in the liver by the stimulation of IL-6. The IL-6 has been demonstrated to accelerate angiogenesis and inhibit ferroptosis, thus promoting the progression and metastasis of tumours [31,60]. Cancer cells can produce a number of chemokines and cytokines, which in turn results in an increase in the serum CRP level [60]. Moreover, a higher prevalence of tobacco use, alcohol consumption, and poor oral hygiene has been observed in patients with HNC, which has also been demonstrated to elevate CRP levels [61–63]. The role of the CRP is of significance, given that it has been identified as a potential mediator of carcinogenesis and cancer progression via the activation of the FcgRs/MAPK/ERK, FcgRs/NF-kB/NLRP3, and FcgRs/IL-6/AKT/STAT3 pathways [60].

This is the first study of its kind to attempt to find a correlation in HNC patients prior to cancer diagnosis using pre-diagnosis CRP data. The Weibull regression results indicate that elevated CRP levels (>15 mg/L) are associated with an increased hazard of HNC. Serum CRP level can be measured in a simple and repeatable manner, and the associated financial cost is reasonable. Therefore, it could be regarded as a routine clinical marker in patients with HNC [59,60]. Salivary CRP has recently attracted considerable attention from the scientific community. A large body of research has demonstrated the correlation between CRP levels in blood and saliva, suggesting that salivary CRP can serve as a reliable surrogate marker for serum CRP as an indicator of oral inflammation. The advent of novel technologies capable of detecting CRP in saliva underscores its potential as a diagnostic instrument for various oral inflammatory and immune conditions, including periodontitis and HNC, given the non-invasive nature of saliva collection [63].

In addition to CRP, renal function, measured via eGFR, was investigated for its potential role in modulating cancer risk, given its known association with systemic inflammation. To date, no article has specifically examined the association between reduced eGFR and HNC. The present article therefore constitutes pioneering work in this regard. As demonstrated by other researchers, the incidence rate of cancer was found to be comparatively higher in patients suffering from chronic kidney disease than in the general population [64,65]. In contrast, Wong et al. did not observe an association between reduced renal function and the overall risk of cancer [66]. Whilst no significant correlation was identified in the present study with eGFR, this may be attributable to the complex relationship between renal function and systemic inflammation.

In accordance with the projected changes in population growth and ageing, and assuming that overall cancer rates remain constant, Bray et al. predict that over 35 million new cancer cases will occur in the year 2050. This figure signifies a 77% increase on the 20 million cases estimated in 2022 [40]. The authors therefore consider research into the identification of risk factors and early detection to be of the utmost importance. Such research may contribute to a reduction in the alarming figure in the future.

3.1. Future Directions

The present study identified the presence of embedded and impacted teeth (K01) as a significant oral health variable associated with an elevated HNC risk. This finding signifies a previously unreported association in the literature and could serve as a basis for further research.

Elevated levels of CRP (>15 mg/L) have also been demonstrated to be associated with an increased hazard of HNC. However, further investigation is required to ascertain whether elevated CRP levels occur prior to the biological onset of cancer or whether they act as a risk factor for its onset.

The absence of a substantial correlation between eGFR and the variables under investigation in the present study may be ascribed to the complex relationship between renal function and systemic inflammation. Further research is required to ascertain the underlying mechanisms.

3.2. Strengths and Limitations

A key strength of this study is its utilisation of real-world clinical data collected over a seven-year follow-up period, allowing for a comprehensive evaluation of the natural history and progression of HNC. This study benefits from a substantial sample size, which enhances statistical power and reliability. Additionally, the inclusion of biomarker data obtained through laboratory assessments provides an objective measure of systemic health, strengthening the validity of associations examined. Diagnoses were established by physicians based on ICD-10 codes, ensuring clinical applicability and diagnostic accuracy. The longitudinal study design facilitated the assessment of temporal and potential causal relationships, while the implementation of robust time-to-event analytical models enhanced methodological rigour by accounting for both systemic and intraoral health parameters.

However, several limitations should be acknowledged. The dataset originated from a single-centre clinical registry, which may restrict the external validity and generalizability of findings to broader or more diverse populations. Moreover, while ICD-10 coding ensures diagnostic precision, it lacks granularity regarding disease severity or staging, limiting a more detailed assessment of disease burden. The absence of key socioeconomic and demographic variables precluded an in-depth evaluation of their potential confounding effects. Smoking, which is a well-documented risk factor for HNC, was not accounted for due to the unavailability of smoking status within the clinical database, which might have influenced the robustness of the observed associations. This study is also a retrospective observational study; therefore, inherent biases related to data availability, selection, and recording could not entirely be excluded. Lastly, given the observational nature of this study, unmeasured confounding variables remain a potential source of bias, necessitating cautious interpretation of causal inferences.

4. Materials and Methods

This study involving human participants was approved by the Ethics Committee at the University of Debrecen (Approval Number: 6054-2022) on 20 April 2022. This study was conducted in accordance with relevant local laws and institutional guidelines. It utilised secondary analysis of pre-anonymized, de-identified hospital records, ensuring that no identifiable personal information was accessible during the research process.

4.1. Data Cleaning and Processing

This study utilised a retrospective longitudinal design based on hospital records collected between 2007 and 2022 by the University Hospital of Debrecen in Hungary. The initial dataset consisted of 37,164 hospital records, representing visits of participants between years 2007 and 2022 diagnosed with various conditions by physicians using ICD-10 codes. To improve data quality and focus on the study period with the most complete and precise records, all data prior to 2015 were excluded. This exclusion was justified by the minimal number of HNC cases recorded before 2015 and the widespread missing laboratory data. After this step, 23,742 records remained, covering visits from 2015 to 2022.

Participants with a diagnosis of HNC (ICD-10 codes: C00–C10, C14) at or prior to baseline were excluded to ensure all individuals were free of the failure event at the start of follow-up. Head and neck cancer cases were identified using the variable "HNC", defined as present when any of the following ICD-10 codes were recorded: C00, C01, C02, C03, C04, C05, C06, C07, C08, C09, C10, or C14. For periodontitis, a binary variable "periodontitis" was created based on the presence of any of the following ICD-10 codes: K05, K05.2, K05.3, K05.4, K05.5, or K05.6.

Participants were eligible for inclusion if they had at least one documented dental or oral health-related diagnosis within the hospital's electronic health records (EHRs) between 2015 and 2022, ensuring the availability of longitudinal data for oral–systemic health assessment. Only individuals with confirmed head and neck cancer (HNC; ICD-10: C00–C14) diagnosed after the baseline dental visit were included in survival analyses to mitigate immortal time bias. To ensure data completeness and feasibility of time-to-event modelling, participants were required to have a minimum of two years of follow-up within the hospital database. Only adult patients ($\geq$18 years old) were eligible, as paediatric and adolescent populations follow distinct clinical trajectories in both oral pathology and malignancy risk.

Exclusion criteria encompassed individuals with a prevalent diagnosis of HNC at baseline, ensuring that all participants were free from the failure event at study entry. Participants with insufficient follow-up data (i.e., lost to follow-up before two years) were excluded to minimise bias from incomplete risk estimation. Given the study's focus on periodontitis and its systemic implications, patients with only secondary or incidental dental diagnoses (e.g., traumatic dental injuries without evidence of chronic inflammation) were excluded to maintain clinical homogeneity. Additionally, participants with systemic inflammatory conditions unrelated to oral health (e.g., autoimmune diseases, systemic lupus erythematosus, rheumatoid arthritis) were excluded to prevent confounding effects on systemic inflammatory markers. Finally, patients with multiple healthcare system transfers or fragmented EHRs were excluded to minimise selection and information bias due to incomplete medical histories.

The follow-up time for each participant was calculated as the difference between their baseline year and either the year of diagnosis (for those who developed oral cancer) or the study's endpoint in 2022 (for censored observations). The follow-up exit year was defined as the diagnosis year for participants with oral cancer and as 2022 for those who did not develop the event. For participants with no recorded event year, follow-up time was adjusted to ensure a minimum value of 1 year.

Key systemic health markers were categorised to facilitate analysis. C-reactive protein (CRP) levels were dichotomized into two categories, high (>15 mg/L) and normal ($\leq$15 mg/L), with this binary variable created only for participants with available CRP data. Similarly, estimated glomerular filtration rate (eGFR) was classified based on clinical guidelines as normal ($\geq$60 mL/min/1.73 m^2) or indicative of kidney disease (<60 mL/min/1.73 m^2).

After processing, the dataset consisted of 4833 unique participants and 23,742 hospital records, all free from HNC at baseline and with complete follow-up information for survival analysis. This robust dataset, based on physician-diagnosed ICD-10 codes, enabled accurate categorization and reliable analysis of systemic and oral health conditions.

As this study relies on retrospective hospital records, the potential for unmeasured confounding cannot be excluded, particularly regarding lifestyle factors such as smoking and alcohol consumption, which were not systematically recorded in the database. Additionally, while ICD-10 coding provides standardised diagnostic criteria, it lacks granularity in disease severity and progression, potentially leading to residual misclassification bias. To minimise these methodological constraints, we applied strict inclusion criteria, used validated time-to-event models, and accounted for key systemic health markers in the analysis to improve internal validity.

4.2. Statistical Analysis
4.2.1. Baseline Characteristics

The baseline characteristics of the study population were summarised to describe the cohort at the start of follow-up. Continuous variables, such as age, were presented as means and standard deviations, in addition to medians and interquartile ranges. Categorical variables, including periodontitis, systemic health markers like CRP and eGFR, and demographic factors like sex, along with other covariates included in the analysis, were summarised as counts and percentages.

4.2.2. Log-Rank Tests and Kaplan–Meier Survival Analysis

The first step in survival analysis involved evaluating differences in survival probabilities across categories of key predictors using the Kaplan–Meier method [67]. Survival curves were generated to visually estimate and compare survival probabilities over time. The figures generated were limited to those covariates that exhibited statistical significance in the final model. The Kaplan–Meier survival function is defined as follows:

$$\hat{s}(t) = \prod_{t_i \leq t} \left(1 - \frac{d_i}{n_i} \right) \tag{1}$$

where t_i represents observed event times, d_i is the number of events at t_i, and n_i is the number of participants at risk immediately before t_i. Differences between survival curves were assessed using the log-rank test [68], which evaluates the null hypothesis that survival distributions across groups are identical. The test statistic is calculated as follows:

$$X^2 = \sum \frac{(O_i - E_i)^2}{E_i} \tag{2}$$

where O_i represents the observed number of events in group i, and E_i is the expected number of events if the survival times were identical across groups. Significant results suggest differences in survival curves between groups. Statistical significance was determined at a threshold of $p < 0.05$, indicating differences in survival probabilities.

4.2.3. Cumulative Hazard Estimation

Cumulative hazard plots were generated using the Nelson–Aalen estimator, a non-parametric method that provides an estimate of the cumulative hazard over time [69,70]. This estimator is defined as follows:

$$\hat{H}(t) = \sum_{t_i \leq t} \frac{d_i}{n_i} \tag{3}$$

where $\hat{H}(t)$ is the cumulative hazard at time t, d_i represents the number of events (failures) at time t_i, and n_i denotes the number of individuals at risk immediately before t_i. The cumulative hazard function is a useful tool to visualise the aggregate risk of the event over the study period, providing a complementary perspective to Kaplan–Meier survival curves by focusing on the accumulation of risk rather than the probability of survival. The plots incorporated 95% confidence intervals to account for variability in the hazard estimates.

4.2.4. Weibull Regression

To quantify the effect of systemic and oral health markers on oral cancer risk, parametric survival regression models were employed, with the Weibull distribution chosen for the primary analysis [71,72]. The Weibull model was selected for its flexibility in accommodating hazard rates that change over time, allowing for hazards that either increase or decrease as a function of time. The hazard function for the Weibull model is expressed as follows:

$$h(t|X) = \lambda \gamma t^{\gamma-1} \exp(X'\beta) \tag{4}$$

where $h(t|X)$ represents the hazard rate at time t, λ is the scale parameter, γ is the shape parameter, X is the vector of covariates, and β represents the coefficients. The corresponding survival function is as follows:

$$S(t|X) = \exp\left(-\lambda t^{\gamma} \exp(X'\beta)\right) \tag{5}$$

The parameters γ and λ were estimated along with covariate effects β to determine the hazard ratios (HRs) and their 95% confidence intervals (CIs). Significant predictors of oral cancer included systemic health markers, such as high CRP levels and impaired eGFR, as well as periodontitis, all of which were retained in the final model.

The Weibull model estimates the hazard of HNC at any given time based on systemic and oral health markers, allowing for time-dependent risk assessment rather than assuming a constant hazard over follow-up. Unlike the Cox proportional hazards model, which assumes a constant hazard ratio over time (proportional hazards assumption), the Weibull model accommodates varying hazard rates by incorporating a shape parameter (γ). This flexibility enables the identification of whether HNC risk increases ($\gamma > 1$) or decreases ($\gamma < 1$) over time, making it particularly suitable for diseases with evolving risk dynamics, such as inflammation-driven malignancies.

4.2.5. Model Comparison

To validate the choice of the Weibull distribution, alternative parametric models were fitted, including the exponential, log-normal, and log-logistic distributions. Additionally, the Weibull model was compared to the semi-parametric Cox proportional hazards model. Model fit was assessed using the Akaike Information Criterion (AIC) [73] and Bayesian Information Criterion (BIC) [74]. Lower AIC and BIC values indicated better model fit. Among the models evaluated, the Weibull regression demonstrated the lowest AIC and BIC values, confirming its suitability for the data.

4.2.6. Model Validation

The predictive performance of the Weibull model was evaluated using receiver operating characteristic (ROC) curves [75] and Harrell's C-index [76]. The ROC curve quantified the ability of the model to distinguish between participants who developed oral cancer and those who did not. The area under the curve (AUC) provided a summary measure of discrimination, with higher values indicating better predictive ability. Harrell's C-index, a concordance measure for survival data, assessed the agreement between predicted and observed survival times.

A two-tailed significance threshold of $p < 0.05$ was used for all statistical tests and models throughout the analysis. All statistical analyses and visualisations were performed using Intercooled Stata v18 [77].

5. Conclusions

Our findings suggest elevated CRP as a potential adjunctive biomarker for stratifying HNC risk, particularly in individuals with chronic oral inflammatory conditions. The significantly increased hazard observed among patients with periodontitis suggests that this population should be reclassified as high-risk, warranting targeted surveillance and earlier diagnostic interventions. Given CRP's cost-effectiveness and accessibility in routine laboratory panels, its integration into multifactorial HNC risk prediction models could enhance early detection, particularly in resource-limited settings where comprehensive oncologic screening is unfeasible.

The observed association between structural oral abnormalities, including embedded or impacted teeth, and HNC risk outlines the necessity of comprehensive dental assessments as part of oncologic risk stratification. These findings reinforce the imperative of interdisciplinary collaboration between dental and medical professionals, advocating for the inclusion of systemic inflammatory markers and periodontal health assessments in routine clinical evaluations. Future research should validate CRP's predictive utility in prospective cohort studies and explore its potential integration with other systemic and molecular biomarkers to refine early detection strategies.

Author Contributions: Conceptualization: A.S.G. and A.C.N.; Methodology: A.S.G.; Validation: A.S.G.; Formal Analysis: A.S.G.; Data Curation: A.S.G.; Writing—Original Draft Preparation: A.S.G., A.C.N., K.S. and Á.T.; Writing—Review and Editing: A.S.G., A.C.N., K.S. and Á.T.; Visualisation: A.S.G.; Supervision: A.C.N. and A.S.G. All authors have read and agreed to the published version of the manuscript.

Funding: This paper was supported by the EKÖP-24-3 University Research Scholarship Program of the Ministry for Culture and Innovation from the source of the National Research, Development and Innovation Fund. This paper was supported by the János Bolyai Research Scholarship of the Hungarian Academy of Sciences.

Institutional Review Board Statement: The research involving human participants received approval from the Ethics Committee of the University of Debrecen. All procedures were carried out in compliance with local laws and institutional guidelines. Ethical approval was granted by the Ethics Committee of the University of Debrecen under protocol number 6054-2022, dated 20 April 2022.

Informed Consent Statement: Patient consent was waived due to secondary data analysis using de-identified data.

Data Availability Statement: The data presented in this study are available on request from the corresponding author. The data are not publicly available due to institutional restrictions.

Conflicts of Interest: The authors declare no conflicts of interest.

References

1. Oral Health. Available online: https://www.who.int/news-room/fact-sheets/detail/oral-health (accessed on 21 February 2025)
2. Wong, T.; Wiesenfeld, D. Oral Cancer. *Aust. Dent. J.* **2018**, *63*, S91–S99. [CrossRef]
3. Lip and Oral Cavity Cancer Treatment—NCI. Available online: https://www.cancer.gov/types/head-and-neck/patient/adult/lip-mouth-treatment-pdq (accessed on 21 February 2025).
4. Chaurasia, A.; Alam, S.I.; Singh, N. Oral cancer diagnostics. *Natl. J. Maxillofac. Surg.* **2021**, *12*, 324–332. [CrossRef]
5. Lestón, J.S.; Dios, P.D. Diagnostic clinical aids in oral cancer. *Oral Oncol.* **2010**, *46*, 418–422. [CrossRef]
6. Montero, P.H.; Patel, S.G. Cancer of the Oral Cavity. *Surg. Oncol. Clin. N. Am.* **2015**, *24*, 491–508. [CrossRef]
7. Heller, M.A.; Nyirjesy, S.C.; Balsiger, R.; Talbot, N.; VanKoevering, K.K.; Haring, C.T.; Old, M.O.; Kang, S.Y.; Seim, N.B. Modifiable risk factors for oral cavity cancer in non-smokers: A systematic review and meta-analysis. *Oral Oncol.* **2023**, *137*, 106300 [CrossRef]
8. Kwon, T.; Lamster, I.B.; Levin, L. Current Concepts in the Management of Periodontitis. *Int. Dent. J.* **2021**, *71*, 462–476. [CrossRef]
9. Gasmi, A.; Benahmed, A.G.; Noor, S.; Mujawdiya, P. Porphyromonas Gingivalis in the Development of Periodontitis: Impact on Dysbiosis and Inflammation. *Arch. Razi Inst.* **2022**, *77*, 1539–1551. [CrossRef]
10. Ghanem, A.S.; Németh, O.; Móré, M.; Nagy, A.C. Role of oral health in heart and vascular health: A population-based study *PLoS ONE* **2024**, *19*, e0301466. [CrossRef]
11. Irani, S.; Barati, I.; Badiei, M. Periodontitis and oral cancer-current concepts of the etiopathogenesis. *Oncol. Rev.* **2020**, *14*, 23–34 [CrossRef] [PubMed]
12. Dhingra, K.; Vandana, K.L. Indices for measuring periodontitis: A literature review. *Int. Dent. J.* **2020**, *61*, 76–84. [CrossRef]
13. Flemmig, T.F. Periodontitis. *Ann. Periodontol.* **1999**, *4*, 32–37. [CrossRef]
14. Chattopadhyay, I.; Verma, M.; Panda, M. Role of Oral Microbiome Signatures in Diagnosis and Prognosis of Oral Cancer. *Technol. Cancer Res. Treat.* **2019**, *18*, 1533033819867354. [CrossRef]
15. Su, S.-C.; Chang, L.-C.; Huang, H.-D.; Peng, C.-Y.; Chuang, C.-Y.; Chen, Y.-T.; Lu, M.-Y.; Chiu, Y.-W.; Chen, P.-Y.; Yang, S.-F. Oral microbial dysbiosis and its performance in predicting oral cancer. *Carcinogenesis* **2020**, *42*, 127–135. [CrossRef]
16. Knight, E.T.; Liu, J.; Seymour, G.J.; Faggion, C.M.; Cullinan, M.P. Risk factors that may modify the innate and adaptive immune responses in periodontal diseases. *Periodontology 2000* **2016**, *71*, 22–51. [CrossRef]
17. Yasuno, T.; Tada, K.; Takahashi, K.; Watanabe, M.; Ito, K.; Arima, H.; Masutani, K. Dysbiosis of oral bacteria in patients with chronic kidney disease. *Ren. Replace. Ther.* **2024**, *10*, 63. [CrossRef]
18. Santacroce, L.; Passarelli, P.C.; Azzolino, D.; Bottalico, L.; Charitos, I.A.; Cazzolla, A.P.; Colella, M.; Topi, S.; Godoy, F.G.; D'addona, A. Oral microbiota in human health and disease: A perspective. *Exp. Biol. Med.* **2023**, *248*, 1288–1301. [CrossRef]
19. Asteriou, E.; Gkoutzourelas, A.; Mavropoulos, A.; Katsiari, C.; Sakkas, L.I.; Bogdanos, D.P. Curcumin for the Management of Periodontitis and Early ACPA-Positive Rheumatoid Arthritis: Killing Two Birds with One Stone. *Nutrients* **2018**, *10*, 908 [CrossRef]
20. Tuominen, H.; Rautava, J. Oral Microbiota and Cancer Development. *Pathobiology* **2020**, *88*, 116–126. [CrossRef]
21. Aggarwal, B.B. Nuclear factor-κB. *Cancer Cell* **2004**, *6*, 203–208. [CrossRef]
22. Ma, Q.; Hao, S.; Hong, W.; Tergaonkar, V.; Sethi, G.; Tian, Y.; Duan, C. Versatile function of NF-κB in inflammation and cancer *Exp. Hematol. Oncol.* **2024**, *13*, 68. [CrossRef]
23. Usui, M.; Onizuka, S.; Sato, T.; Kokabu, S.; Ariyoshi, W.; Nakashima, K. Mechanism of alveolar bone destruction in periodontitis—Periodontal bacteria and inflammation. *Jpn. Dent. Sci. Rev.* **2021**, *57*, 201–208. [CrossRef]
24. Machado, V.; Botelho, J.; Escalda, C.; Hussain, S.B.; Luthra, S.; Mascarenhas, P.; Orlandi, M.; Mendes, J.J.; D'aiuto, F. Serum C-Reactive Protein and Periodontitis: A Systematic Review and Meta-Analysis. *Front. Immunol.* **2021**, *12*, 706432. [CrossRef]
25. Pandey, A. C-Reactive Protein (CRP) and its Association with Periodontal Disease: A Brief Review. *J. Clin. Diagn. Res.* **2014**, *8*, ZE21. [CrossRef]
26. Zhang, D.; Sun, M.; Samols, D.; Kushner, I. STAT3 Participates in Transcriptional Activation of the C-reactive Protein Gene by Interleukin-6. *J. Biol. Chem.* **1996**, *271*, 9503–9509. [CrossRef]
27. Ngwa, D.N.; Pathak, A.; Agrawal, A. IL-6 regulates induction of C-reactive protein gene expression by activating STAT3 isoforms. *Mol. Immunol.* **2022**, *146*, 50–56. [CrossRef]
28. Kramer, F.; Torzewski, J.; Kamenz, J.; Veit, K.; Hombach, V.; Dedio, J.; Ivashchenko, Y. Interleukin-1β stimulates acute phase response and C-reactive protein synthesis by inducing an NFκB- and C/EBPβ-dependent autocrine interleukin-6 loop. *Mol. Immunol.* **2008**, *45*, 2678–2689. [CrossRef]
29. Hart, P.C.; Rajab, I.M.; Alebraheem, M.; Potempa, L.A. C-Reactive Protein and Cancer—Diagnostic and Therapeutic Insights. *Front. Immunol.* **2020**, *11*, 595835. [CrossRef]
30. Antonelli, A.; Battaglia, A.M.; Sacco, A.; Petriaggi, L.; Giorgio, E.; Barone, S.; Biamonte, F.; Giudice, A. Ferroptosis and oral squamous cell carcinoma: Connecting the dots to move forward. *Front. Oral Health* **2024**, *5*, 1461022. [CrossRef]

31. Li, M.; Jin, S.; Zhang, Z.; Ma, H.; Yang, X. Interleukin-6 facilitates tumor progression by inducing ferroptosis resistance in head and neck squamous cell carcinoma. *Cancer Lett.* **2021**, *527*, 28–40. [CrossRef] [PubMed]

32. Zhou, L.; Wu, Y.; Ying, Y.; Ding, Y. Current knowledge of ferroptosis in the pathogenesis and prognosis of oral squamous cell carcinoma. *Cell. Signal.* **2024**, *119*, 111176. [CrossRef] [PubMed]

33. Wen, Z.; Zhang, Y.; Gao, B.; Chen, X. Baicalin induces ferroptosis in oral squamous cell carcinoma by suppressing the activity of FTH1. *J. Gene Med.* **2024**, *26*, e3669. [CrossRef] [PubMed]

34. Kenessey, I.; Nagy, P.; Polgár, C. The Hungarian situation of cancer epidemiology in the second decade of the 21st century. *Magy Onkol* **2022**, *66*, 175–184.

35. O'sullivan, A.; Kabir, Z.; Harding, M. Lip, Oral Cavity and Pharyngeal Cancer Burden in the European Union from 1990–2019 Using the 2019 Global Burden of Disease Study. *Int. J. Environ. Res. Public Health* **2022**, *19*, 6532. [CrossRef]

36. Dhull, A.K.; Atri, R.; Dhankhar, R.; Chauhan, A.K.; Kaushal, V. Major Risk Factors in Head and Neck Cancer: A Retrospective Analysis of 12-Year Experiences. *World J. Oncol.* **2018**, *9*, 80–84. [CrossRef]

37. Irani, S. New Insights into Oral Cancer—Risk Factors and Prevention: A Review of Literature. *Int. J. Prev. Med.* **2020**, *11*, 202. [CrossRef]

38. OECD. EU Country Cancer Profile: Hungary 2023. In *EU Country Cancer Profiles*; OECD: Paris, France, 2023. [CrossRef]

39. Hussein, A.A.; Helder, M.N.; de Visscher, J.G.; Leemans, C.R.; Braakhuis, B.J.; de Vet, H.C.; Forouzanfar, T. Global incidence of oral and oropharynx cancer in patients younger than 45 years versus older patients: A systematic review. *Eur. J. Cancer* **2017**, *82*, 115–127. [CrossRef]

40. Bray, F.; Laversanne, M.; Sung, H.; Ferlay, J.; Siegel, R.L.; Soerjomataram, I.; Jemal, A. Global cancer statistics 2022: GLOBOCAN estimates of incidence and mortality worldwide for 36 cancers in 185 countries. *CA Cancer J. Clin.* **2024**, *74*, 229–263. [CrossRef]

41. Ghanem, A.S.; Móré, M.; Nagy, A.C. Assessing the impact of sociodemographic and lifestyle factors on oral health: A cross-sectional study in the Hungarian population. *Front. Public Health* **2023**, *11*, 1276758. [CrossRef]

42. Nemes, J.A.; Redl, P.; Boda, R.; Kiss, C.; Márton, I.J. Oral Cancer Report from Northeastern Hungary. *Pathol. Oncol. Res.* **2008**, *14*, 85–92. [CrossRef] [PubMed]

43. Tezal, M.; Scannapieco, F.A.; Wactawski-Wende, J.; Meurman, J.H.; Marshall, J.R.; Rojas, I.G.; Stoler, D.L.; Genco, R.J. Dental Caries and Head and Neck Cancers. *Arch. Otolaryngol. Neck Surg.* **2013**, *139*, 1054–1060. [CrossRef] [PubMed]

44. Mensch, K.; Nagy, G.; Nagy, Á.; Bródy, A. A szájüreg leggyakoribb bakteriális eredetű kórképeinek jellegzetességei, diagnosztikája és kezelése. *Orvosi Hetil.* **2019**, *160*, 739–746. [CrossRef] [PubMed]

45. Gaffen, S.; Hajishengallis, G. A new inflammatory cytokine on the block: Re-thinking periodontal disease and the Th1/Th2 paradigm in the context of Th17 cells and IL-17. *J. Dent. Res.* **2008**, *87*, 817–828. [CrossRef] [PubMed]

46. Gaur, P.; Singh, A.K.; Shukla, N.K.; Das, S.N. Inter-relation of Th1, Th2, Th17 and Treg cytokines in oral cancer patients and their clinical significance. *Hum. Immunol.* **2014**, *75*, 330–337. [CrossRef]

47. Anvar, M.T.; Rashidan, K.; Arsam, N.; Rasouli-Saravani, A.; Yadegari, H.; Ahmadi, A.; Asgari, Z.; Vanan, A.G.; Ghorbaninezhad, F.; Tahmasebi, S. Th17 cell function in cancers: Immunosuppressive agents or anti-tumor allies? *Cancer Cell Int.* **2024**, *24*, 355. [CrossRef]

48. Huang, N.; Dong, H.; Luo, Y.; Shao, B. Th17 Cells in Periodontitis and Its Regulation by A20. *Front. Immunol.* **2021**, *12*, 742925. [CrossRef] [PubMed]

49. Liu, J.; Ouyang, Y.; Zhang, Z.; Wen, S.; Pi, Y.; Chen, D.; Su, Z.; Liang, Z.; Guo, L.; Wang, Y. The role of Th17 cells: Explanation of relationship between periodontitis and COPD? *Inflamm. Res.* **2022**, *71*, 1011–1024. [CrossRef]

50. Gaur, P.; Shukla, N.K.; Das, S.N. Phenotypic and Functional Characteristics of Th17 (CD4$^+$IL17A$^+$) Cells in Human Oral Squamous Cell Carcinoma and Its Clinical Relevance. *Immunol. Investig.* **2017**, *46*, 689–702. [CrossRef]

51. Brauer, M.; A Roth, G.; Aravkin, A.Y.; Zheng, P.; Abate, K.H.; Abate, Y.H.; Abbafati, C.; Abbasgholizadeh, R.; Abbasi, M.A.; Abbasian, M.; et al. Global burden and strength of evidence for 88 risk factors in 204 countries and 811 subnational locations, 1990–2021: A systematic analysis for the Global Burden of Disease Study 2021. *Lancet* **2024**, *403*, 2162–2203. [CrossRef]

52. Komlós, G.; Csurgay, K.; Horváth, F.; Pelyhe, L.; Németh, Z. Periodontitis as a risk for oral cancer: A case–control study. *BMC Oral Health* **2021**, *21*, 640. [CrossRef]

53. Surlari, Z.; Virvescu, D.I.; Baciu, E.-R.; Vasluianu, R.-I.; Budală, D.G. The Link between Periodontal Disease and Oral Cancer—A Certainty or a Never-Ending Dilemma? *Appl. Sci.* **2021**, *11*, 12100. [CrossRef]

54. Elebyary, O.; Barbour, A.; Fine, N.; Tenenbaum, H.C.; Glogauer, M. The Crossroads of Periodontitis and Oral Squamous Cell Carcinoma: Immune Implications and Tumor Promoting Capacities. *Front. Oral Health* **2021**, *1*, 584705. [CrossRef] [PubMed]

55. Gonde, N.; Rathod, S.; Kolte, A.; Lathiya, V.; Ughade, S. Association between tooth loss and risk of occurrence of oral cancer—A systematic review and meta-analysis. *Dent. Res. J.* **2023**, *20*, 4. [CrossRef]

56. Singh, N.; Baby, D.; Rajguru, J.P.; Patil, P.B.; Thakkannavar, S.S.; Pujari, V.B. Inflammation and cancer. *Ann. Afr. Med.* **2019**, *18*, 121–126. [CrossRef]

57. Adeyemo, W.L. Do pathologies associated with impacted lower third molars justify prophylactic removal? A critical review of the literature. *Oral Surg. Oral Med. Oral Pathol. Oral Radiol. Endodontology* **2006**, *102*, 448–452. [CrossRef]

58. Araújo, J.P.; Kowalski, L.P.; Rodrigues, M.L.; de Almeida, O.P.; Pinto, C.A.L.; Alves, F.A. Malignant transformation of an odontogenic cyst in a period of 10 years. *Case Rep. Dent.* **2014**, *2014*, 1–5. [CrossRef]

59. Salema, H.; Joshi, S.; Pawar, S.; Nair, V.S.; Deo, V.V.; Sanghai, M.M. Evaluation of the Role of C-reactive Protein as a Prognostic Indicator in Oral Pre-malignant and Malignant Lesions. *Cureus* **2024**, *16*, e60812. [CrossRef] [PubMed]

60. Zhang, Y.; Gu, D. Prognostic Impact of Serum CRP Level in Head and Neck Squamous Cell Carcinoma. *Front. Oncol.* **2022**, *12*, 889844. [CrossRef]

61. Volpato, S.; Pahor, M.; Ferrucci, L.; Simonsick, E.M.; Guralnik, J.M.; Kritchevsky, S.B.; Fellin, R.; Harris, T.B. Relationship of alcohol intake with inflammatory markers and plasminogen activator inhibitior-1 in well-functioning older adults. *Circulation* **2004**, *109*, 607–612. [CrossRef] [PubMed]

62. Shiels, M.S.; Katki, H.A.; Freedman, N.D.; Purdue, M.P.; Wentzensen, N.; Trabert, B.; Kitahara, C.M.; Furr, M.; Li, Y.; Kemp, T.J.; et al. Cigarette smoking and variations in systemic immune and inflammation markers. *JNCI J. Natl. Cancer Inst.* **2014**, *106*, dju294. [CrossRef]

63. AlShammari, A.; AlSaleh, S.; AlKandari, A.; AlSaqabi, S.; AlJalahmah, D.; AlSulimmani, W.; AlDosari, M.; AlHazmi, H.; AlQaderi, H. The association between dental caries and serum crp in the us adult population: Evidence from NHANES 2015–2018. *BMC Public Health* **2024**, *24*, dju294. [CrossRef]

64. Malyszko, J.; Tesarova, P.; Capasso, G.; Capasso, A. The link between kidney disease and cancer: Complications and treatment. *Lancet* **2020**, *396*, 277–287. [CrossRef] [PubMed]

65. Hu, M.; Wang, Q.; Liu, B.; Ma, Q.; Zhang, T.; Huang, T.; Lv, Z.; Wang, R. Chronic Kidney Disease and Cancer: Inter-Relationships and Mechanisms. *Front. Cell Dev. Biol.* **2022**, *10*, 868715. [CrossRef] [PubMed]

66. Wong, G.; Staplin, N.; Emberson, J.; Baigent, C.; Turner, R.; Chalmers, J.; Zoungas, S.; Pollock, C.; Cooper, B.; Harris, D.; et al. Chronic kidney disease and the risk of cancer: An individual patient data meta-analysis of 32,057 participants from six prospective studies. *BMC Cancer* **2016**, *16*, 488. [CrossRef]

67. Kaplan, E.L.; Meier, P. Nonparametric Estimation from Incomplete Observations. *J. Am. Stat. Assoc.* **1958**, *53*, 457–481. [CrossRef]

68. Mantel, N. Evaluation of survival data and two new rank order statistics arising in its consideration. *Cancer Chemother. Rep.* **1966**, *50*, 163–170.

69. Nelson, W. Hazard Plotting for Incomplete Failure Data. *J. Qual. Technol.* **1969**, *1*, 27–52. [CrossRef]

70. McKeague, I.W. Introduction to Aalen (1978) Nonparametric Inference for a Family of Counting Processes. In *Breakthroughs in Statistics*; Springer Series in Statistics; Kotz, S., Johnson, N.L., Eds.; Springer: New York, NY, USA, 1997; pp. 347–367. [CrossRef]

71. Weibull, W. A Statistical Distribution Function of Wide Applicability. *J. Appl. Mech.* **1951**, *18*, 293–297. [CrossRef]

72. Collett, D. *Modelling Survival Data in Medical Research*; Taylor & Francis: London, UK, 2015. [CrossRef]

73. Akaike, H. A new look at the statistical model identification. *IEEE Trans. Autom. Control* **1974**, *19*, 716–723. [CrossRef]

74. Schwarz, G. Estimating the Dimension of a Model. *Ann. Stat.* **1978**, *6*, 461–464. [CrossRef]

75. Treisman, M. Book Review: Signal Detection Theory and ROC Analysis. *Q. J. Exp. Psychol.* **1977**, *29*, 745–746. [CrossRef]

76. Harrell, F.E.; Califf, R.M.; Pryor, D.B.; Lee, K.L.; Rosati, R.A. Evaluating the Yield of Medical Tests. *JAMA J. Am. Med. Assoc.* **1982**, *247*, 2543. [CrossRef]

77. StataCorp. *Stata Statistical Software: Release 18*; StataCorp LLC: College Station, TX, USA, 2023.

International Journal of
Molecular Sciences

Article

Transcriptional Expression of SLC2A3 and SDHA Predicts the Risk of Local Tumor Recurrence in Patients with Head and Neck Squamous Cell Carcinomas Treated Primarily with Radiotherapy or Chemoradiotherapy

Mercedes Camacho [1], Silvia Bagué [2], Cristina Valero [3], Anna Holgado [3], Laura López-Vilaró [2], Ximena Terra [4], Francesc-Xavier Avilés-Jurado [5,†] and Xavier León [3,6,*,†]

[1] Genomics of Complex Diseases, Research Institute Hospital Sant Pau, IIB Sant Pau, 08041 Barcelona, Spain; mcamacho@santpau.cat
[2] Pathology Department, Hospital de la Santa Creu i Sant Pau, Universitat Autònoma de Barcelona, 08041 Barcelona, Spain; sbague@santpau.cat (S.B.); llopezv@santpau.cat (L.L.-V.)
[3] Otorhinolaryngology Department, Hospital de la Santa Creu i Sant Pau, Universitat Autònoma de Barcelona, 08041 Barcelona, Spain; cvalero@santpau.cat (C.V.); aholgado@santpau.cat (A.H.)
[4] MoBioFood Research Group, Biochemistry and Biotechnology Department, Universitat Rovira i Virgili, Campus Sescelades, 43007 Tarragona, Spain; ximena.terra@urv.cat
[5] Head Neck tumors Unit, Hospital Clínic Barcelona, Universitat de Barcelona, IDIBAPS, 08036 Barcelona, Spain; faviles@clinic.cat
[6] Centro de Investigación Biomédica en Red de Bioingeniería, Biomateriales y Nanomedicina (CIBER-BBN), 28029 Madrid, Spain
* Correspondence: xleon@santpau.cat; Tel.: +34-93-5565679
† These authors contributed equally to this work.

Abstract: Reprogramming of metabolic pathways is crucial to guarantee the bioenergetic and biosynthetic demands of rapidly proliferating cancer cells and might be related to treatment resistance. We have previously demonstrated the deregulation of the succinate pathway in head and neck squamous cell carcinoma (HNSCC) and its potential as a diagnostic and prognostic marker. Now we aim to identify biomarkers of resistance to radiotherapy (RT) by analyzing the expression of genes related to the succinate pathway and nutrient flux across the cell membrane. We determined the transcriptional expression of succinate receptor 1 (SUCNR1), succinate dehydrogenase A (SDHA), and the solute carrier (SLC) superfamily transporters responsible for the influx or efflux of a wide variety of nutrients (SLC2A3 and SLC16A3) in tumoral tissue from 120 HNSCC patients treated with RT or chemoradiotherapy (CRT). Our results indicated that the transcriptional expression of the glucose transporter SLC2A3 together with SDHA had the best predictive capacity for local response after treatment with RT or CRT. High SLC2A3 and SDHA expression predicted poor outcomes after RT or CRT, with these patients having a 4.2 times higher risk of local recurrence compared to the rest of the patients. These results might indicate that tumors that shifted toward a higher glucose influx and a higher oxidation of succinate via mitochondrial complex II present an ideal environment for radioresistance development. Patients with a high transcriptional expression of both SLC2A3 and SDHA had a significantly higher risk of local recurrence after treatment with RT or CRT.

Keywords: SLC2A3; GLUT3; SLC16A; SUCNR1; SDHA; biomarker; Warburg effect; head and neck squamous cell carcinoma

1. Introduction

Radiotherapy (RT) is widely used in patients with head and neck squamous cell carcinoma (HNSCC), either alone or in multimodality therapeutic strategies including surgery and chemotherapy. Currently, there are no genomic or molecular markers other than p16 or human papillomavirus (HPV) status that can effectively predict the outcome of RT in patients with HNSCC. Accordingly, the identification of molecular markers that predict response to RT would be an important milestone in head and neck oncology and would help move toward the development of personalized treatments to maximize survival and minimize morbidity [1].

Aberrant glucose metabolism has been recognized as a common feature in cancer for nearly a century [2]. This phenomenon, known as the Warburg effect, consists of a significantly elevated rate of glucose consumption and lactate excretion that is largely insensitive to oxygen availability [3]. There is significant evidence to demonstrate the metabolic reprogramming of tumor cells, as an addition to excessive uptake and metabolism of key nutrients to support rapid proliferation and invasion capacities. The transport of key nutrients for cancer metabolism, such as glucose and amino acids, is of particular interest for their roles in tumor progression and metastasis. Solute carrier (SLC) superfamily transporters are responsible for the influx or efflux of a wide variety of nutrients that are needed for the cells to function. To meet the increased demand for nutrients and energy, SLC transporters are frequently deregulated in cancer cells. For example, elevated SLC2A1 (GLUT1) and SLC2A3 (GLUT3), which facilitate the transport of glucose across the plasmatic membrane, have been associated with increased cancer metabolism [4]. Specifically, the role of glucose transporters in oral squamous cell carcinoma has been recently reviewed by Botha et al. [5]. The authors concluded that SLC2A1 and SLC2A3 have a role in the pathophysiology of oral squamous cell carcinoma and represent valuable biomarkers in assessing diagnosis and prognosis. SLC16A family lactate transporters have also been reported to be deregulated, in part to maintain intracellular pH for continued growth with increased lactate production [6].

The tricarboxylic acid (TCA) cycle is strategically situated at the center of cellular metabolism, where it serves as an anabolic hub for the synthesis of macromolecules such as fatty acids, cholesterol, and amino acids that are crucial to support rapidly growing tumors. The TCA cycle must be constantly replenished by carbon atoms, a process named anaplerosis [7]. The current dogma in biochemistry is that the main anaplerotic substrates are pyruvate (from glucose), which is converted into oxalacetate, and glutamate (from glutamine), which is metabolized to succinate, with lesser contributions from precursors of propionyl-CoA, such as odd-chain fatty acids, amino acids, and C5 ketone bodies that also feed into succinate [8].

Our group has previously demonstrated that the circulating succinate levels are elevated in HNSCC, identifying this oncometabolite as a potentially valuable non-invasive biomarker for HNSCC diagnosis [9]. Furthermore, we demonstrated an important role of the succinate-related pathway in tumor development and response to treatment in patients with HNSCC, where high succinate receptor 1 (SUCNR1) together with high succinate dehydrogenase A (SDHA) expression predicts poor locoregional disease-free survival in a sub-group of patients treated with RT or chemoradiotherapy (CRT) [9].

The close relationship between the transporters of anaplerotic substrates and the succinate pathway in the TCA cycle as an anabolic hub for the synthesis of macromolecules to support rapidly growing tumors made us hypothesize that the potential of SUCNR1/SDHA as prognostic biomarkers could be enhanced including in the model transporters of key nutrients that are crucial for cancer metabolism.

The present study aims to evaluate the prognostic potential of the transcriptional expression of genes encoding solute carrier transporters (SLC2A3 and SLC16A3) and to develop a predictive model that integrates these genes with others associated with the succinate pathway (SUCNR1 and SDHA), thereby enhancing prognostic accuracy. The analysis was conducted using a cohort of patients with HNSCC treated with RT or CRT.

2. Results

We analyzed 120 patients with squamous cell carcinomas located in the oral cavity, oropharynx, hypopharynx, or larynx treated primarily with RT or CRT. During the follow-up period, 39 patients (32.5%) had a local recurrence of the tumor, 14 (11.7%) had a regional recurrence, and 15 (12.5%) had distant metastases. We defined local recurrence as the persistence or recurrence of carcinoma at the primary tumor site following completion of radiotherapy or chemoradiotherapy.

2.1. Transcriptional Expression of SUCNR1, SLC2A3, SLC16A3, and SDHA

When analyzing the transcriptional expression according to clinical variables, we observed significant differences according to sex in the expression of SLC2A3. SLC2A3 expression was higher in female patients. There were no significant differences in the expression of the genes analyzed according to the history of toxics consumption or local extension (cT) or regional extension (cN) of the tumor. In the case of patients with oropharyngeal carcinomas, SUCNR1 expression was higher in patients with HPV-positive tumors.

There were significant differences in the SLC2A3 transcriptional expression according to the local tumor control after RT or CRT. Patients with local recurrence had significantly higher SLC2A3 expression values than patients who had local control after treatment (p = 0.043). Patients with a local recurrence tended to have higher SDHA expression, but without reaching statistical significance (p = 0.085). Figure S1 of the Supplementary Materials shows the distribution of SLC2A3 expression according to local control. Table S1 of the Supplementary Materials shows the median transcriptional expression values of the genes analyzed according to the different variables studied.

2.2. Results of the Recursive Partitioning Analysis (RPA)

We individually evaluated with an RPA the relationship between the expression values of each gene and local disease control after treatment. We found a significant relationship between the high expression of SLC2A3, SLC16A3, and SDHA and a higher rate of local recurrence. No relationship was found between SUCNR1 expression and local disease control. Table 1 shows the distribution of patients according to the cut-off values obtained with the RPA, as well as the 5-year local recurrence-free survival for each of the categories obtained with these cut-off values. Figure S2 of the Supplementary Materials shows the local recurrence-free survival curves according to the transcriptional expression categories of SLC2A3, SLC16A3, and SDHA obtained with the RPA.

When we jointly analyzed the transcriptional expression values of the three genes that were related to local disease control, the RPA model classified patients into three categories, with the first level of classification based on SLC2A3 expression and the second level of classification for patients with high SLC2A3 expression based on SDHA expression. A classification tree with three terminal nodes was obtained (Figure 1): patients with low SLC2A3 expression (n = 64, percentage of local recurrence 21.9%), patients with high SLC2A3 and low SDHA expression (n = 21, percentage of local recurrence 14.3%), and patients with high SLC2A3 and high SDHA expression (n = 35, percentage of local recurrence 62.9%). We then proceeded to group the two terminal nodes with lower local recurrence rates, classifying patients into two groups: Group 1, patients with low SLC2A3 expression and patients with high SLC2A3 and low SDHA expression (n = 85, local recurrence rate 20.0%), and Group 2, patients with high SLC2A3 and high SDHA expression (n = 35, local recurrence rate 62.9%).

Table 1. Distribution of patients according to the cut-off points obtained with the classification and regression tree and 5-year local recurrence-free survival (LRFS) for each of the categories obtained with these cut-off points.

	Cut-Off	N	5–Year LRFS (95% CI)	*p*
SLC2A3	Low ($\leq$19.16)	64	77.2% (66.6–87.8%)	0.005
	High (>19.16)	56	54.0% (40.7–67.3%)	
SLAC16A3	Low ($\leq$40.83)	50	77.0% (65.0–89.0%)	0.025
	High (>40.83)	70	58.2% (46.2–70.2%)	
SDHA	Low ($\leq$29.13)	51	79.5% (68.1–90.9%)	0.009
	High (>29.13)	69	56.5% (44.5–68.5%)	

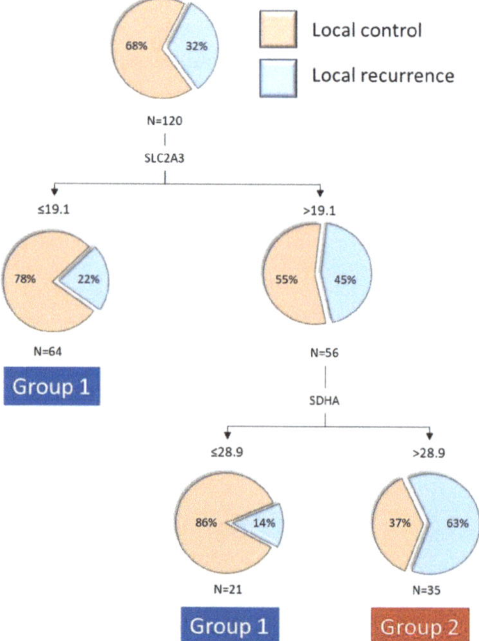

Figure 1. Classification and regression tree according to the transcriptional expression values of SLC2A3 and SDHA considering the local control after treatment with radiotherapy or chemoradiotherapy as the dependent variable.

Supplementary Table S2 presents the distribution of patients according to the SLC2A3-SDHA expression category in relation to clinical variables. Significant differences were observed only in the distribution of patients based on the extension of the primary tumor. Specifically, the frequency of patients with early-stage tumors (cT1-2) was higher in Group 1 compared to Group 2 (78.6% vs. 21.4%, *p* = 0.041).

2.3. Survival According to the SLC2A3-SDHA Expression Category

Figure 2 shows the local recurrence-free survival according to the transcriptional SLC2A3-SDHA group. Five-year local recurrence-free survival for Group 1 was 79.1% (95% CI:70.3–87.9%), and for Group 2, it was 35.1% (95% CI:18.7–51.5%) (*p* = 0.0001).

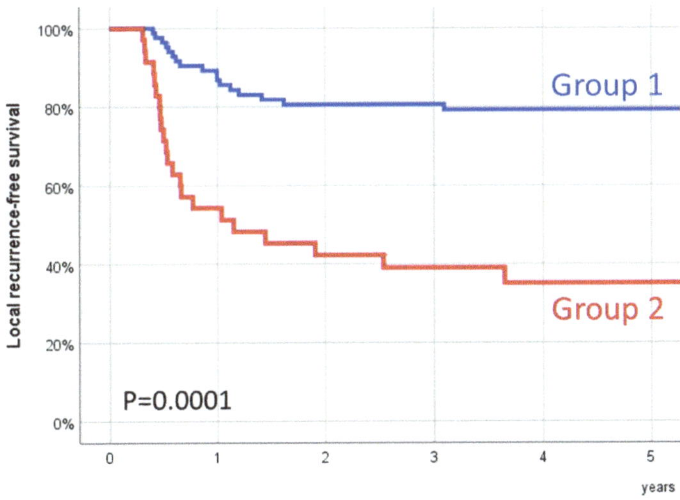

Figure 2. Local recurrence-free survival after treatment with radiotherapy or chemoradiotherapy according to SLC2A3 and SDHA transcriptional expression. Group 1, patients with low SLC2A3 expression or patients with high SLC2A3 and low SDHA expression; Group 2, patients with high SLC2A3 and high SDHA expression.

The observed differences in local disease control were independent of the treatment modality. For patients treated with RT (n = 54), 5-year local recurrence-free survival for Group 1 (n = 42) was 87.7% (95% CI: 77.7–97.7%), and for Group 2 (n = 12), it was 41.7% (95% CI: 13.9–69.5%) (p = 0.0001). For patients treated with CRT (n = 66), 5-year local recurrence-free survival for Group 1 (n = 43) was 70.6% (95% CI: 56.5–84.7%), and for Group 2 (n = 23), it was 31.4% (95% CI: 11.0–51.8%) (p = 0.001). The local recurrence-free survival curves, stratified by the SLC2A3-SDHA transcriptional group and treatment type, are shown in Figure S3 of the Supplementary Materials.

Similarly, we observed differences in local disease control regardless of the local extension of the primary tumor. For patients with early tumors (cT1-T2, n = 70), 5-year local recurrence-free survival for Group 1 (n = 55) was 85.0% (95% CI: 75.4–94.6%), and for Group 2 (n = 15), it was 53.3% (95% CI: 28.0–78.6%) (p = 0.005). In patients with locally advanced tumors (cT3-T4, n = 50), 5-year local recurrence-free survival for Group 1 (n = 30) was 68.3% (95% CI: 51.1–85.5%), and for Group 2 (n = 20), it was 21.4% (95% CI: 2.0–40.8%) (p = 0.001). Figure S4 in the Supplementary Materials shows the local recurrence-free survival curves according to the SLC2A3-SDHA transcriptional group depending on the local tumor extension category.

2.4. Multivariable Analysis

Table 2 shows the result of a multivariable analysis in which local recurrence-free survival was considered as the dependent variable. The only variable that was significantly associated with local control was the category of SLC2A3-SDHA expression. Relative to patients in Group 1, patients in Group 2 (high SLC2A3 and high SDHA expression) had a 4.24-fold increased risk of local recurrence (95% CI: 2.07–8.69, p = 0.0001).

Table 2. Results of the multivariable analysis considering local recurrence-free survival as the dependent variable.

		HR (CI95%)	*p*
Age		1.02 (0.98–1.06)	0.234
Sex	Male	1	
	Female	0.74 (0.25–2.22)	0.602
Toxics consumption	No	1	
	Moderate	0.44 (0.07–2.57)	0.367
	Severe	0.45 (0.10–1.99)	0.296
Location	Oral cavity	1	
	Oropharynx	0.84 (0.12–5.86)	0.865
	Hypopharynx	2.14 (0.25–17.99)	0.481
	Larynx	1.50 (0.19–11.33)	0.693
Local extension	cT1-T2	1	
	cT3-T4	2.29 (0.94–5.58)	0.066
Regional extension	cN0	1	
	cN+	1.04 (0.38–2.88)	0.929
Histological grade	Well differentiated	1	
	Moderately differentiated	2.99 (0.47–19.01)	0.244
	Poorly differentiated	0.89 (0.08–8.95)	0.925
Treatment	Radiotherapy	1	
	Chemoradiotherapy	2.20 (0.56–8.55)	0.253
SLC2A3-SDHA	Group 1	1	
	Group 2	4.24 (2.07–8.69)	0.0001

When analyzing patients with oropharyngeal carcinomas according to HPV status, we observed that in HPV-negative tumors, the advantage in local recurrence-free survival was maintained for patients in Group 1 (5-year local recurrence-free survival for patients in Group 1 66.9% versus 12.9% for patients in Group 2), although the differences did not reach statistical significance ($p = 0.123$). For patients with HPV-positive tumors, the 5-year local recurrence-free survival for patients in Group 1 (n = 9) was 88.9%, and for patients in Group 2 (n = 3), it was 100%.

2.5. Local Recurrence-Free Survival According to SLC2A3 and SDHA Transcriptional Expression

Table 3 shows the 5-year local recurrence-free survival rates according to whether patients had high or low expression of SLC2A3 and SDHA according to the cut-off points obtained in the individual analysis of each gene. A significant decrease in local recurrence-free survival was observed only in the combination of elevated SLC2A3 and elevated SDHA expression. Notably, for patients with low SLC2A3 expression (n = 64), no differences in 5-year local recurrence-free survival were seen according to the SDHA expression category (75.8% for patients with low SDHA expression vs. 78.6% for those with high SDHA expression, $p = 0.826$). Figure S5 in the Supplementary Materials shows the local recurrence-free survival of patients according to whether they had high or low SLC2A3 and SDHA expression.

Table 3. Five-year local recurrence-free survival (LRFS) according to SLC2A3 and SDHA transcriptional expression.

Transcriptional SLC2A3/SDHA Expression	5-Year LRFS (95% CI)
High SLC2A3/low SDHA (n = 21)	85.2% (69.7–100%)
High SLC2A3/high SDHA (n = 35)	35.1% (18.6–51.6%)
Low SLC2A3/low SDHA (n = 30)	75.5% (59.6–91.4%)
Low SLC2A3/high SDHA (n = 34)	78.6% (64.5–92.7%)

2.6. Regional Recurrence and Distant Metastasis-Free Survival According to SLC2A3 and SDHA Transcriptional Expression

Regional recurrence-free survival for Group 1 was significantly higher than for Group 2 (5-year regional recurrence-free survival: 93.9%, 95% CI: 88.8–99.0% versus 71.5%, 95% CI: 55.0–88.0%, $p = 0.001$). Distant metastasis-free survival for Group 1 was also higher than Group 2, although in this case, the differences did not reach statistical significance (5-year distant metastasis-free survival: 88.7%, 95% CI: 81.6–95.8% versus 80.3%, 95% CI: 66.0–94.6%, $p = 0.249$). Group 1 patients had significantly better disease-specific survival. Five-year disease-specific survival for Group 1 was 75.6% (95% CI: 65.8–85.4%), and for Group 2, it was 51.2% (95% CI: 33.2–69.2%) ($p = 0.005$).

2.7. Results of the External Validation Study with the TCGA Data

Significant differences in SLC2A3 transcriptional expression were observed among patients in the TCGA cohort based on tumor status ($p = 0.044$). Patients who were tumor-free at the last contact or time of death (n = 340, 70.8%) exhibited significantly lower SLC2A3 expression compared to those with active tumors (n = 140, 29.2%). The distribution of SLC2A3 expression by tumor status is presented in Figure S6 of the Supplementary Materials. In contrast, SDHA expression did not differ significantly based on tumor status ($p = 0.144$).

When analyzing the transcriptional expression of SLC2A3 and SDHA with an RPA considering tumor status as the dependent variable, we obtained a classification tree with three terminal nodes, with SLC2A3 expression as the primary classifier and SDHA expression further stratifying patients with low SLC2A3 levels. Among patients with high SLC2A3 expression (n = 55), 49.1% had an active tumor at the last contact or time of death. In contrast, the proportion of patients with active tumors was 32.3% among those with low SLC2A3 and high SDHA expression (n = 189) and 22.3% among those with low SLC2A3 and low SDHA expression (n = 229). These differences were statistically significant ($p = 0.0001$). The classification and regression tree based on SLC2A3 and SDHA expression are presented in Figure S7 of the Supplementary Materials.

3. Discussion

According to our results, the joint assessment of SLC2A3 and SDHA transcriptional expression allowed us to define a group of patients with an elevated risk of local tumor recurrence after treatment with RT or CRT. Patients with elevated expression of both SLC2A3 and SDHA had a 4.24-fold increased risk of local recurrence relative to all other patients. The predictive capacity of SLC2A3-SDHA expression was consistent, regardless of the treatment modality (RT or CRT) or the local extension of the tumor (cT1–T2 or cT3–T4).

Elevated expression of glucose transporters is associated with decreased survival in most cancer models, including HNSCCs [4,10]. In a study of patients with surgically treated oral cavity carcinomas, Ayala et al. [11] found that patients with high immunohistochemical expression of SLC2A3 had a significantly increased risk of recurrence and decreased survival in both uni- and multivariable analyses. In another study carried out in patients with oral cavity carcinomas treated with surgery, Estilo et al. [12] found a significant association between elevated SLC2A3 transcriptional expression and depth of invasion, pathologic staging, and risk of recurrence. Similarly, in patients with laryngeal carcinomas treated with surgery and/or radiotherapy, Baer et al. [13] found a significant association between the immunohistochemical expression of SLC2A3 and survival. In patients with advanced laryngeal carcinomas treated with surgery, Starska et al. [14] found an increase in the transcriptional and immunohistochemical expression of SLC2A3 in tumor tissue relative to adjacent normal laryngeal tissue. Tumors with elevated SLC2A3 expression tended to have worse survival, but the differences did not reach statistical significance. In addition, SLC2A3 transcriptional expression appears in genetic signatures associated with prognosis and response to treatment in patients with HNSCC [15,16].

Some authors have found a relationship between SLC2A1 expression and response to RT in HNSCC patients. Kunkel et al. found that HNSCC with a high immunohisto-chemical expression of SLC2A1 had increased resistance to RT [17]. Chen et al. observed a significant decrease in disease-free survival and disease-specific survival in patients with p16-negative oropharyngeal or hypopharyngeal carcinomas treated with RT or CRT and a high immunohistochemical expression of SLC2A1 [18]. Nonetheless, to our knowledge, there are no studies that have analyzed the relationship between SLC2A3 expression and response to RT.

SDHA is a mitochondrial TCA cycle enzyme that converts succinate to fumarate. Mutations in the gene encoding SDHA that result in altered SDHA function have been associated with the appearance of tumors such as paragangliomas and pheochromocy-tomas, gastrointestinal stromal tumors, renal cell carcinomas, and pituitary adenomas [19]. However, there is no evidence that SDHA itself is mutated or dysfunctional in HNSCCs.

In a previous study from our group in a cohort of 41 patients with HNSCC independent of the cohort of the present study, we were able to demonstrate that SDHA with elevated transcriptional expression was associated with the loco-regional control of the disease in patients treated with RT or CRT [9]. In this study, we performed an external validation of the prognostic capacity of SDHA expression in patients treated with RT or CRT.

The number of studies that have analyzed the involvement of SDHA in the process of carcinogenesis is limited. Chattopadhyay et al. [20] found that high expression of SDHA was associated with the risk of metastatic spread and poor clinical outcome in patients with uveal melanoma, and Olszewski et al. found inhibition of tumor growth in a patient-derived xenograft model of SDHA-deficient tumors [21]. On the other hand, SDHA has been described as a tumor suppressor because it reduces the accumulation and secretion of succinate, considered an oncometabolite [22]. Li et al. found that SDHA was frequently downregulated in hepatocellular carcinoma tissues and that this downregulation was associated with poor prognosis [23]. By analyzing the data included in The Cancer Genome Atlas (TCGA), it can be observed that the prognostic capacity of SDHA transcriptional expression depends on the tumor type [24]. High expression of SDHA was associated with significantly reduced overall survival in breast cancer patients, whereas it was associated with increased survival in renal cancer patients. For patients with HNSCC, high SDHA expression was associated with a decrease in overall survival, but the differences did not reach statistical significance ($p = 0.16$). It should be considered that patients included in TCGA are mainly patients with oral cavity carcinomas and a large majority of patients treated with surgery, which significantly differs from the patients analyzed in our study.

Dysregulation in SDH may exert effects on many metabolic pathways [25]. Recently, Schöpf et al. [26] studied the rewiring of metabolism in prostate cancer. Their results reveal a shift toward higher oxidation of succinate, which is associated with deleterious mutations in mitochondrial Complex I genes, and a rewired expression of mitochondrial metabolic enzymes. Their study on oxidative phosphorylation capacities in prostate cancer tissues uncovered increased oxidation of succinate via Complex II as compensation for a decreased capacity to oxidize substrates via Complex I. In our cohort of HNSCC patients, the increased SDHA expression might also be associated with an increased oxidative phosphorylation capacity which might confer radiation resistance. Nonetheless, SDH enzymes are characterized by fine regulatory mechanisms including regulation of mRNA expression, post-translational modification, and endogenous inhibition [27]. Further studies are needed to investigate the role of this enzyme in both the carcinogenesis and progression of HNSCC.

The nutrient supply in tumor cells may originate from cellular metabolism or may be imported from the external microenvironment across the plasma membrane via transporters, fueling TCA for the synthesis of metabolic precursors, and OXPHOS for ATP production in mitochondria. Therefore, on one hand, a high expression of SLC2A3 might enhance aerobic glycolysis and the production of lactate, and on the other hand, an increased expression of SDH might favor (I) the conversion of succinate to fumarate as part of the tricarboxylic acid cycle and (II) the oxidative phosphorylation, delivering reducing equivalents into the electron transport chain via FADH2 and finally producing ATP. This shift seems an ideal environment for cancer development because it allows sparing of bio precursors for other needs, otherwise used for NADH production, while high succinate oxidation is maintained for efficient ATP production using FADH2 as an electron donor. This hypothesis agrees with the fact that in our results, when SLC2A3 expression is low, SDHA (high or low expression levels) does not affect the outcomes. This result highlights the importance of glucose and its influx in the metabolism of HNSCC cells [8]. In contrast, when SLC2A3 is elevated, the potential enhanced glucose influx is not enough to induce a radioresistant phenotype if the SDHA levels are low. Recently, Olszewski et al. [22] demonstrated that both reduced glucose entrance into the cell and low TCA enzyme expression are needed to reduce tumor growth. They showed that inhibiting class I glucose transporters was effective in inhibiting tumor growth in patient-derived xenograft models of SDHA-deficient tumors.

Analysis of SLC2A3 and SDHA transcriptional expression in patients with HNSCC from the TCGA dataset revealed trends consistent with those observed in our patients. Specifically, higher transcriptional expression of SLC2A3 and SDHA was associated with an increased likelihood of having an active tumor at the last contact or time of death. However, it is important to note that the characteristics of patients in the TCGA dataset differ significantly from those in our study, with a higher proportion of oral cavity carcinomas and a predominant use of surgical treatment.

There are several limitations in our study that need to be considered when analyzing the results. This is a retrospective study in which a limited number of genes related to glucose metabolism were studied. Moreover, we performed analysis only of the transcriptional expression, without providing information on the signaling pathways activated, the local concentration of substrates, or the existence of post-translational regulatory mechanisms. The limited number of patients with oral cavity carcinomas in this study may affect the generalizability of our findings to this specific sub-group. Additionally, while it would have been valuable to assess the expression of SLC2A and SDHA in recurrent tumors, we were unable to perform this analysis due to the lack of sufficient samples.

External validation studies are necessary before SLC2A3/SDHA transcriptional expression can be considered as a reliable biomarker for response to RT or CRT in patients with HNSCC. If confirmed in future studies, the predictive role of SLC2A3 and SDHA expression could contribute to more personalized treatment strategies, particularly in identifying patients at higher risk of local recurrence following radiotherapy. Furthermore, these findings may provide a foundation for future research into targeted therapies for patients with elevated SLC2A3 and SDHA expression.

4. Materials and Methods

4.1. Patients

The present study was performed retrospectively from biopsies obtained from the primary location of the tumor prior to any type of oncologic treatment in 120 patients with histologically confirmed squamous cell carcinomas located in the oral cavity, oropharynx, hypopharynx, or larynx treated with RT or CRT during the period 2008–2016. The patients included in the present study were diagnosed and treated in a different center than the one in which the initial determinations of the relationship between the succinate pathway and prognosis in patients with HNSCC were carried out [9]. Clinical information was obtained from a database that prospectively collects data related to the clinical characteristics, treatment, and follow-up of all patients with malignant head and neck tumors treated at our center. All patients included in the study were evaluated by a multidisciplinary tumor board, who proposed treatment with RT or CRT according to the institutional treatment protocols. In general, treatment was RT alone for patients with early-stage tumors (stages I–II) and CRT for advanced-stage tumors (stages III–IV), depending on the clinical characteristics of the patient. Table 4 shows the characteristics of the patients included in the study.

Table 4. Characteristics of the patients included in the study.

		N (%)
Age	Mean 62.59 years (rank 38.1–98.7 years)	
Sex	Male	106 (88.3%)
	Female	14 (11.7%)
Toxics consumption	No	13 (10.8%)
	Moderate	17 (14.2%)
	Severe	90 (75.0%)
Location	Oral cavity	6 (5.0%)
	Oropharynx	50 (41.7%)
	Hypopharynx	15 (12.5%)
	Larynx	49 (40.8%)
Local extension	cT1	23 (19.2%)
	cT2	47 (39.2%)
	cT3	35 (29.2%)
	cT4	15 (12.5%)
Regional extension	cN0	71 (59.2%)
	cN1	14 (11.7%)
	cN2	33 (27.5%)
	cN3	2 (1.7%)
Stage	I	22 (18.3%)
	II	26 (21.7%)
	III	28 (23.3%)
	IV	44 (36.7%)
Histological grade	Well differentiated	11 (9.2%)
	Moderately differentiated	97 (80.8%)
	Poorly differentiated	12 (10.0%)
Treatment	Radiotherapy	54 (45.0%)
	Chemoradiotherapy	66 (55.0%)

Given the interaction between tobacco and alcohol consumption, a new combined variable of toxics consumption was created with 3 categories: no consumption; moderate consumption (<20 cigarettes/day and/or <80 g alcohol/day); and severe consumption ($\geq$20 cigarettes/day or $\geq$80 g alcohol/day). For patients with oropharyngeal carcinomas, information regarding the HPV status of the tumor was available for 42 of the patients included in the study. The presence of viral DNA was determined with RT-PCR with the SPF-10 PCR/DEIA/LiPA25 system until 2012 and with the PCR/CLART HPV2 system thereafter. For all positive HPV-DNA samples, immunohistochemical expression of p16INK4a was evaluated, considering as positive those specimens with intense and diffuse staining of more than 70% of the tumor tissue. HPV-related tumors (HPV-positive) were considered those with the presence of viral DNA together with immunopositivity to p16INK4a. Twelve oropharyngeal tumors (28.6%) were considered HPV-positive.

RT treatment given was 70 Gy at the primary tumor and morphologically and/or metabolically positive lymph nodes and 50 Gy in the lymph node areas at risk of microscopic disease, according to international consensus guidelines. CT treatment consisted in the administration of two to three cycles of cisplatin at a dose of 100 mg/m^2 every 21 days (n = 59) or carboplatin administered weekly at a dose of 1.5 AUC (n = 7), administered concomitantly with RT.

All patients included in the study had a follow-up period of more than 3 years. The mean follow-up period of the patients was 5.2 years (standard deviation of 3.7 years).

4.2. Transcriptional Analysis

The biopsy samples obtained from each patient were immediately enclosed in RNA-later (Quiagen GmbH, Hilden, Germany) to prevent mRNA degradation and stored at -80 °C until processing. Total RNA was extracted using Trizol (Invitrogen, Carlsbad, CA, USA) according to the manufacturer's instructions. The cDNA was obtained by reverse transcription of 1 µg RNA with High-Capacity cDNA Archive Kit (Applied Biosystems, Foster City, CA, USA), and transcriptional expression of SUCNR1, SLC2A3, SLC16A3, SDHA, and Beta-actin as endogenous control were assessed by RT-PCR on an ABI Prism 7000 using pre-designed validated assays (TaqMan Gene Expression Assays; Applied Biosystems).

4.3. External Validation Study: The Cancer Genome Atlas Database

We conducted an external validation study to assess the prognostic significance of SLC2A3 and SDHA transcriptional expression using data from The Cancer Genome Atlas (TCGA) [24]. Patients with available tumor status information were included in the analysis. Tumor status was defined as either tumor-free or with an active tumor at the last contact or time of death. Among the 473 patients with tumor status data, 334 (70.6%) were tumor-free, while 139 (29.4%) had active disease. The characteristics of these patients, stratified by tumor status, are presented in Table S3 of the Supplementary Materials.

4.4. Statistical Analysis

We compared the transcriptional expression levels of SUCNR1, SLC2A3, SLC16A3, and SDHA according to sex, toxics consumption, location of the primary tumor, clinical local (cT) and regional (cN) extension of the tumor, and local control of the tumor after treatment with RT or CRT. The distribution of the transcriptional expression of the genes analyzed did not meet the criteria of normality, so we used the non-parametric Mann–Whitney U or Kruskal–Wallis tests in the comparisons of the transcriptional expression values.

The relationship between the transcriptional expression for each of the genes analyzed and local disease control after treatment was assessed with a recursive partitioning analysis (RPA) using the classification and regression tree model. If a relationship between local disease control and transcriptional expression was present, the RPA identified the transcriptional expression cut-off value with the highest prognostic capacity. Subsequently, genes that were associated with local disease control when analyzed individually were then included jointly in another RPA, considering the local control as the dependent variable. Local recurrence-free survival analysis was performed according to the categories obtained with the RPA using the Kaplan–Meier method. Differences between survival curves were analyzed with the log-rank test. A multivariable analysis was carried out considering local recurrence-free survival as the dependent variable, including as one of the independent variables the categories derived from the RPA.

The transcriptional expression values of SLC2A3 and SDHA in TCGA patients did not follow a normal distribution; therefore, we used the non-parametric Mann–Whitney U tests in the comparison of the transcriptional expression values according to tumor status. Subsequently, we performed an RPA, using SLC2A3 and SDHA expression values as predictors and tumor status as the dependent variable.

5. Conclusions

SLC2A3/SDHA transcriptional expression was significantly associated with local control in HNSCC patients treated with RT or CRT. Patients with a high transcriptional expression of both SLC2A3 and SDHA had a significantly higher risk of local recurrence after treatment with RT or CRT.

Supplementary Materials: The following supporting information can be downloaded at https://www.mdpi.com/article/10.3390/ijms26062451/s1.

Author Contributions: Conceptualization, M.C. and X.L.; methodology, F.-X.A.-J.; software, C.V.; validation, A.H. and S.B.; formal analysis, X.L.; investigation, S.B., L.L.-V. and A.H.; writing—original draft preparation, X.L. and X.T.; writing—review and editing, M.C. and C.V.; supervision, F.-X.A.-J.; funding acquisition, X.L. All authors have read and agreed to the published version of the manuscript.

Funding: This research was funded by grants from Plan Estatal de I + D + I of the Instituto de Salud Carlos III (FIS PI18/0844 to F.A. and PI24/01605 to X.L.), Fondo Europeo de Desarrollo Regional (FEDER), A Way to Build Europe, and Asociación Española Contra el Cáncer (LABAE18025AVIL to F.A.).

Institutional Review Board Statement: This study was approved by the Institutional Review Board of our center (IIBSP-CCC-14-93) and followed the principles outlined in the Declaration of Helsinki.

Informed Consent Statement: Informed consent was obtained from all subjects involved in the study.

Data Availability Statement: Data are available from the corresponding author upon reasonable request.

Conflicts of Interest: The authors declare no conflicts of interest related to this study.

Abbreviations

The following abbreviations are used in this manuscript:

HNSCC	Head and neck squamous cell carcinoma
RT	Radiotherapy
SUCNR1	Succinate receptor 1
SDHA	Succinate dehydrogenase A
SLC	Solute carrier
SLC2A3	Solute carrier family 2 member 3
SLC16A3	Solute carrier family 16 member 3
CRT	Chemoradiotherapy

HPV	Human papillomavirus
GLUT1	Glucose transporter type 1
SLC2A1	Solute carrier family 2 member 1
GLUT3	Glucose transporter type 3
TCA	Tricarboxylic acid cycle
RPA	Recursive partitioning analysis
TCGA	The Cancer Genome Atlas

References

1. Pardo-Reoyo, S.; Roig-Lopez, J.L.; Yang, E.S. Potential biomarkers for radiosensitivity in head and neck cancers. *Ann. Transl. Med.* **2016**, *4*, 524. [CrossRef] [PubMed]
2. Warburg, O.; Wind, F.; Negelein, E. The metabolism of tumors in the body. *J. Gen. Physiol.* **1927**, *8*, 519–530. [CrossRef] [PubMed]
3. Potter, M.; Newport, E.; Morten, K.J. The Warburg effect: 80 years on. *Biochem. Soc. Trans.* **2016**, *44*, 1499–1505. [CrossRef] [PubMed]
4. Ancey, P.B.; Contat, C.; Meylan, E. Glucose transporters in cancer—From tumor cells to the tumor microenvironment. *FEBS J.* **2018**, *285*, 2926–2943. [CrossRef]
5. Botha, H.; Farah, C.S.; Koo, K.; Cirillo, N.; McCullough, M.; Paolini, R.; Celentano, A. The role of glucose transporters in oral squamous cell carcinoma. *Biomolecules* **2021**, *11*, 1070. [CrossRef]
6. Felmlee, M.A.; Jones, R.S.; Rodriguez-Cruz, V.; Follman, K.E.; Morris, M.E. Monocarboxylate transporters (SLC16): Function, regulation, and role in health and disease. *Pharmacol. Rev.* **2020**, *72*, 466–485. [CrossRef]
7. Owen, O.E.; Kalhan, S.C.; Hanson, R.W. The key role of anaplerosis and cataplerosis for citric acid cycle function. *J. Biol. Chem.* **2002**, *277*, 30409–30412. [CrossRef]
8. Brunengraber, H.; Roe, C.R. Anaplerotic molecules: Current and future. *J. Inherit. Metab. Dis.* **2006**, *29*, 327–331. [CrossRef]
9. Terra, X.; Ceperuelo-Mallafré, V.; Merma, C.; Benaiges, E.; Bosch, R.; Castillo, P.; Flores, J.C.; León, X.; Valduvieco, I.; Basté, N.; et al. Succinate pathway in head and neck squamous cell carcinoma: Potential as a diagnostic and prognostic marker. *Cancers* **2021**, *13*, 1653. [CrossRef]
10. Chen, X.; Lu, P.; Zhou, S.; Zhang, L.; Zhao, J.H.; Tang, J.H. Predictive value of glucose transporter-1 and glucose transporter-3 for survival of cancer patients: A meta-analysis. *Oncotarget* **2017**, *8*, 13206–13213. [CrossRef]
11. Ayala, F.R.; Rocha, R.M.; Carvalho, K.C.; Carvalho, A.L.; da Cunha, I.W.; Lourenço, S.V.; Soares, F.A. GLUT1 and GLUT3 as potential prognostic markers for oral squamous cell carcinoma. *Molecules* **2010**, *15*, 2374–2387. [CrossRef] [PubMed]
12. Estilo, C.L.; O-charoenrat, P.; Talbot, S.; Socci, N.D.; Carlson, D.L.; Ghossein, R.; Williams, T.; Yonekawa, Y.; Ramanathan, Y.; Boyle, J.O.; et al. Oral tongue cancer gene expression profiling: Identification of novel potential prognosticators by oligonucleotide microarray analysis. *BMC Cancer* **2009**, *9*, 11. [CrossRef]
13. Baer, S.; Casaubon, L.; Schwartz, M.R.; Marcogliese, A.; Younes, M. Glut3 expression in biopsy specimens of laryngeal carcinoma is associated with poor survival. *Laryngoscope* **2002**, *112*, 393–396. [CrossRef]
14. Starska, K.; Forma, E.; Jóźwiak, P.; Bryś, M.; Lewy-Trenda, I.; Brzezińska-Błaszczyk, E.; Krześlak, A. Gene and protein expression of glucose transporter 1 and glucose transporter 3 in human laryngeal cancer-the relationship with regulatory hypoxia-inducible factor-1α expression, tumor invasiveness, and patient prognosis. *Tumour Biol.* **2015**, *364*, 2309–2321. [CrossRef] [PubMed]
15. Han, Y.; Wang, X.; Xia, K.; Su, T. A novel defined hypoxia-related gene signature to predict the prognosis of oral squamous cell carcinoma. *Ann. Transl. Med.* **2021**, *9*, 1565. [CrossRef]
16. Lu, W.; Wu, Y.; Huang, S.; Zhang, D. A Ferroptosis-related gene signature for predicting the prognosis and drug sensitivity of head and neck squamous cell carcinoma. *Front. Genet.* **2021**, *12*, 755486. [CrossRef]
17. Kunkel, M.; Moergel, M.; Stockinger, M.; Jeong, J.H.; Fritz, G.; Lehr, H.A.; Whiteside, T.L. Overexpression of GLUT-1 is associated with resistance to radiotherapy and adverse prognosis in squamous cell carcinoma of the oral cavity. *Oral. Oncol.* **2007**, *43*, 796–803. [CrossRef] [PubMed]
18. Chen, R.Y.; Lin, Y.C.; Chen, S.W.; Lin, T.Y.; Hsieh, T.C.; Yen, K.Y.; Liang, J.A.; Yang, S.N.; Wang, Y.C.; Chen, Y.H.; et al. Immunohistochemical biomarkers and volumetric parameters for predicting radiotherapy-based outcomes in patients with p16-negative pharyngeal cancer. *Oncotarget* **2017**, *8*, 72342–72351. [CrossRef]
19. MacFarlane, J.; Seong, K.C.; Bisambar, C.; Madhu, B.; Allinson, K.; Marker, A.; Warren, A.; Park, S.M.; Giger, O.; Challis, B.G.; et al. A review of the tumour spectrum of germline succinate dehydrogenase gene mutations: Beyond phaeochromocytoma and paraganglioma. *Clin. Endocrinol.* **2020**, *93*, 528–538. [CrossRef]
20. Chattopadhyay, C.; Oba, J.; Roszik, J.; Marszalek, J.R.; Chen, K.; Qi, Y.; Eterovic, K.; Robertson, A.G.; Burks, J.K.; McCannel, T.A.; et al. Elevated endogenous SDHA drives pathological metabolism in highly metastatic uveal melanoma. *Investig. Ophthalmol. Vis. Sci.* **2019**, *60*, 4187–4195. [CrossRef]

21. Olszewski, K.; Barsotti, A.; Feng, X.J.; Momcilovic, M.; Liu, K.G.; Kim, J.I.; Morris, K.; Lamarque, C.; Gaffney, J.; Yu, X.; et al. Inhibition of glucose transport synergizes with chemical or genetic disruption of mitochondrial metabolism and suppresses TCA cycle-deficient tumors. *Cell Chem. Biol.* **2022**, *29*, 423–435.e10. [CrossRef] [PubMed]

22. King, A.; Selak, M.A.; Gottlieb, E. Succinate dehydrogenase and fumarate hydratase: Linking mitochondrial dysfunction and cancer. *Oncogene* **2006**, *25*, 4675–4682. [CrossRef]

23. Li, J.; Liang, N.; Long, X.; Zhao, J.; Yang, J.; Du, X.; Yang, T.; Yuan, P.; Huang, X.; Zhang, J.; et al. SDHC-related deficiency of SDH complex activity promotes growth and metastasis of hepatocellular carcinoma via ROS/NFκB signaling. *Cancer Lett.* **2019**, *461*, 44–55. [CrossRef] [PubMed]

24. Available online: https://tcga-data.nci.nih.gov/tcga (accessed on 15 June 2023).

25. Sant'Anna-Silva, A.C.B.; Perez-Valencia, J.A.; Sciacovelli, M.; Lalou, C.; Sarlak, S.; Tronci, L.; Nikitopoulou, E.; Meszaros, A.T. Frezza, C.; Rossignol, R.; et al. Succinate anaplerosis has an onco-driving potential in prostate cancer cells. *Cancers* **2021**, *13*, 1727. [CrossRef] [PubMed]

26. Schöpf, B.; Weissensteiner, H.; Schäfer, G.; Fazzini, F.; Charoentong, P.; Naschberger, A.; Rupp, B.; Fendt, L.; Bukur, V.; Giese, I.; et al. OXPHOS remodeling in high-grade prostate cancer involves mtDNA mutations and increased succinate oxidation. *Nat Commun.* **2020**, *11*, 1487. [CrossRef]

27. Dalla Pozza, E.; Dando, I.; Pacchiana, R.; Liboi, E.; Scupoli, M.T.; Donadelli, M.; Palmieri, M. Regulation of succinate dehydrogenase and role of succinate in cancer. *Semin. Cell Dev. Biol.* **2020**, *98*, 4–14. [CrossRef]

International Journal of
Molecular Sciences

Article

Immunohistochemical Profiling of IDO1 and IL4I1 in Head and Neck Squamous Cell Carcinoma: Interplay for Metabolic Reprogramming?

Benedikt Schmidl [1,*], Maren Lauterbach [1], Fabian Stögbauer [2], Carolin Mogler [2], Julika Ribbat-Idel [3], Sven Perner [3] and Barbara Wollenberg [1]

[1]   Department of Otolaryngology Head and Neck Surgery, Technical University Munich,
       81675 Munich, Germany; marenwallesch@gmx.de (M.L.); barbara.wollenberg@tum.de (B.W.)
[2]   Institute of General and Surgical Pathology, TUM School of Medicine and Health, Technical University
       of Munich, 81675 Munich, Germany; carolin.mogler@tum.de (C.M.)
[3]   Department of Pathology, University of Lübeck, 23562 Lübeck, Germany;
       sven.perner1972@googlemail.com (S.P.)
*   Correspondence: benedikt.schmidl@tum.de

Abstract: Head and neck squamous cell carcinoma (HNSCC) is a heterogeneous and malignant disease with a limited number of biomarkers and insufficient targeted therapies. The current therapeutic landscape is challenged by low response rates, underscoring the need for new therapeutic targets. The success of immunotherapy in HNSCC has highlighted the importance of the immune microenvironment, and since metabolic reprogramming, especially altered tryptophan metabolism, is an important aspect in immune evasion, the interplay of the two enzymes IDO1 and IL4I1 was investigated in HNSCC to assess their immunosuppressive roles and potential as prognostic biomarkers. The immunohistochemical expression of IDO1 and IL4I1 was evaluated by an experienced head and neck pathologist in a tissue microarray (TMA) of 402 patients with HNSCC. Clinical and pathological data were retrieved, and the overall survival of the patients was calculated. In this study, IDO1 and IL4I1 were expressed by HNSCC tumor cells in the TMA of 402 patients. The overall survival analysis of the clinical data of the patients revealed that high IL4I1 expression was significantly associated with worse OS ($p = 0.0073$), while IDO1 expression did not reach statistical significance ($p = 0.087$). The combination of both markers led to a clinically significant stratification of patients. Especially p16-negative OPSCC with a high IL4I1 expression demonstrated poor survival. Immunologic differences between IDO1 and IL4I1 were detected in a TMA of 403 patients, with IDO1 and IL4I1 being expressed by HNSCC. A low IL4I1 expression in HNSCC led to a significantly better OS in this study, while IDO1 expression did not have a significant effect. Additional studies are necessary to investigate the complex interplay in the metabolic reprogramming of tumor cells.

Keywords: IDO1; IL4I1; immunohistochemistry; TMA; metabolic reprogramming

1. Introduction

Head and neck squamous cell carcinoma (HNSCC) is a heterogeneous and malignant disease that originates from the mucosa of the upper aerodigestive tract and is associated with a poor prognosis and a lack of preoperative and prognostic biomarkers [1]. The heterogeneity of the disease impairs the use of targeted therapy and is further complicated by an insufficient knowledge of the interplay of the immune system and tumor cells of the tumor microenvironment (TME) [2]. The currently approved immunotherapy agents for HNSCC target the PD-1 receptor on lymphocytes and thereby block ligands that could deactivate them [3]. PD-1 expression is an important mechanism contributing to the exhausted effector T cell phenotype and the expression of PD-1 on effector T cells, and PD-L1 on neoplastic cells enables tumor cells to evade anti-tumor immunity [4]. Currently, this treatment is approved for the recurrent/metastatic HNSCC using an IgG4 humanized antibody against programmed cell death 1 (PD1). While the introduction of PD-1 inhibition has revolutionized the therapeutic landscape, there is still a response rate of only 20% for HNSCC [5,6]. For that reason, there is a need to investigate novel targets for (immune-)therapy that can be used to develop drugs either as a standalone therapy or in combination with existing immunotherapy agents [7,8].

Since metabolic reprogramming is one of the most prominent features of HNSCC, with amino acid metabolism as the most significantly altered one, elucidating aberrant metabolic profiles might be the key to understanding the mechanisms of tumor immune escape [9].

One of these potential candidates is therefore indoleamine 2, 3-dioxygenase 1 (IDO/IDO1/INDO), a rate-limiting enzyme that metabolizes the essential amino acid, tryptophan (Trp), into downstream kynurenines (Kyn) [10]. IDO1 has a potential immunosuppressive role with IDO and/or tryptophan dioxygenase (TDO)-mediated depletion of Trp and/or the accumulation of Kyn, which is associated with the suppression of immune effector cells and the upregulation, activation, and/or induction of tolerogenic immune cells [11,12]. A high expression of IDO1 leads to a high rate of tryptophan conversion and depletion. This induces cell cycle arrest and/or anergy in the effector cytotoxic lymphocyte (CTL) compartment. This also leads to the activation/maturation of regulatory T cells (Treg) in association with CTLA4-mediated CD80/CD86 co-inhibition. Kynurenine (Kyn) also directly induces the apoptosis of CTL. TGF-β signaling results in the phosphorylation of the IDO1 intrinsic immunoreceptor tyrosine-based inhibitory motifs (ITIM), leading to non-canonical NF-κB activation and autocrine reinforcement of IDO1 and TGF-β expression [12–14]. There is a growing number of clinical trials focused on IDO1, with many studies coupling multiple substances to test the combinatorial benefit [12,15].

Another promising target in the metabolic landscape of HNSCC is interleukin 4-induced gene 1 (IL4I1), an amino acid-catabolizing enzyme, that is mainly secreted in the synaptic cleft and expressed by antigen-presenting cells ([3]). As IDO1, IL4I1 also generates bioactive metabolites from tryptophan. In scRNAseq data, there is an overlapping expression pattern in myeloid cells of the tumor microenvironment, suggesting the two enzymes control a network of tryptophan-specific metabolic events [16]. IL4I1 activates the aryl hydrocarbon receptor through the generation of indole metabolites and kynurenic acid and is associated with reduced survival in glioma patients. IL4I1 can suppress T cell proliferation and promote cancer cell motility and suppress adaptive immunity in chronic lymphocytic leukemia (CLL) in mice. It has been suggested that since IDO1 inhibitors do not block IL4I1, IL4I1 may be the reason for the prior failure of clinical studies combining immunotherapy with IDO1 inhibition [17].

Only a few studies have investigated the role of IDO1 and IL4I1 in HNSCC so far, with limited insights into the interplay of IDO1 and IL4l1, but they have demonstrated that IDO1 expression in HNSCC is positively correlated with several immune-related molecules, and that IL4I1 is a metabolic immune checkpoint that activates the aryl hydrocarbon receptor (AHR) and promotes tumor progression [18,19]. The objective of this study was therefore to evaluate the expression of IDO and IL4I1 in a tissue microarray of HNSCC patients to assess a functional and prognostic role and to discover the dynamic immune landscape within HNSCC.

2. Results

For the analysis, the TMA cores were evaluated by an experienced head and neck pathologist and the mean expression of IDO1 and IL4I1 in the tumor was calculated and used for the subsequent analysis. The TMA cohort had an even distribution of early (52.8%) and advanced HNSCC (47.2%). Distant metastasis was present in 12.9% and regional lymph node metastasis in 37.9% of the cases. There was an even distribution of anatomical sites, with 26.0% oropharyngeal squamous cell carcinoma (OPSCC) and 29.5% laryngeal carcinoma as the most common sites. A total of 53.1% of the 128 OPSCC demonstrated a positive p16 staining (68 vs. 60). A total of 87.9% of the patients were smokers (313 out of 356). Regular alcohol consumption was reported in 43.0% of the patients. The overall clinical and pathological characteristics are depicted in Figure 1.

		Overall TMA	
		n	%
T	T1-2	188	52,8%
	T3-4	168	47,2%
N	N0	221	62,1%
	N+	135	37,9%
M	M0	310	87,1%
	M1	46	12,9%
p16	Negative	255	71,6%
	Positive	101	28,4%
Smoking	Non Smoker	43	12,1%
	Smoker	313	87,9%
Sex	Male	273	76,7%
	Female	83	23,3%
C2	Non C2	203	57,0%
	C2	153	43,0%
Location	Larynx	105	29,5%
	Hypopharynx	45	12,6%
	Oral Cavity	78	21,9%
	Oropharynx p16-	60	16,9%
	Oropharynx p16+	68	19,1%

Figure 1. Clinical and pathological data of the HNSCC TMA Cohort of 402 patients. Depicted is the percentage of the total number of the category. Abbreviations: C2 = Alcohol Consumption, CUP = Cancer of Unknown Primary.

2.1. The Expression of IDO1 and IL4I1 and Association with Clinicopathologic Characteristics

Both IDO1 and IL4I1 were expressed by HNSCC tumor cells in the TMA. Exemplary images of the immunohistochemical staining of IDO1 and IL4I1 are depicted in Supplementary Figure S1.

To assess the optimal cutoff of the expression of IDO1 and IL4I1, maximally selected rank statistics were calculated using the "maxstat method" in R. To differentiate high and low expression groups, a cutoff of 14% of IDO1 positivity was calculated, while for IL4I1 any expression was considered positive/high expression. For IDO1, 68 patients (19.1%) were classified as low expression, while 288 patients (80.9%) were classified as high expression. The mean expression for IDO1 was 10.5%, with a standard deviation of 8.5%, reflecting variability primarily due to the wide range of expression levels in the high-expression group. For IL4I1, 236 patients (66.3%) were classified as low expression, and 120 patients (33.7%) were classified as high expression. The mean expression for IL4I1 was 2.3%, with a standard deviation of 3.0%, indicating that while most samples had low expression, there was still considerable variation in the high-expression group (Supplementary Figure S2).

The associations were tested only in patients with complete clinical data, resulting in a cohort of 356 patients. The comparison of the clinical data and the immunohistochemical expression using Fisher's exact test revealed no association of IDO1 expression with T-stage (p = 0.058), nor with lymphatic metastasis (p = 0.091). When looking at the different anatomical locations of the HNSCC, the majority of hypopharyngeal carcinoma showed a high expression of IDO1, as well as p16-negative and p16-positive oropharyngeal squamous cell carcinoma (OPSCC). Distant metastasis, gender, p16 expression in general, and recurrence did not have significant associations, whereas smokers were more often in the high-expression group (p = 0.023) (Figure 2).

(A)

		IDO1 Expression		
		Low	High	p-value
T	T1-2	40	148	0.058
	T3-4	28	140	
N	N0	40	181	0.091
	N+	28	107	
M	M0	61	249	0.551
	M1	7	39	
p16	Negative	45	210	0.296
	Positive	23	78	
Nicotin	Non Smoker	14	29	**0.023**
	Smoker	54	259	
Sex	Male	53	220	0.874
	Female	15	68	
Rec.	No Recurrence	50	219	0.642
	Recurrence	18	69	
Alcohol Con.	Non C2	44	159	0.174
	C2	24	129	
Location	Larynx	21	84	0.089
	Hypopharynx	5	40	
	Oral Cavity	19	59	
	Oropharynx p16-	6	54	
	Oropharynx p16+	17	51	
Total		68	288	

(B)

		IL4I1 Expression		
		Low	High	p-value
T	T1-2	121	67	0.064
	T3-4	115	53	
N	N0	139	82	**0.021**
	N+	97	38	
M	M0	207	103	0.673
	M1	29	17	
p16	Negative	164	91	0.217
	Positive	72	29	
Nicotin	Non Smoker	28	15	0.865
	Smoker	208	105	
Sex	Male	186	87	0.188
	Female	50	33	
Rec.	No Recurrence	184	85	0.152
	Recurrence	52	35	
Alcohol Con.	Non C2	130	73	0.311
	C2	106	47	
Location	Larynx	60	45	0.057
	Hypopharynx	33	12	
	Oral Cavity	51	27	
	Oropharynx p16-	39	21	
	Oropharynx p16+	53	15	
Total		236	120	

Figure 2. Association of IDO1 and IL4L1 Expression and clinical and pathological data (**A**) Expression of IDO1 and. (**B**) Expression of IL4L1 and association with clinical and pathological data. Fishers Exact Test was used to calculate the statistical significance. A p-value of 0.05 was considered significant.

For IL4I1, there was a significant association with lymphatic metastasis (p = 0.021), while T-stage, M-stage, p16 expression, smoking, gender, recurrence, and alcohol consumption were not associated significantly (Figure 2). Hypopharyngeal carcinoma, p16-negative OPSCC, and p16-positive OPSCC demonstrated low Il4I1 expressions more often.

2.2. Prognostic Impact of the Expression of IDO1 and IL4I1

Next, the clinical data of the 402 patients in the TMA were used to calculate Kaplan–Meier survival curves and log-rank testing was applied to calculate statistical significance. A high expression of IDO1 in the cohort of all patients was associated with poor OS, but did not reach significance (p = 0.087), while in the different subgroups stratified by the anatomical location, there was a no significantly improved survival in the different subgroups. Since the expression of p16 is the most important prognostic factor in OPSCC, the cases were also stratified as p16-positive or p16-negative, but there was no statistically significant differentiation of survival based on IDO1 expression.

At the same time, the expression of IL4I1 in the cohort of all patients was associated with a significantly worse overall survival (p = 0.0073), whereas in the different subgroups stratified by anatomical location, there was a significantly worse OS in the group of p16-negative OPSCC (p = 0.018). The other subgroups did not achieve a significant patient stratification.

In the next step, IDO1 expression and IL4I1 expression were combined in univariate survival analysis, resulting in a significant stratification of patients with the group of high IDO1, and high IL4I1 expression resulted in the worst overall survival, since the group of low IDO1 expression and high IL4I only involved 17 patients. The group of low IDO1 and low IL4I1 expression had the best overall survival of the cohort (Figure 3).

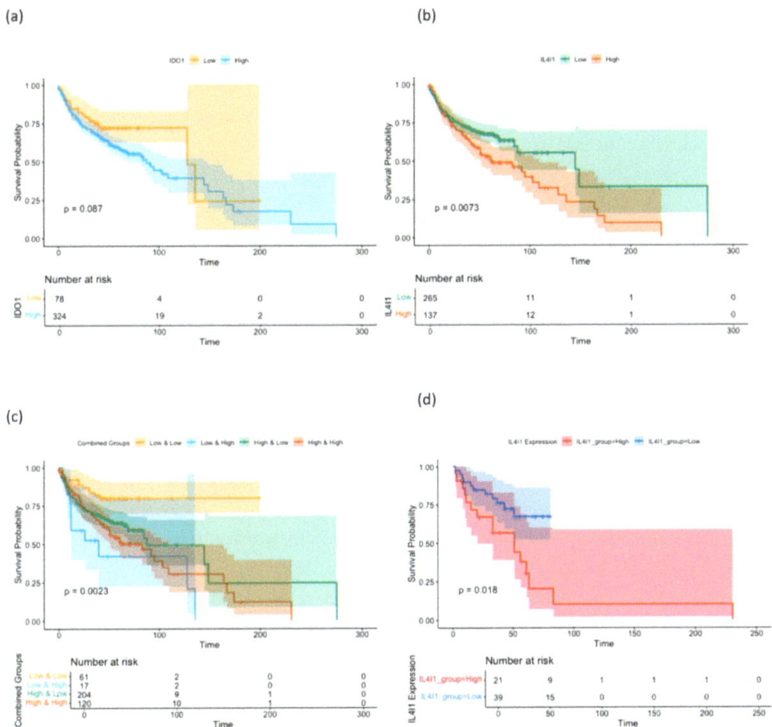

Figure 3. Kaplan-Meier survival curves stratified by (**a**) IDO1 and (**b**) IL4I1 and (**c**) combined expression levels in head and neck squamous cell carcinoma (HNSCC) patients. (**d**) Subanalysis of p16-negative OPSCC stratified by IL4I1 expression. Patients were divided into high and low expression groups based on the optimal cutoffs for IL4I1 and IDO1 expression. The curves illustrate survival probabilities over time and the number at risk. Statistical significance was assessed using the log-rank test.

In the cohort of 356 patients, multivariable Cox proportional hazards regression analysis demonstrated that a lower expression of IDO1 (hazard ratio = 1.4278, p = 0.3286) and IL4I1 (hazard ratio = 0.7953, p = 0.2264) did not significantly reduce the risk of events individually. However, the interaction between low IDO and low IL4I1 expression achieved a slightly significant decrease in hazard (hazard ratio = 0.3301, p = 0.0263).

3. Discussion

IDO1 has been presented as a novel immune-related gene in oral squamous cell carcinoma (OSCC) in the mRNA sequencing data of the TCGA dataset [20], potentially associated with tumor progression, immune evasion, and suppression, and IDO1 inhibitors have already been used in clinical trials for cancer immunotherapy in other carcinomas [21]. At the same time, some of the first clinical trials have already failed in highly immunogenic tumors such as melanoma [22]. There are data that suggest that immune checkpoint blockade (ICB) might induce the expression of IL4I1, an amino acid-catabolizing enzyme that is also involved in metabolic reprogramming, and thereby compromise the effect of IDO inhibition [17].

The objective of this study was therefore to investigate the controversial role of IDO1 and IL4I1 in the largest cohort so far of HNSCC to lay the foundation for a potential use of a combined therapeutic approach in the future. The analysis of the expression of IDO and IL4I1 in this study revealed no significant impact of the expression of IDO1 on the overall survival of patients, while there was a significant association of low expression of IL4I1 leading to a better survival. Both markers were expressed by HNSCC cells in the TMA, and the expression of IDO1 was associated with smoking, whereas tumors with low expression of IL4I1 had less lymphatic metastasis. When analyzing the subgroups stratified by anatomical localization, there is a significantly worse survival of p16-negative OPSCC with a high IL4I1 expression. When both IDO1 and IL4I1 were combined, the survival of the low IDO1 and low IL4I1 was significantly worse than the other groups.

The results for IDO1 in this study elaborate the findings of a bioinformatic study showing a significantly higher expression of IDO1 in HNSCC in the TCGA dataset, especially in HPV + HNSCC compared to healthy control tissue [18], but no prognostic role for overall and disease-free survival [18]. Another study did not focus on the IDO1 expression in tumor cells, but found IDO1 expressing immune cells, especially macrophages, to be more abundant in advanced stages of OSCC and a reduced progression-free survival [19]. Simultaneously, there are studies in other tumors showing that a high IDO1 expression might reflect the presence of a T cell-inflamed phenotype, associated with good prognosis and potentially a response to immunotherapy and chemotherapy [23]. IDO1 was even proposed as a surrogate marker for a more robust spontaneous antitumor immune response in some cancers [24], backed by studies evaluating the IDO1 expression in circulating tumor cells (CTCs) at baseline and after completion of chemoradiotherapy, the finding of a significant overexpression at baseline compared with the post-treatment counterparts, and that mRNA expression at baseline is associated with better survival in terms of progression-free survival. Controversially, in the same study, post-treatment IDO1 mRNA levels were correlated with unfavorable prognosis in terms of overall survival [23].

For Il4I1 there is only a single study so far that investigated the role of IL4I1 in head–neck cancer-derived mesenchymal stromal cells (MSC) in the microenvironment. A microarray gene revealed that HNSCC-MSCs in response to IFN-γ and TNF-α express IL4I1, which is then able to suppress T cell proliferation in vitro [25]. A clinical neoadjuvant study of pembrolizumab for oral tongue squamous cell carcinoma found IL4I1 as a differentially expressed fatty acid metabolism-related gene and a high expression in anti-PD1 therapy responders [26]. Unfortunately, there was no clinical correlation and validation of this bioinformatic result, which might have been able to back the theory of increased IL4I1 expression in patients with anti-PD1 therapy [17]. The data from our study highlight the importance of IL4I1 and IDO1 expression in HNSCC and suggest a potential interplay between these enzymes, which is especially important since it was suggested that IL4I1 may be the reason for the prior failure of clinical studies combining immunotherapy with IDO1 inhibition [17]. Specifically, the combination of low IDO1 and low IL4I1 expression in this study of HNSCC tissue was associated with a significant decrease in hazard, indicating a possible synergistic effect on overall survival. While our study is based on immunohistochemical expression analysis, these findings support the hypothesis that IDO1 and IL4I1 may cooperatively influence the tumor immune microenvironment, possibly through metabolic reprogramming and immune evasion mechanisms. Although direct functional assays are necessary to elucidate this interplay further, our results provide a strong foundation for future investigations, including metabolic profiling and mechanistic studies, to investigate the precise biological interactions between these two immunosuppressive enzymes.

While the results of this study highlight the increasing importance of the interplay of IDO and IL4I1 in HNSCC, there are a few limitations. Since HNSCC are quite heterogeneous, the small tissue cores that are used to generate a TMA might not fully represent this, especially when the marker expression is unevenly distributed across the tumor tissue. This might even lead to over or underestimation of the marker expression [27,28]. Another important aspect is that the expression of IDO and IL4I1 in this study represents only a single-time-point snapshot. In the future, biopsies from recurrence, or a correlation with liquid-biopsy samples, might show an even more dynamic change in the two markers [29,30]. Additionally, the visual assessment of the IDO and IL4I1 expression relies on the interpretation of a pathologist. For this reason, scoring criteria were predefined and the pathologist was blinded. Another limitation was the lack of some of the clinical data resulting in lower patient numbers for the testing of associations, which might have resulted in the worse survival data for the relatively small group of low IDO1 and high IL4I1 expression (n = 17). Since automatic cutoff calculation was used in this study due to the novelty of the markers, clinically more meaningful cutoffs might lead to different results in the future. Additionally, the character of this study does not allow the establishment of causal relationships or mechanisms but lays the foundation for future studies investigating the inhibition of both markers in the clinical setting.

4. Materials and Methods

4.1. Patient Cohort

This study included a well-characterized cohort (Lübeck cohort) of 402 patients with an HNSCC, who were treated according to local treatment guidelines at the Department for Otorhinolaryngology of the University Hospital Schleswig-Holstein, Campus Lübeck, Germany, between 2001 and 2016. Clinical data, survival data, and patient characteristics were obtained from the Department for Otorhinolaryngology of the University Hospital Schleswig-Holstein, Campus Lübeck, Germany. To ensure patient confidentiality, the clinical data were anonymized before being shared with the researchers, making patient identification impossible. This study was approved by the ethics committee of the Technical University of Munich The characteristics of the patient cohort are depicted in Figure 1.

4.2. Immunohistochemistry and the Assessment of the IDO1 and IL4I1 Expression

Formalin-fixed paraffin-embedded (FFPE) tumor tissues were retrieved from the archives of the Institute of Pathology of the University Hospital Schleswig-Holstein, Campus Lübeck, Germany and tissue microarrays (TMA) were constructed from representative tumor areas. Three 0.6 mm cores of each tumor specimen were assembled into TMA blocks. The FFPE tissue was cut with a microtom into 2–3 µm thick sections and deparaffinized at 65 °C. Immunohistochemical staining was conducted with the IDO1 primary antibody (1:100, Thermo Fisher, Salem, MA, USA) and IL4I1 primary antibody (1:100, Thermo Fisher, Salem, MA, USA). Slides were counterstained with hematoxylin. After dehydration by immersion in the ethanol series and xylol (2 min each) the slides were examined under light microscopy.

The immunohistochemically stained slides were evaluated by an experienced pathologist (F.S.) using a standardized scoring protocol to ensure reproducibility and minimize observer bias. For each case, five high-power fields (HPF) were selected from representative tumor areas, and the percentage of tumor cells exhibiting positive cytoplasmic staining for IDO1 or IL4I1 was assessed separately for each HPF. The average percentage across these HPFs was calculated and used for further analysis. To determine the optimal cutoff values for high versus low expression of IDO1 and IL4I1, we employed the "maxstat" function in R, which performs maximally selected rank statistics to identify the threshold that best stratifies the data based on outcome measures [31]. The cutoff value yielding the highest Youden Index was selected to maximize sensitivity and specificity. This approach ensures an objective, data-driven threshold for classification. To account for multiple testing, p-values were adjusted using the Bonferroni correction method, thereby reducing the risk of false positive results. A p-value of less than 0.05 was considered statistically significant. The summary data of the immunohistochemistry scoring, including the distribution of high and low expression cases, mean expression levels, and correlation with clinical parameters, are provided in the results section.

4.3. Statistical Analysis

Kaplan–Meier survival analysis and log-rank testing were used to compare survival rates for different patient groups and clinical characteristics. Associations were tested with Fisher's exact test and Bonferroni correction. At $p < 0.05$ the null hypothesis was rejected, and the result was considered statistically significant. Statistical calculation was performed using Prism (Version 10.4.2) (GraphPad Software, La Jolla, CA, USA) and R (Version 2.15.2, R Core Team, 2024) using the survival, ggplot, and coxph packages.

5. Conclusions

This study is the first highlighting the potential of a combination of two markers of the tryptophan pathway and to assess the interplay of these markers for the progression of HNSCC and the impact on overall survival. The results demonstrate a complex interplay with IL4I1 being shown as a prognostic and important factor in HNSCC, laying the foundation for further studies into analyzing the impact of dual therapy of these markers in vitro for HNSCC.

Supplementary Materials: The following supporting information can be downloaded at: https://www.mdpi.com/article/10.3390/ijms26083719/s1.

Author Contributions: Conceptualization, B.W. and B.S.; formal analysis, B.S., M.L., F.S., S.P. and J.R.-I.; methodology, B.S. and M.L.; validation, F.S. and C.M.; writing—review and editing, B.S., B.W. and F.S. All authors have read and agreed to the published version of the manuscript.

Funding: This research received no external funding.

Institutional Review Board Statement: The study was conducted in accordance with the Declaration of Helsinki, and approved by the Ethics Committee of TUM (Reference: 285/20 S-KH, Approval 2022).

Informed Consent Statement: Patient consent was waived due to the retrospective and descriptive nature of the study.

Data Availability Statement: The original contributions presented in this study are included in the article/Supplementary Material. Further inquiries can be directed to the corresponding author(s).

Conflicts of Interest: The authors declare no conflict of interest.

References

1. Johnson, D.E.; Burtness, B.; Leemans, C.R.; Lui, V.W.Y.; Bauman, J.E.; Grandis, J.R. Head and neck squamous cell carcinoma. Nature reviews. *Dis. Primers* **2020**, *6*, 92. [CrossRef] [PubMed]
2. Fasano, M.; Della Corte, C.M.; Di Liello, R.; Viscardi, G.; Sparano, F.; Iacovino, M.L.; Paragliola, F.; Piccolo, A.; Napolitano, S.; Martini, G.; et al. Immunotherapy for head and neck cancer: Present and future. *Crit. Rev. Oncol.* **2022**, *174*, 103679. [CrossRef]
3. Zeitler, L.; Murray, P.J. IL4i1 and IDO1: Oxidases that control a tryptophan metabolic nexus in cancer. *J. Biol. Chem.* **2023**, *299*, 104827. [CrossRef] [PubMed]

4.	Kwok, G.; Yau, T.C.C.; Chiu, J.W.; Tse, E.; Kwong, Y.-L. Pembrolizumab (Keytruda). *Hum. Vaccines Immunother.* **2016**, *12*, 2777–2789. [CrossRef]

5.	Bhatia, A.; Burtness, B. Treating Head and Neck Cancer in the Age of Immunotherapy: A 2023 Update. *Drugs* **2023**, *83*, 217–248. [CrossRef] [PubMed]

6.	Johnson, S.B.; King, A.J.; Warner, E.L.; Aneja, S.; Kann, B.H.; Bylund, C.L. Using ChatGPT to evaluate cancer myths and misconceptions: Artificial intelligence and cancer information. *JNCI Cancer Spectr.* **2023**, *7*, pkad015. [CrossRef]

7.	Le, X.; Ferrarotto, R.; Wise-Draper, T.; Gillison, M. Evolving Role of Immunotherapy in Recurrent Metastatic Head and Neck Cancer. *J. Natl. Compr. Cancer Netw.* **2020**, *18*, 899–906. [CrossRef]

8.	Lee, Y.S.; Johnson, D.E.; Grandis, J.R. An update: Emerging drugs to treat squamous cell carcinomas of the head and neck. *Expert Opin. Emerg. Drugs* **2018**, *23*, 283–299. [CrossRef]

9.	Zhang, X.; Shi, J.; Jin, S.; Wang, R.; Li, M.; Zhang, Z.; Yang, X.; Ma, H. Metabolic landscape of head and neck squamous cell carcinoma informs a novel kynurenine/Siglec-15 axis in immune escape. *Cancer Commun.* **2024**, *44*, 670–694. [CrossRef]

10.	Zhai, L.; Bell, A.; Ladomersky, E.; Lauing, K.L.; Bollu, L.; Sosman, J.A.; Zhang, B.; Wu, J.D.; Miller, S.D.; Meeks, J.J.; et al. Immunosuppressive IDO in Cancer: Mechanisms of Action, Animal Models, and Targeting Strategies. *Front. Immunol.* **2020**, *11*, 1185. [CrossRef]

11.	Zhai, L.; Ladomersky, E.; Lenzen, A.; Nguyen, B.; Patel, R.; Lauing, K.L.; Wu, M.; A Wainwright, D. IDO1 in cancer: A Gemini of immune checkpoints. *Cell. Mol. Immunol.* **2018**, *15*, 447–457. [CrossRef] [PubMed]

12.	Zhai, L.; Spranger, S.; Binder, D.C.; Gritsina, G.; Lauing, K.L.; Giles, F.J.; Wainwright, D.A. Molecular Pathways: Targeting IDO1 and Other Tryptophan Dioxygenases for Cancer Immunotherapy. *Clin. Cancer Res.* **2015**, *21*, 5427–5433. [CrossRef] [PubMed]

13.	Qin, Y.; Wang, N.; Zhang, X.; Han, X.; Zhai, X.; Lu, Y. IDO and TDO as a potential therapeutic target in different types of depression. *Metab. Brain Dis.* **2018**, *33*, 1787–1800. [CrossRef]

14.	Ye, Z.; Yue, L.; Shi, J.; Shao, M.; Wu, T. Role of IDO and TDO in Cancers and Related Diseases and the Therapeutic Implications. *J. Cancer* **2019**, *10*, 2771–2782. [CrossRef]

15.	Zakharia, Y.; McWilliams, R.R.; Rixe, O.; Drabick, J.; Shaheen, M.F.; Grossmann, K.F.; Kolhe, R.; Pacholczyk, R.; Sadek, R.; Tennant, L.L.; et al. Phase II trial of the IDO pathway inhibitor indoximod plus pembrolizumab for the treatment of patients with advanced melanoma. *J. Immunother. Cancer* **2021**, *9*, e002057. [CrossRef]

16.	Cheng, S.; Li, Z.; Gao, R.; Xing, B.; Gao, Y.; Qin, S.; Zhang, L.; Ouyang, H.; Du, P.; Jiang, L.; et al. A pan-cancer single-cell transcriptional atlas of tumor infiltrating myeloid cells. *Cell* **2021**, *184*, 792–809.e23. [CrossRef]

17.	Sadik, A.; Patterson, L.F.S.; Öztürk, S.; Mohapatra, S.R.; Panitz, V.; Secker, P.F.; Pfänder, P.; Loth, S.; Salem, H.; Prentzell, M.T.; et al. IL4I1 Is a Metabolic Immune Checkpoint that Activates the AHR and Promotes Tumor Progression. *Cell* **2020**, *182*, 1252–1270.e34. [CrossRef] [PubMed]

18.	Gkountana, G.V.; Wang, L.; Giacomini, M.; Hyytiäinen, A.; Juurikka, K.; Salo, T.; Al-Samadi, A. IDO1 correlates with the immune landscape of head and neck squamous cell carcinoma: A study based on bioinformatics analyses. *Front. Oral Heal.* **2024**, *5*, 1335648. [CrossRef]

19.	Struckmeier, A.-K.; Radermacher, A.; Fehrenz, M.; Bellin, T.; Alansary, D.; Wartenberg, P.; Boehm, U.; Wagner, M.; Scheller, A.; Hess, J.; et al. IDO1 is highly expressed in macrophages of patients in advanced tumour stages of oral squamous cell carcinoma. *J. Cancer Res. Clin. Oncol.* **2022**, *149*, 3623–3635. [CrossRef]

20.	Hasegawa, M.; Amano, Y.; Kihara, A.; Matsubara, D.; Fukushima, N.; Takahashi, H.; Chikamatsu, K.; Nishino, H.; Mori, Y.; Yoshida, N.; et al. Guanylate binding protein 5 is an immune-related biomarker of oral squamous cell carcinoma: A retrospective prognostic study with bioinformatic analysis. *Cancer Med.* **2024**, *13*, e7431. [CrossRef] [PubMed]

21.	Tang, K.; Wu, Y.-H.; Song, Y.; Yu, B. Indoleamine 2,3-dioxygenase 1 (IDO1) inhibitors in clinical trials for cancer immunotherapy. *J. Hematol. Oncol.* **2021**, *14*, 68. [CrossRef] [PubMed]

22.	Van den Eynde, B.J.; van Baren, N.; Baurain, J.-F. Is There a Clinical Future for IDO1 Inhibitors After the Failure of Epacadostat in Melanoma? *Annu. Rev. Cancer Biol.* **2020**, *4*, 241–256. [CrossRef]

23.	Economopoulou, P.; Kladi-Skandali, A.; Strati, A.; Koytsodontis, G.; Kirodimos, E.; Giotakis, E.; Maragoudakis, P.; Gagari, E.; Maratou, E.; Dimitriadis, G.; et al. Prognostic impact of indoleamine 2,3-dioxygenase 1 (IDO1) mRNA expression on circulating tumour cells of patients with head and neck squamous cell carcinoma. *ESMO Open* **2020**, *5*, e000646. [CrossRef]

24.	Munn, D.H.; Mellor, A.L. IDO in the Tumor Microenvironment: Inflammation, Counter-Regulation, and Tolerance. *Trends Immunol.* **2016**, *37*, 193–207. [CrossRef]

25.	Mazzoni, A.; Capone, M.; Ramazzotti, M.; Vanni, A.; Locatello, L.G.; Gallo, O.; De Palma, R.; Cosmi, L.; Liotta, F.; Annunziato, F.; et al. IL4I1 Is Expressed by Head–Neck Cancer-Derived Mesenchymal Stromal Cells and Contributes to Suppress T Cell Proliferation. *J. Clin. Med.* **2021**, *10*, 2111. [CrossRef]

26.	Tan, X.; Li, G.; Deng, H.; Xiao, G.; Wang, Y.; Zhang, C.; Chen, Y. Obesity enhances the response to neoadjuvant anti-PD1 therapy in oral tongue squamous cell carcinoma. *Cancer Med.* **2024**, *13*, e7346. [CrossRef]

27. Hoos, A.; Cordon-Cardo, C. Tissue Microarray Profiling of Cancer Specimens and Cell Lines: Opportunities and Limitations. *Mod. Pathol.* **2001**, *81*, 1331–1338. [CrossRef] [PubMed]
28. Jawhar, N.M. Tissue Microarray: A rapidly evolving diagnostic and research tool. *Ann. Saudi Med.* **2009**, *29*, 123–127. [CrossRef]
29. Ferrandino, R.M.; Chen, S.; Kappauf, C.; Barlow, J.; Gold, B.S.; Berger, M.H.; Westra, W.H.; Teng, M.S.; Khan, M.N.; Posner, M.R.; et al. Performance of Liquid Biopsy for Diagnosis and Surveillance of Human Papillomavirus–Associated Oropharyngeal Cancer. *Arch. Otolaryngol. Neck Surg.* **2023**, *149*, 971–977. [CrossRef]
30. Huber, L.T.; Kraus, J.M.; Ezić, J.; Wanli, A.; Groth, M.; Laban, S.; Hoffmann, T.K.; Wollenberg, B.; Kestler, H.A.; Brunner, C. Liquid biopsy: An examination of platelet RNA obtained from head and neck squamous cell carcinoma patients for predictive molecular tumor markers. *Explor. Target. Anti-tumor Ther.* **2023**, *4*, 422–446. [CrossRef]
31. Hothorn, T.; Zeileis, A. Generalized Maximally Selected Statistics. *Biometrics* **2008**, *64*, 1263–1269. [CrossRef] [PubMed]

Article

A Pilot Immunohistochemical Study Identifies Hedgehog Pathway Expression in Sinonasal Adenocarcinoma

Matko Leović [1,†], Antonija Jakovčević [2,†], Ivan Mumlek [3], Irena Zagorac [4], Maja Sabol [5,*] and Dinko Leović [6]

1 Clinical Hospital Center Zagreb, Kišpatićeva 12, 10000 Zagreb, Croatia; matko.leovic@kbc-zagreb.hr
2 Department of Pathology, Cllinical Hospital Center Zagreb, Kišpatićeva 12, 10000 Zagreb, Croatia; antonia.jakovcevic@kbc-zagreb.hr
3 Department of Maxillofacial and Oral Surgery, Clinical Hospital Center Osijek, Josipa Huttlera 4, 31000 Osijek, Croatia; mumlek.ivan@kbco.hr
4 Department of Pathology, Clinical Hospital Center Osijek, Josipa Huttlera 4, 31000 Osijek, Croatia; zagorac.irena@kbco.hr
5 Laboratory for Hereditary Cancer, Division of Molecular Medicine, Ruđer Bošković Institute, Bijenička Cesta 54, 10000 Zagreb, Croatia
6 Maxillofacial Surgery Unit, Department of Otorhinolaryngology and Head and Neck Surgery, Clinical Hospital Center Zagreb, Kišpatićeva 12, 10000 Zagreb, Croatia; dinko.leovic@kbc-zagreb.hr
* Correspondence: maja.sabol@irb.hr; Tel.: +38-514-560-997
† These authors contributed equally to this work.

Abstract: Tumors of the head and neck, more specifically the squamous cell carcinoma, often show upregulation of the Hedgehog signaling pathway. However, almost nothing is known about its role in the sinonasal adenocarcinoma, either in intestinal or non-intestinal subtypes. In this work, we have analyzed immunohistochemical staining of six Hedgehog pathway proteins, sonic Hedgehog (SHH), Indian Hedgehog (IHH), Patched1 (PTCH1), Gli family zinc finger 1 (GLI1), Gli family zinc finger 2 (GLI2), and Gli family zinc finger 3 (GLI3), on 21 samples of sinonasal adenocarcinoma and compared them with six colon adenocarcinoma and three salivary gland tumors, as well as with matching healthy tissue, where available. We have detected GLI2 and PTCH1 in the majority of samples and also GLI1 in a subset of samples, while GLI3 and the ligands SHH and IHH were generally not detected. PTCH1 pattern of staining shows an interesting pattern, where healthy samples are mostly positive in the stromal compartment, while the signal shifts to the tumor compartment in tumors. This, taken together with a stronger signal of GLI2 in tumors compared to non-tumor tissues, suggests that the Hedgehog pathway is indeed activated in sinonasal adenocarcinoma. As Hedgehog pathway inhibitors are being tested in combination with other therapies for head and neck squamous cell carcinoma, this could provide a therapeutic option for patients with sinonasal adenocarcinoma as well.

Keywords: hedgehog; sinonasal adenocarcinoma; tumor–stroma interaction

1. Introduction

The Hedgehog signaling pathway regulates cell proliferation, differentiation, and tissue polarity during embryonic development. The gradient of Hedgehog ligands is involved in these processes at both short- and long distances [1]. Hedgehog signaling regulates the development of various tissues in the craniofacial region, including teeth, lips, palate, and salivary glands [2–4], and its dysregulation during development results in craniofacial deformations, cleft lip and palate, holoprosencephaly, cyclopia, and tooth dysplasia [5]. It is also crucial for the development of epithelial tissues, such as the epidermis, touch domes, hair follicles, sebaceous glands, mammary glands, teeth, nails, and gastric and intestinal epithelium [6]. In adult tissues, its activity is limited to somatic stem cell maintenance and tissue repair [7] and is often upregulated in tumors, where it provides the same signals as during embryogenesis, guiding cell proliferation and differentiation [8].

In humans, three ligands Sonic Hedgehog (SHH), Indian Hedgehog (IHH), and Desert Hedgehog (DHH) act as pathway activation signals in a tissue-specific way. SHH is the most widespread ligand, while DHH is specific for the reproductive system and IHH for bone, cartilage, and digestive tract [9]. The ligands are released from the producing cells, and they can stimulate either the cells that produced them (autocrine activation), the neighboring cells (paracrine activation), or remote tissues if they are distributed long distances by lipid vesicles or as a freely diffusible molecule [10]. This is crucial for the maintenance of the tumor microenvironment, as tumor–stroma interactions can affect tumor cell survival and response/resistance to therapy. The reception of the pathway is regulated by the PTCH1 protein, which is the main receptor for all three Hedgehog proteins. PTCH1 is a tumor suppressor, it keeps control of the Hedgehog pathway through the autoregulative loop, as it is the transcriptional target of the pathway itself. As long as the ligand is present, PTCH1 will translocate from the primary cilia and enable translocation of SMO to the ciliary tip, which will enable activation of GLI transcription factors (GLI1, GLI2, and GLI3). However, when the ligand is not present, PTCH1 will stay localized in the cilia, preventing the accumulation of SMO and therefore preventing the activation of GLI transcription factors [11]. The presence of PTCH1 on the cell membrane signifies that the cell is in a ready state to receive the Hedgehog ligand(s).

The Hedgehog pathway has been found upregulated in head and neck tumors in both in vitro and in vivo models. We and others have identified overexpression of Hedgehog pathway components in head and neck squamous cell carcinoma (HNSCC) [12,13]. Activation of the pathway in HNSCC has been associated with invasion into the bone [14], lymph node invasion [15], radioresistance [16], and worse overall survival [17]. Squamous cell carcinomas are the most frequent tumor types in the upper aerodigestive tract, but other neoplasms can also be found, such as salivary gland-type tumors and sinonasal adenocarcinoma [18]. Sinonasal adenocarcinomas most frequently occur in the ethmoidal sinus and nasal cavum. According to the recent WHO Classification of Tumors, sinonasal adenocarcinomas are considered as tumors of epithelial origin and are classified into two categories: intestinal type adenocarcinomas (ITACs) and non-intestinal type adenocarcinomas (non-ITACs). ITAC is the second most common type of sinonasal adenocarcinoma (after squamous cell carcinoma), which usually has an aggressive clinical presentation, invading the bone and surrounding tissues: orbit, anterior and middle cranial fossa, etc. Non-ITACs are mostly of low-grade malignant potential, without bone destruction [19]. It is considered that ITAC is developed through the process of intestinal metaplasia of sinonasal olfactory epithelium. ITAC histologically resembles colonic adenocarcinoma and can occur in five histological subtypes: colonic, papillary, solid, mucinous, and mixed. Its development is mostly connected with hard-wood dust exposure and leather manufacturing. The incidence of ITAC among these individuals is 500–1000 times higher compared to the non-exposed population [20,21]. Chronic exposure and irritation by organic dust lead to chronic inflammation, which can stimulate tumor initiation [22,23]. Sinonasal ITAC and "true" intestinal adenocarcinoma share similar immunohistochemical profiles. In both, expressions of CK20, CDX-2, vilin, and MUC2 are present. On the other hand, these markers are not expressed in non-ITACs. Contrary to colorectal carcinoma, *KRAS* and *BRAF* mutations are rare in ITAC. Expression of EGFR protein is stronger in patients with ITAC exposed to wood dust than in those exposed to leather dust. In the subgroup of patients who were not exposed to organic dust, EGFR expression is absent [20,21]. Metastatic spread of intestinal adenocarcinomas to the sinonasal tract is rare [24]. Surgery is the primary method of treatment for these tumors. Data considering survival vary depending on the type of treatment, design of study cohorts, and methodology of adverse events calculation. In the study by Cantu et al., the 5-year and 10-year cause-specific mortality for ITACs was reported as 44% and 53%, respectively [25]. A population-based study of 848 patients with ITAC revealed a 5-year relative survival rate of 63 ± 2.1% [26]. A study of 169 endoscopically treated patients revealed a 5-year overall survival of 68.9% and event-free survival of 63.3% [27]. A recent study of 535 patients, including those treated by radiotherapy and chemotherapy, reports a 5-year overall survival of 52% [28].

In this study, we investigated the expression of six Hedgehog pathway proteins (PTCH1, GLI1, GLI2, GLI3, SHH, and IHH) on 21 formalin-fixed paraffin-embedded (FFPE) samples of sinonasal adenocarcinoma: 18 ITAC and 3 non-ITACs. Six colon adenocarcinoma samples were used as controls to compare to the ITAC subtype. As the Hedgehog signaling pathway is also implicated in intestinal development, homeostasis, and colon adenocarcinoma, and IHH is the Hedgehog ligand relevant for the colon adenoma formation, this ligand was included in addition to the most prevalent SHH [29–31]. Additionally, three samples of salivary tumors were used as an outlier subtype of upper aerodigestive tract tumors, where involvement of the Hedgehog pathway has also been demonstrated [32]. To our knowledge, this is the first study to identify Hedgehog pathway upregulation in sinonasal adenocarcinoma.

2. Results

In total, 30 FFPE samples were stained for Hedgehog pathway proteins GLI1, GLI2, GLI3, PTCH1, SHH, and IHH. Immunohistochemical staining showed positive staining of these signaling pathway proteins in sinonasal adenocarcinoma (Figure 1, Table 1).

Overall, the most frequently detected protein was GLI2; it was found in 90% of all tumors, and 66.7% of tumor stromal tissues on average. It was also frequently detected in healthy tissues, 88.3% on average. In all cases, GLI2 staining was detected in the cytoplasm, and some samples also showed some nuclear staining. All three GLI proteins are transcription factors, which can be detected in the cytoplasm when the ligand is not present and in the nucleus when the binding of the ligand activates the signaling cascade.

The second most frequently detected protein was PTCH1, with 53.3% positive tumor samples and 76.6% positive tumor stroma. Interestingly, in the healthy tissues, the stromal compartment was more often positive for PTCH1 than the epithelial compartment ($p = 0.0025$), while this was not the case when comparing the tumor and the tumor stroma compartment (Figure 2). For GLI2 scores, there is no difference between healthy tissues, while for the tumor samples, there is an increase in scores in the tumor tissues compared to the tumor stroma compartment ($p = 0.0009$). Interestingly, in the salivary tumors, the expression pattern of PTCH1 differs from the other analyzed tumor types: Healthy epithelium shows PTCH1 positivity, whereas none of the other tissues showed this pattern.

Some positivity was also detected for GLI1 protein, 16.7% of tumor samples on average, while other examined proteins of the Hedgehog signaling pathway were mostly undetectable (Table 1). GLI1 was not detected in the colon adenocarcinoma, while some positive samples were detected for the ITAC and non-ITAC and the salivary tumor. The distribution of IHC scores for GLI1, GLI2, and PTCH1 for all samples and all tumor types is presented in Supplementary Figure S1.

The focus of further analyses was on the two most abundant proteins, GLI2 and PTCH1. First, the staining scores of ITAC and non-ITAC subtypes of sinonasal adenocarcinoma were compared. As the non-ITAC subtype is very rare compared to the intestinal subtype, only three samples of this type were collected in our cohort. The staining score of GLI2 in the non-ITAC subtype seems to be higher than the ITAC subtype, even though this may be the consequence of the small number of non-ITAC samples and needs to be verified on a larger cohort ($p = 0.0049$) (Figure 3A). This does not seem to be the case for PTCH1 staining (Figure 3C). There were also no differences between the two subtypes in the stromal compartment for these two proteins (Figure 3B,D). Therefore, in our further analysis, we have grouped the ITAC and non-ITAC subtypes into a single group of sinonasal adenocarcinoma.

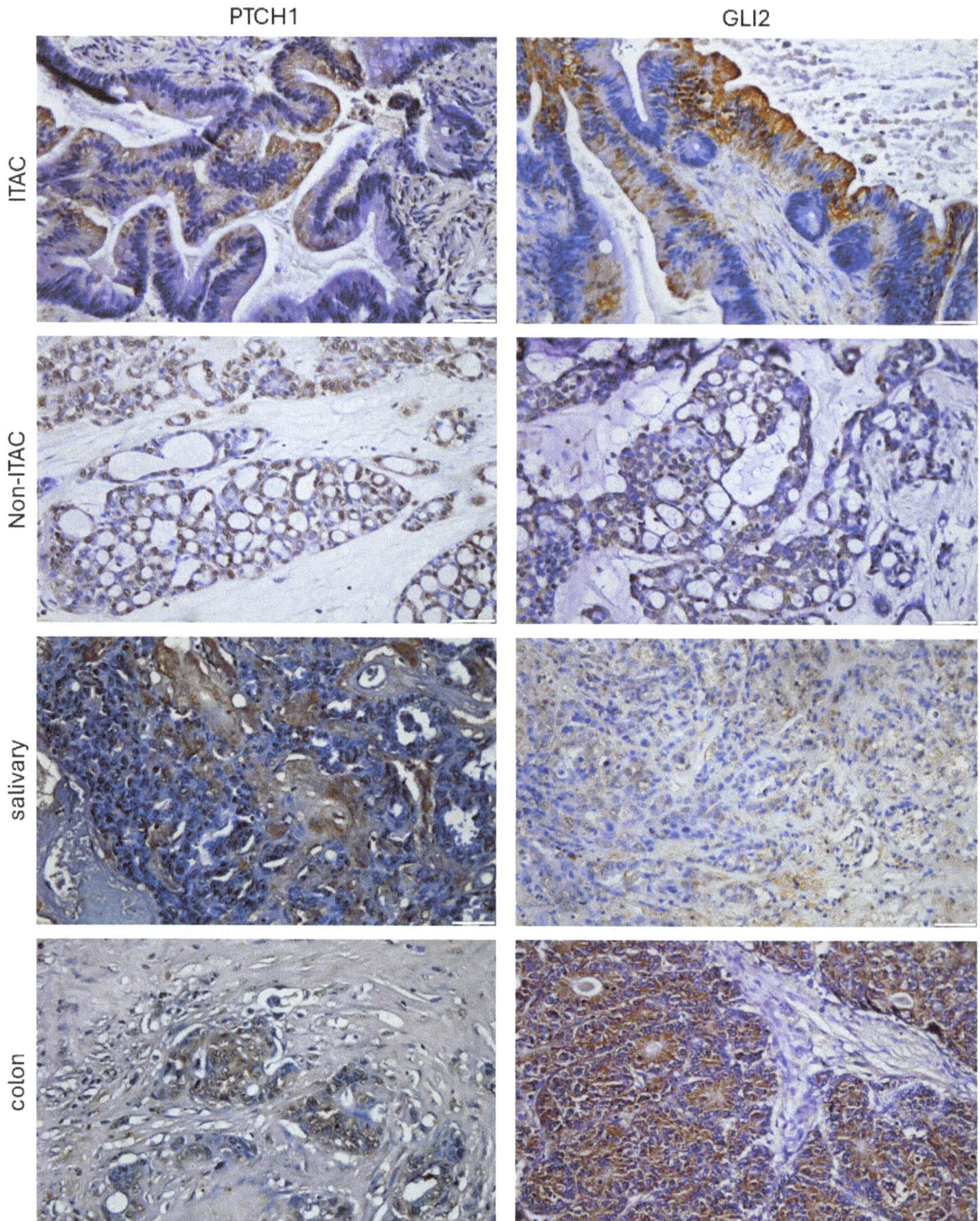

Figure 1. Examples of IHC staining for GLI2 and PTCH1 proteins in ITAC (cytoplasmatic staining in tumor cells), non-ITAC (cytoplasmatic and nuclear staining in tumor cells), salivary (cytoplasmatic staining in stromal cells), and colon (cytoplasmatic staining in tumor cells) adenocarcinomas. Scale bar = 50 μm.

Table 1. Summary table of the number and percentage of samples with positive staining for the HH-GLI pathway proteins GLI1, GLI2, GLI3, PTCH1, SHH, and IHH in four tumor types (colon adenocarcinoma, sinonasal adenocarcinoma of the intestinal type, sinonasal adenocarcinoma of the non-intestinal type, and salivary adenocarcinoma) for four different regions (tumor mass, tumor stroma, healthy epithelium, and healthy stroma).

Protein	Tumor Regions	Total n	All	Colon (n = 6)	Intestinal (n = 18)	Non-Intestinal (n = 3)	Salivary (n = 3)
GLI1	Tumor	30	5 (16.7%)	0 (0%)	4 (22.2%)	1 (33.3%)	0 (0%)
	Tumor stroma	30	5 (16.7%)	0 (0%)	4 (22.2%)	1 (33.3%)	0 (0%)
GLI2	Tumor	30	27 (90%)	5 (83.3%)	16 (88.9%)	3 (100%)	3 (100%)
	Tumor stroma	30	20 (66.7%)	4 (66.7%)	14 (77.8%)	1 (33.3%)	1 (33.3%)
GLI3	Tumor	30	1 (3.3%)	0 (0%)	1 (5.5%)	0 (0%)	0 (0%)
	Tumor stroma	30	1 (3.3%)	0 (0%)	0 (0%)	0 (0%)	1 (33.3%)
PTCH1	Tumor	30	16 (53.3%)	4 (66.7%)	8 (44.4%)	2 (66.7%)	2 (66.7%)
	Tumor stroma	30	23 (76.7%)	5 (83.3%)	14 (77.8%)	2 (66.7%)	2 (66.7%)
SHH	Tumor	30	0 (0%)	0 (0%)	0 (0%)	0 (0%)	0 (0%)
	Tumor stroma	30	0 (0%)	0 (0%)	0 (0%)	0 (0%)	0 (0%)
IHH	Tumor	30	0 (0%)	0 (0%)	0 (0%)	0 (0%)	0 (0%)
	Tumor stroma	30	0 (0%)	0 (0%)	0 (0%)	0 (0%)	0 (0%)
	Healthy Regions	**Total n**	**All**	**Colon (n = 6)**	**Intestinal (n = 9)**	**Non-Intestinal (n = 1)**	**Salivary (n = 2)**
GLI1	Healthy epithelium	18	1 (5.5%)	0 (0%)	0 (0%)	0 (0%)	1 (50%)
	Healthy stroma	18	2 (11.4%)	0 (0%)	2 (22.2%)	0 (0%)	0 (0%)
GLI2	Healthy epithelium	18	15 (83.3%)	6 (100%)	7 (77.8%)	1 (100%)	1 (50%)
	Healthy stroma	18	15 (83.3%)	6 (100%)	7 (77.8%)	1 (100%)	1 (50%)
GLI3	Healthy epithelium	18	1 (5.5%)	0 (0%)	0 (0%)	0 (0%)	1 (50%)
	Healthy stroma	18	1 (5.5%)	0 (0%)	1 (11.1%)	0 (0%)	0 (0%)
PTCH1	Healthy epithelium	18	6 (33.3%)	2 (33.3%)	2 (22.2%)	0 (0%)	2 (100%)
	Healthy stroma	18	16 (88.9%)	5 (83.3%)	9 (100%)	1 (100%)	1 (50%)
SHH	Healthy epithelium	18	3 (16.7%)	0 (0%)	2 (22.2%)	0 (0%)	1 (50%)
	Healthy stroma	18	3 (16.7%)	0 (0%)	2 (22.2%)	0 (0%)	1 (50%)
IHH	Healthy epithelium	18	0 (0%)	0 (0%)	0 (0%)	0 (0%)	0 (0%)
	Healthy stroma	18	0 (0%)	0 (0%)	0 (0%)	0 (0%)	0 (0%)

When analyzing staining scores of sinonasal adenocarcinoma with the two control groups, the colon adenocarcinoma and the salivary adenocarcinoma, no differences in distribution were found between the three tumor subtypes regarding PTCH1 and GLI2 staining scores (Supplementary Figure S1).

The involvement of the tumor stroma compartment in tumor biology was further examined on the sinonasal adenocarcinoma subgroup, as other groups were used as referent groups, and had too few samples for a meaningful statistical analysis. The trend seen in all samples is again seen here in the sinonasal adenocarcinoma subgroup, with PTCH1 expression significantly different between the healthy epithelium and healthy stroma ($p < 0.0001$), and GLI2 expression significantly different between the tumor and tumor stroma ($p = 0.030$) (Figure 4).

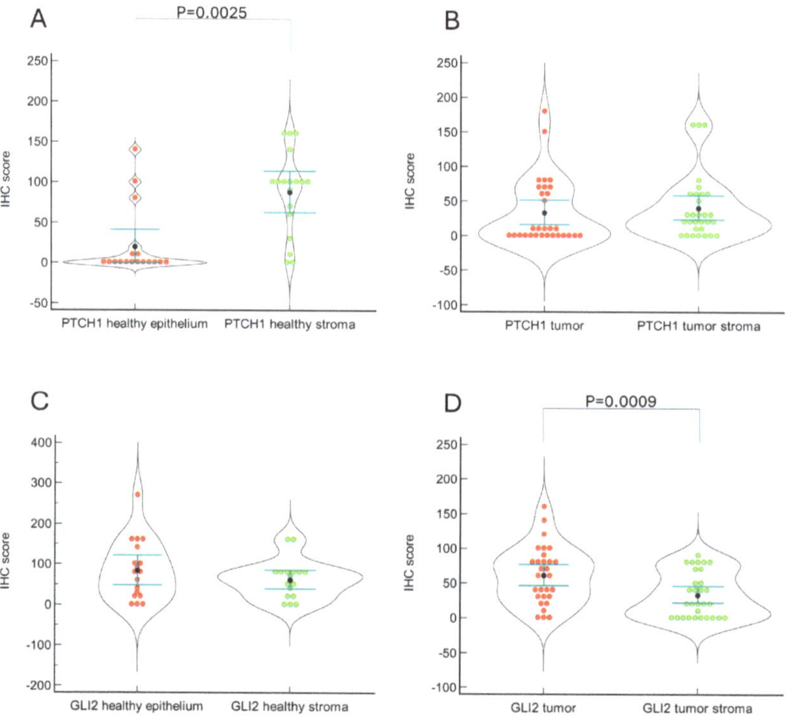

Figure 2. Staining scores of PTCH1 and GLI2 in healthy tissues and tumor tissues for all analyzed samples. PTCH1 staining differs in the healthy tissues, where the stromal compartment shows stronger staining compared to the epithelial compartment ($p = 0.0001$). GLI2 staining differs in the tumor samples, where the tumor tissues show stronger staining compared to the tumor stromal compartment ($p = 0.0052$). (**A**): PTCH1 staining in healthy tissues, (**B**): PTCH1 staining in tumor tissues, (**C**): GLI2 staining in healthy tissues, (**D**): GLI2 staining in tumor tissues.

The absence of either ligand suggests that the pathway is not activated by an autocrine mechanism in these tumor samples. Rather, the presence of PTCH1 and GLI proteins may signify a ready state to receive the ligand signal from remote tissues. Alternatively, it may suggest that their expression is the result of non-canonical activation of the pathway resulting in the expression of the two known downstream targets, PTCH1 and GLI2.

Based on the analysis of the different sample regions, it can be deduced that the PTCH1 protein is preferentially expressed in the stromal compartment of both sinonasal and referent colon adenocarcinoma. This supports the paracrine signaling model in these tumor types, where the stromal cells express HH-GLI pathway receptors and activate downstream signaling, while the activation signal does not originate in the stroma but rather in the tumor (not the case here) or in remote tissues. It is likely that the stromal cells can, upon reception of the signal, activate the pathway and produce various growth factors and cytokines that can support tumor growth. This analysis, however, is beyond the scope of this work.

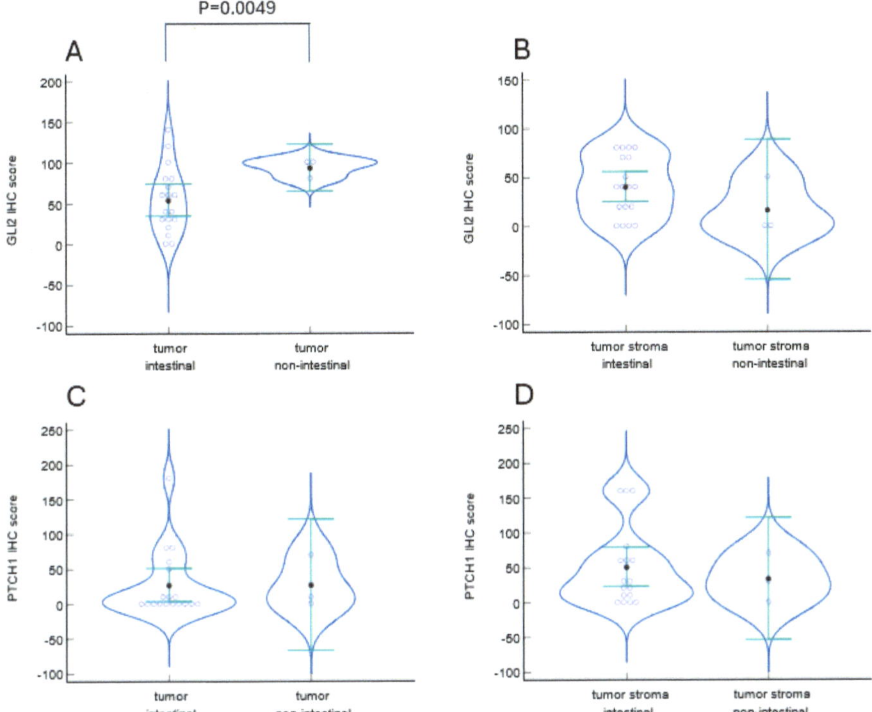

Figure 3. Comparison of the GLI2 and PTCH1 staining scores for the intestinal and non-intestinal subtypes of sinonasal adenocarcinoma. GLI2 score is slightly higher in the non-intestinal tumors compared to the intestinal ones ($p = 0.0049$), while for PTCH1, there are no differences in staining between the two subtypes. (**A**): GLI2 staining in the tumor tissue for ITAC and non-ITAC, (**B**): GLI2 staining in the stroma for ITAC and non-ITAC, (**C**): PTCH1 staining in the tumor tissue for ITAC and non-ITAC, (**D**): PTCH1 staining in the stroma for ITAC and non-ITAC.

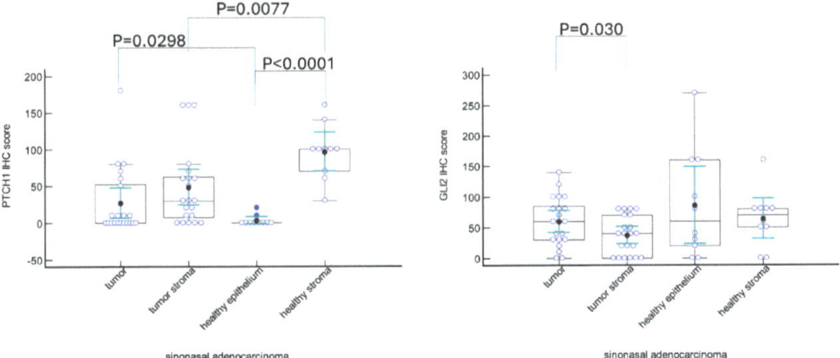

Figure 4. IHC scores for PTCH1 and GLI2 proteins in sinonasal adenocarcinoma. PTCH1 expression is the highest in the healthy stromal cells, and it is downregulated in the tumor stroma ($p = 0.0077$). On the other hand, healthy epithelium shows very weak or no expression of PTCH1, which is increased in tumor tissue ($p = 0.0298$). GLI2 expression is mostly uniform throughout the groups, with stronger staining of tumor tissue compared to its stromal compartment.

3. Discussion

As the Hedgehog signaling pathway regulates the development of the craniofacial structures and epithelium, it is not surprising that its aberrant activation can be detected in tumors developing from these tissues. The involvement of Hedgehog signaling in squamous cell carcinoma of the head and neck region has been well-documented [33]. The Hedgehog signaling pathway has been implicated in salivary gland neoplasms. Vidal et al. have demonstrated positive SHH and GLI1 staining of salivary glands and their neoplasms [32]. In our study, we did not detect SHH in the salivary glands or salivary tumors, but we did detect both GLI2 and PTCH1, which Vidal et al. did not include in their study, confirming the activation of the Hedgehog pathway in the salivary gland tumors. Olfactory neuroblastoma (ONB), another rare tumor of the epithelium in the nasal cavity, also showed activation of the Hedgehog pathway (IHC staining of PTCH1, GLI1, and GLI2), as well as inhibition by Hedgehog pathway inhibitor cyclopamine in vitro on two ONB cell lines [34]. However, there is no information on sinonasal adenocarcinoma. As sinonasal adenocarcinoma is most frequently of the ITAC subtype, which resembles colorectal tumors, we included six colon adenocarcinoma samples to compare to the ITAC.

The activation of the Hedgehog pathway in colon adenocarcinoma epithelial cells has been demonstrated 15 years ago [35]. At the same time, it has been demonstrated that Hedgehog signaling is also involved in the differentiation and renewal of colon epithelial lining as PTCH1, GLI1, and GLI2 were detected in the crypts [36]. In our study, the results were similar, with GLI2 detected in both healthy colon and tumors. However, Alinger et al. also detected SHH and DHH expression in the epithelium of their samples, while we did not. As the Hedgehog pathway is often activated in wound healing and tissues that require constant renewal, such as colon and oral epithelium, it is not surprising to detect its expression in healthy tissues [37]. Another possible explanation for the presence of pathway proteins is the sampling of the control tissues, which were taken from the same individual, and may therefore be affected by unseen processes during field cancerization, which is very common in tumors of the head and neck [38]. Mazumdar et al. have tested two Hedgehog pathway inhibitors on colon cancer cell lines and concluded that inhibition works better at the level of GLI proteins rather than at the level of membrane component SMO, suggesting that GLI activation, possibly through non-canonical pathways, is the contributing factor in colon carcinogenesis [39].

Tumor–stroma communication is extremely important in the maintenance of the favorable microenvironment for cancer progression. For example, in the healthy intestine, the ligand IHH is produced in the epithelial cells and received by the mesenchymal cells [29]. In the mouse model, downstream components of the signaling pathway are confined to the stromal compartment. Furthermore, the same study showed that the expression of downstream components GLI1, GLI2, and GLI3 is disconnected from the expression of the ligands and receptors. Activation of Hedgehog signaling in the stromal compartment resulted in the induction of epithelial differentiation markers and restriction of colonic stem cell markers [40]. In oral squamous cell carcinoma, SHH and GLI1 are both found in the stromal compartment and could be the source of the ligand for both paracrine and autocrine activation in the tumor cells [41]. In fact, SHH can contribute to therapeutic resistance in HNSCC and is a predictor of shorter overall survival and disease-free survival in patients treated with cisplatin [42]. Hypopharyngeal squamous cell carcinoma cell line FaDu can be inhibited by Hedgehog inhibitor JK184 when cells are implanted into the maxillary sinus of nude mice [43]. This, however, does not reflect the potential biology of the sinonasal cavity, but rather the HNSCC cell line itself, and it has been demonstrated that these cells respond to Hedgehog pathway inhibition [44].

When comparing the literature data and the staining patterns of sinonasal adenocarcinoma in our study to the colon and HNSCC tumors, sinonasal adenocarcinoma shows more similarities to the colon than to the HNSCC, primarily due to the lack of co-localization of the upstream and downstream components of the signaling pathway. This makes Hedgehog signaling pathway activation in the sinonasal adenocarcinoma more similar to the paracrine model seen in colon adenocarcinoma than to the autocrine or mixed autocrine plus paracrine model seen in HNSCC. What is especially interesting is the shift of PTCH1 expression, which changes from the healthy stroma to the tumor compartment, and this is matched with the stronger GLI2 staining in the tumor as well. This may indicate that the tumor cells become the receiving cells for the outside signal, resulting in their proliferation.

Sinonasal adenocarcinoma, when inoperable, is often subjected to radiotherapy as part of the treatment protocol. However, radiotherapy can result in unforeseen effects. When HNSCCs are irradiated, there is the activation of GLI1 in the stroma, which contributes to the repopulation of the tumor after therapy and radioresistance [16]. The same is true for colon cancer cells, where SHH and GLI1 expression are increased after irradiation and contribute to tumor repopulation after radiotherapy [45]. Hedgehog pathway inhibitor vismodegib can sensitize HNSCC cell lines to radiation therapy [46]. In the salivary gland, irradiation induces cellular senescence, leading to impaired salivary gland function and dry mouth. It has been demonstrated in both mouse and pig models that the re-introduction of SHH can preserve salivary gland function [47,48]. Therefore, an investigation into Hedgehog pathway expression pre- and post-irradiation would be very informative for these tumors, and it may reflect on patient quality of life and survival. This should be investigated in a separate study with a larger cohort.

The GLI transcription factors were in most cases detected in the cytoplasm of the cells, with occasional nuclear staining, which may signify that the pathway is not activated, but rather poised and ready for activation. This is supported by the fact that ligands required to activate the pathway (SHH and IHH) were not detected in our samples; therefore, there is no signal for translocation of the transcription factors to the nucleus. It is surprising that the Hedgehog ligands, SHH and IHH, have not been detected in our samples. The question remains whether SHH ligand can be delivered from a remote tissue, and if such SHH-expressing cells exist in this region. According to some recent studies, SHH can be detected in the nasal mucus and is decreased in patients with hyposmia [49]. The same was demonstrated by the same group for the parotid saliva and patients with taste dysfunction [50]. Maurya et al. hypothesize that the SHH protein present in the nasal mucus is necessary for the activation of the Hedgehog pathway in the olfactory cilia, as they have demonstrated that the Hedgehog pathway is required for olfactory perception in a mouse model [51]. Therefore, it is possible that this can also be the source of the ligands for the sinonasal adenocarcinoma, but this needs to be investigated further.

4. Materials and Methods

In this pilot retrospective study, archival samples of 18 intestinal adenocarcinoma of the sinus, 3 non-intestinal adenocarcinoma of the sinus, 3 salivary gland tumors, and 6 colon adenocarcinoma FFPE samples were collected from the Department of Otorynolaryngology and Head and Neck Surgery, Clinical Hospital Centre Zagreb and the Department of Oral and Maxillofacial Surgery, Clinical Hospital Osijek. For 18 samples, accompanying healthy tissue controls from the same patient were also available. Ethical approval was granted by the Ethical Committee of the Clinical Hospital Centre Zagreb (no. 02/21 AG) on 25 November 2019. Due to the retrospective nature of the study, patient consent was not required.

4.1. Immunohistochemical Staining

FFPE slides (thickness 4–5 μm) were deparaffinized in Bioclear (Biognost, Zagreb, Croatia), rehydrated in 100%, 90%, and 70% ethanol, and finally in water. Rehydrated slides were warmed to 100 °C in citrate Target retrieval solution pH 6 (Agilent, Santa Clara, CA, USA) and left to cool to room temperature before proceeding with blocking of the endogenous peroxidase activity by 3% hydrogen peroxide in methanol (Kemika, Zagreb, Croatia). Slides were washed in TBST buffer and blocked with serum-free Protein block (Agilent, Santa Clara, CA, USA), followed by incubation with the following antibodies overnight at 4 °C: anti-PTCH1 (1:100, 1750-1-AP, ProteinTech, Planegg-Martinsried, Germany), anti-GLI1 (1:100, NB600-600, Novus Biologicals, Centennial, CO, USA), anti-GLI2 (1:100, sc-271786, Santa Cruz Biotechnology, Dallas, TX, USA), anti-GLI3 (1:200, GTX104362, GeneTex, Irvine, CA, USA), anti-SHH (1:100, sc-365112, Santa Cruz Biotechnology, Dallas, TX, USA), and anti-IHH (1:100, sc-271101, Santa Cruz Biotechnology, Dallas, TX, USA). For negative control, the primary antibody was replaced with 2%BSA/TBST. Following incubation, the detection was performed using the LSAB2 universal kit (Agilent, Santa Clara, CA, USA), the signal was visualized with DAB (Agilent, Santa Clara, CA, USA), and the slides were counterstained with hematoxylin (Biognost, Zagreb, Croatia). Slides were dehydrated using 70%, 90%, and 100% ethanol solutions and Bioclear and fixed in the Biomount medium (Biognost, Zagreb, Croatia).

4.2. Slide Analysis

Slides were examined by an expert pathologist and scored by assessing the signal intensity (0–3) and percentage of positive cells (0–100%) for both the tumor mass and the tumor stroma. Additionally, where available, the accompanying healthy tissues were scored in the same way, examining the healthy epithelium and healthy stroma. Final scores were generated by multiplying the staining intensity with the percentage of positive cells for each of the four examined regions separately. Images were taken using the Olympus BX51 microscope with OLYMPUS stream Essentials 2.4 licensed software. The data were analyzed in MedCalc for Windows (v.22-021) using the Kruskal–Wallis test for multiple group comparison, paired samples *t*-test for comparisons between scores between two regions within the sample, and independent samples *t*-test for comparisons of scores between tumor types.

5. Conclusions

Based on our preliminary results on a small cohort of sinonasal adenocarcinoma samples, we conclude that Hedgehog pathway activation can be detected in these samples. The most frequently detected proteins were PTCH1, the pathway receptor, and GLI2, which seems to be the dominant pathway activator among the three GLI proteins. GLI1 expression can also be detected in a smaller fraction of samples. No expression of ligands, SHH and IHH, was detected, suggesting that the ligands are not produced in these tissues but rather delivered from remote tissues, possibly the nasal mucus. The signal detected in the stromal compartment suggests that the mode of Hedgehog signal transduction is paracrine in sinonasal adenocarcinoma. Based on these findings, the sinonasal adenocarcinoma shows more similarities to the colon adenocarcinoma than the HNSCC or salivary gland tumors, which concurs with analyses of other markers by other authors. This may be relevant for the development of future therapies, as upstream inhibitors of Hedgehog signaling might be less effective than those targeting downstream components.

Supplementary Materials: The following supporting information can be downloaded at https://www.mdpi.com/article/10.3390/ijms25094630/s1.

Author Contributions: Conceptualization, D.L. and M.S.; methodology, M.L. and A.J.; formal analysis, M.L., A.J. and M.S.; resources, I.M. and I.Z.; data curation, M.L. and I.M.; writing—original draft preparation, M.L., A.J., M.S. and D.L.; writing—review and editing, M.L., A.J., I.M., I.Z., M.S. and D.L.; visualization, A.J.; supervision, D.L.; funding acquisition, M.S. All authors have read and agreed to the published version of the manuscript.

Funding: This research was funded by the Croatian Science Foundation, grant number HrZZ-IP-2018-01-4889, and donation by the Croatian Society for Head and Neck Tumors.

Institutional Review Board Statement: The study was conducted in accordance with the Declaration of Helsinki and approved by the Ethics Committee of the Clinical Hospital Centre Zagreb (no. 02/21 AG) on 25 November 2019.

Informed Consent Statement: Patient consent was not required due to the archival nature of the samples.

Data Availability Statement: Data are contained within the article and Supplementary Materials.

Conflicts of Interest: The authors declare no conflicts of interest. The funders had no role in the design of the study; in the collection, analyses, or interpretation of data; in the writing of the manuscript; or in the decision to publish the results.

References

1. Gritli-Linde, A.; Lewis, P.; McMahon, A.P.; Linde, A. The Whereabouts of a Morphogen: Direct Evidence for Short- and Graded Long-Range Activity of Hedgehog Signaling Peptides. *Dev. Biol.* **2001**, *236*, 364–386. [CrossRef] [PubMed]
2. Cobourne, M.T.; Green, J.B.A. Hedgehog Signalling in Development of the Secondary Palate. *Front. Oral Biol.* **2012**, *16*, 52–59. [CrossRef]
3. Jaskoll, T.; Leo, T.; Witcher, D.; Ormestad, M.; Astorga, J.; Bringas, P.; Carlsson, P.; Melnick, M. Sonic Hedgehog Signaling Plays an Essential Role during Embryonic Salivary Gland Epithelial Branching Morphogenesis. *Dev. Dyn.* **2004**, *229*, 722–732. [CrossRef] [PubMed]
4. Sagai, T.; Amano, T.; Maeno, A.; Kiyonari, H.; Seo, H.; Cho, S.-W.; Shiroishi, T. SHH Signaling Directed by Two Oral Epithelium-Specific Enhancers Controls Tooth and Oral Development. *Sci. Rep.* **2017**, *7*, 13004. [CrossRef] [PubMed]
5. Abramyan, J. Hedgehog Signaling and Embryonic Craniofacial Disorders. *J. Dev. Biol.* **2019**, *7*, 9. [CrossRef] [PubMed]
6. Zheng, L.; Rui, C.; Zhang, H.; Chen, J.; Jia, X.; Xiao, Y. Sonic Hedgehog Signaling in Epithelial Tissue Development. *Regen. Med. Res.* **2019**, *7*, 3. [CrossRef] [PubMed]
7. Petrova, R.; Joyner, A.L. Roles for Hedgehog Signaling in Adult Organ Homeostasis and Repair. *Development* **2014**, *141*, 3445–3457. [CrossRef] [PubMed]
8. Jing, J.; Wu, Z.; Wang, J.; Luo, G.; Lin, H.; Fan, Y.; Zhou, C. Hedgehog Signaling in Tissue Homeostasis, Cancers, and Targeted Therapies. *Signal Transduct. Target. Ther.* **2023**, *8*, 315. [CrossRef]
9. Pettigrew, C.A.; Asp, E.; Emerson, C.P. A New Role for Hedgehogs in Juxtacrine Signaling. *Mech. Dev.* **2014**, *131*, 137–149. [CrossRef] [PubMed]
10. Ramsbottom, S.A.; Pownall, M.E. Regulation of Hedgehog Signalling Inside and Outside the Cell. *J. Dev. Biol.* **2016**, *4*, 23. [CrossRef]
11. Weiss, L.E.; Milenkovic, L.; Yoon, J.; Stearns, T.; Moerner, W.E. Motional Dynamics of Single Patched1 Molecules in Cilia Are Controlled by Hedgehog and Cholesterol. *Proc. Natl. Acad. Sci. USA* **2019**, *116*, 5550–5557. [CrossRef]
12. Leovic, D.; Sabol, M.; Ozretic, P.; Musani, V.; Car, D.; Marjanovic, K.; Zubcic, V.; Sabol, I.; Sikora, M.; Grce, M.; et al. Hh-Gli Signaling Pathway Activity in Oral and Oropharyngeal Squamous Cell Carcinoma. *Head Neck* **2012**, *34*, 104–112. [CrossRef] [PubMed]
13. Schneider, S.; Thurnher, D.; Kloimstein, P.; Leitner, V.; Petzelbauer, P.; Pammer, J.; Brunner, M.; Erovic, B.M. Expression of the Sonic Hedgehog Pathway in Squamous Cell Carcinoma of the Skin and the Mucosa of the Head and Neck. *Head Neck* **2011**, *33*, 244–250. [CrossRef] [PubMed]
14. Honami, T.; Shimo, T.; Okui, T.; Kurio, N.; Hassan, N.M.M.; Iwamoto, M.; Sasaki, A. Sonic Hedgehog Signaling Promotes Growth of Oral Squamous Cell Carcinoma Cells Associated with Bone Destruction. *Oral Oncol.* **2012**, *48*, 49–55. [CrossRef]
15. Fan, H.-X.; Wang, S.; Zhao, H.; Liu, N.; Chen, D.; Sun, M.; Zheng, J.-H. Sonic Hedgehog Signaling May Promote Invasion and Metastasis of Oral Squamous Cell Carcinoma by Activating MMP-9 and E-Cadherin Expression. *Med. Oncol.* **2014**, *31*, 41. [CrossRef]
16. Gan, G.N.; Eagles, J.; Keysar, S.B.; Wang, G.; Glogowska, M.J.; Altunbas, C.; Anderson, R.T.; Le, P.N.; Morton, J.J.; Frederick, B.; et al. Hedgehog Signaling Drives Radioresistance and Stroma-Driven Tumor Repopulation in Head and Neck Squamous Cancers. *Cancer Res.* **2014**, *74*, 7024–7036. [CrossRef]
17. Wang, Y.-F.; Chang, C.-J.; Lin, C.-P.; Chang, S.-Y.; Chu, P.-Y.; Tai, S.-K.; Li, W.-Y.; Chao, K.S.C.; Chen, Y.-J. Expression of Hedgehog Signaling Molecules as a Prognostic Indicator of Oral Squamous Cell Carcinoma. *Head Neck* **2012**, *34*, 1556–1561. [CrossRef]
18. Stelow, E.B.; Mills, S.E.; Jo, V.Y.; Carlson, D.L. Adenocarcinoma of the Upper Aerodigestive Tract. *Adv. Anat. Pathol.* **2010**, *17*, 262. [CrossRef] [PubMed]
19. Agarwal, A.; Bhatt, A.A.; Bathla, G.; Kanekar, S.; Soni, N.; Murray, J.; Vijay, K.; Vibhute, P.; Rhyner, P.H. Update from the 5th Edition of the WHO Classification of Nasal, Paranasal, and Skull Base Tumors: Imaging Overview with Histopathologic and Genetic Correlation. *Am. J. Neuroradiol.* **2023**, *44*, 1116–1125. [CrossRef]
20. Rampinelli, V.; Ferrari, M.; Nicolai, P. Intestinal-Type Adenocarcinoma of the Sinonasal Tract: An Update. *Curr. Opin. Otolaryngol. Head Neck Surg.* **2018**, *26*, 115–121. [CrossRef]
21. Leivo, I. Intestinal-Type Adenocarcinoma: Classification, Immunophenotype, Molecular Features and Differential Diagnosis. *Head Neck Pathol.* **2017**, *11*, 295–300. [CrossRef]
22. Vivanco, B.; Llorente, J.L.; Perez-Escuredo, J.; Alvarez Marcos, C.; Fresno, M.F.; Hermsen, M.A. Benign Lesions in Mucosa Adjacent to Intestinal-Type Sinonasal Adenocarcinoma. *Pathol. Res. Int.* **2011**, *2011*, 230147. [CrossRef] [PubMed]
23. Hoeben, A.; van de Winkel, L.; Hoebers, F.; Kross, K.; Driessen, C.; Slootweg, P.; Tjan-Heijnen, V.C.G.; van Herpen, C. Intestinal-Type Sinonasal Adenocarcinomas: The Road to Molecular Diagnosis and Personalized Treatment. *Head Neck* **2016**, *38*, 1564–1570. [CrossRef] [PubMed]

24. Zappia, J.J.; Wolf, G.T.; McClatchey, K.D. Signet-Ring Cell Adenocarcinoma Metastatic to the Maxillary Sinus. *Oral Surg. Oral Med. Oral Pathol.* **1992**, *73*, 89–90. [CrossRef] [PubMed]
25. Cantu, G.; Solero, C.L.; Mariani, L.; Lo Vullo, S.; Riccio, S.; Colombo, S.; Pompilio, M.; Perrone, F.; Formillo, P.; Quattrone, P. Intestinal Type Adenocarcinoma of the Ethmoid Sinus in Wood and Leather Workers: A Retrospective Study of 153 Cases. *Head Neck* **2011**, *33*, 535–542. [CrossRef] [PubMed]
26. Turner, J.H.; Reh, D.D. Incidence and Survival in Patients with Sinonasal Cancer: A Historical Analysis of Population-Based Data. *Head Neck* **2012**, *34*, 877–885. [CrossRef]
27. Nicolai, P.; Schreiber, A.; Bolzoni Villaret, A.; Lombardi, D.; Morassi, L.; Raffetti, E.; Donato, F.; Battaglia, P.; Turri–Zanoni, M.; Bignami, M.; et al. Intestinal Type Adenocarcinoma of the Ethmoid: Outcomes of a Treatment Regimen Based on Endoscopic Surgery with or without Radiotherapy. *Head Neck* **2016**, *38*, E996–E1003. [CrossRef] [PubMed]
28. Patel, N.N.; Maina, I.W.; Kuan, E.C.; Triantafillou, V.; Trope, M.A.; Carey, R.M.; Workman, A.D.; Tong, C.C.; Kohanski, M.A.; Palmer, J.N.; et al. Adenocarcinoma of the Sinonasal Tract: A Review of the National Cancer Database. *J. Neurol. Surg. B Skull Base* **2020**, *81*, 701–708. [CrossRef]
29. Buller, N.V.J.A.; Rosekrans, S.L.; Westerlund, J.; van den Brink, G.R. Hedgehog Signaling and Maintenance of Homeostasis in the Intestinal Epithelium. *Physiology* **2012**, *27*, 148–155. [CrossRef]
30. Walton, K.D.; Gumucio, D.L. Hedgehog Signaling in Intestinal Development and Homeostasis. *Annu. Rev. Physiol.* **2021**, *83*, 359–380. [CrossRef]
31. van Dop, W.A.; Heijmans, J.; Büller, N.V.J.A.; Snoek, S.A.; Rosekrans, S.L.; Wassenberg, E.A.; van den Bergh Weerman, M.A.; Lanske, B.; Clarke, A.R.; Winton, D.J.; et al. Loss of Indian Hedgehog Activates Multiple Aspects of a Wound Healing Response in the Mouse Intestine. *Gastroenterology* **2010**, *139*, 1665–1676.e10. [CrossRef]
32. Vidal, M.T.A.; Lourenço, S.V.; Soares, F.A.; Gurgel, C.A.; Studart, E.J.B.; Valverde, L.D.F.; Araújo, I.B.D.O.; Ramos, E.A.G.; Xavier, F.C.D.A.; Dos Santos, J.N. The Sonic Hedgehog Signaling Pathway Contributes to the Development of Salivary Gland Neoplasms Regardless of Perineural Infiltration. *Tumor Biol.* **2016**, *37*, 9587–9601. [CrossRef]
33. Cierpikowski, P.; Leszczyszyn, A.; Bar, J. The Role of Hedgehog Signaling Pathway in Head and Neck Squamous Cell Carcinoma. *Cells* **2023**, *12*, 2083. [CrossRef]
34. Mao, L.; Xia, Y.; Zhou, Y.; Dai, R.; Yang, X.; Wang, Y.; Duan, S.; Qiao, X.; Mei, Y.; Hu, B. Activation of Sonic Hedgehog Signaling Pathway in Olfactory Neuroblastoma. *Oncology* **2009**, *77*, 231–243. [CrossRef]
35. Varnat, F.; Duquet, A.; Malerba, M.; Zbinden, M.; Mas, C.; Gervaz, P.; Ruiz i Altaba, A. Human Colon Cancer Epithelial Cells Harbour Active HEDGEHOG-GLI Signalling That Is Essential for Tumour Growth, Recurrence, Metastasis and Stem Cell Survival and Expansion. *EMBO Mol. Med.* **2009**, *1*, 338–351. [CrossRef]
36. Alinger, B.; Kiesslich, T.; Datz, C.; Aberger, F.; Strasser, F.; Berr, F.; Dietze, O.; Kaserer, K.; Hauser-Kronberger, C. Hedgehog Signaling Is Involved in Differentiation of Normal Colonic Tissue Rather than in Tumor Proliferation. *Virchows Arch.* **2009**, *454*, 369–379. [CrossRef]
37. Das, D.; Fletcher, R.B.; Ngai, J. Cellular Mechanisms of Epithelial Stem Cell Self-Renewal and Differentiation during Homeostasis and Repair. *WIREs Dev. Biol.* **2020**, *9*, e361. [CrossRef]
38. Bansal, R.; Nayak, B.B.; Bhardwaj, S.; Vanajakshi, C.N.; Das, P.; Somayaji, N.S.; Sharma, S. Cancer Stem Cells and Field Cancerization of Head and Neck Cancer—An Update. *J. Fam. Med. Prim. Care* **2020**, *9*, 3178–3182. [CrossRef]
39. Mazumdar, T.; Devecchio, J.; Agyeman, A.; Shi, T.; Houghton, J.A. The GLI Genes as the Molecular Switch in Disrupting Hedgehog Signaling in Colon Cancer. *Oncotarget* **2011**, *2*, 638–645. [CrossRef]
40. Gerling, M.; Büller, N.V.J.A.; Kirn, L.M.; Joost, S.; Frings, O.; Englert, B.; Bergström, Å.; Kuiper, R.V.; Blaas, L.; Wielenga, M.C.B.; et al. Stromal Hedgehog Signalling Is Downregulated in Colon Cancer and Its Restoration Restrains Tumour Growth. *Nat. Commun.* **2016**, *7*, 12321. [CrossRef]
41. Guimaraes, V.S.N.; Vidal, M.T.A.; de Faro Valverde, L.; de Oliveira, M.G.; de Oliveira Siquara da Rocha, L.; Coelho, P.L.C.; Soares, F.A.; de Freitas Souza, B.S.; Bezerra, D.P.; Coletta, R.D.; et al. Hedgehog Pathway Activation in Oral Squamous Cell Carcinoma: Cancer-Associated Fibroblasts Exhibit Nuclear GLI-1 Localization. *J. Mol. Histol.* **2020**, *51*, 675–684. [CrossRef] [PubMed]
42. Noman, A.S.M.; Parag, R.R.; Rashid, M.I.; Rahman, M.Z.; Chowdhury, A.A.; Sultana, A.; Jerin, C.; Siddiqua, A.; Rahman, L.; Shirin, A.; et al. Widespread Expression of Sonic Hedgehog (Shh) and Nrf2 in Patients Treated with Cisplatin Predicts Outcome in Resected Tumors and Are Potential Therapeutic Targets for HPV-Negative Head and Neck Cancer. *Ther. Adv. Med. Oncol.* **2020**, *12*, 1758835920911229. [CrossRef]
43. Liu, J.; Xie, Y.; Wu, S.; Lv, D.; Wei, X.; Chen, F.; Wang, Z. Combined Effects of EGFR and Hedgehog Signaling Blockade on Inhibition of Head and Neck Squamous Cell Carcinoma. *Int. J. Clin. Exp. Pathol.* **2017**, *10*, 9816–9828. [PubMed]
44. Zubčić, V.; Rinčić, N.; Kurtović, M.; Trnski, D.; Musani, V.; Ozretić, P.; Levanat, S.; Leović, D.; Sabol, M. GANT61 and Lithium Chloride Inhibit the Growth of Head and Neck Cancer Cell Lines Through the Regulation of GLI3 Processing by GSK3β. *Int. J. Mol. Sci.* **2020**, *21*, 6410. [CrossRef] [PubMed]
45. Ma, J.; Tian, L.; Cheng, J.; Chen, Z.; Xu, B.; Wang, L.; Li, C.; Huang, Q. Sonic Hedgehog Signaling Pathway Supports Cancer Cell Growth during Cancer Radiotherapy. *PLoS ONE* **2013**, *8*, e65032. [CrossRef]
46. Hehlgans, S.; Booms, P.; Güllülü, Ö.; Sader, R.; Rödel, C.; Balermpas, P.; Rödel, F.; Ghanaati, S. Radiation Sensitization of Basal Cell and Head and Neck Squamous Cell Carcinoma by the Hedgehog Pathway Inhibitor Vismodegib. *Int. J. Mol. Sci.* **2018**, *19*, 2485. [CrossRef]

47.	Zhao, Q.; Zhang, L.; Hai, B.; Wang, J.; Baetge, C.L.; Deveau, M.A.; Kapler, G.M.; Feng, J.Q.; Liu, F. Transient Activation of the Hedgehog-Gli Pathway Rescues Radiotherapy-Induced Dry Mouth via Recovering Salivary Gland Resident Macrophages. *Cancer Res.* **2020**, *80*, 5531–5542. [CrossRef]

48.	Hu, L.; Du, C.; Yang, Z.; Yang, Y.; Zhu, Z.; Shan, Z.; Zhang, C.; Wang, S.; Liu, F. Transient Activation of Hedgehog Signaling Inhibits Cellular Senescence and Inflammation in Radiated Swine Salivary Glands through Preserving Resident Macrophages. *Int. J. Mol. Sci.* **2021**, *22*, 13493. [CrossRef] [PubMed]

49.	Henkin, R.I.; Hosein, S.; Stateman, W.A.; Knoppel, A.B.. Sonic Hedgehog in Nasal Mucus Is a Biomarker for Smell Loss in Patients with Hyposmia. *Cell. Mol. Med. Open Access* **2016**, *2*, 9. [CrossRef]

50.	Henkin, R.I.; Knöppel, A.B.; Abdelmeguid, M.; Stateman, W.A.; Hosein, S. Sonic Hedgehog Is Present in Parotid Saliva and Is Decreased in Patients with Taste Dysfunction. *J. Oral Pathol. Med.* **2017**, *46*, 829–833. [CrossRef]

51.	Maurya, D.K.; Bohm, S.; Alenius, M. Hedgehog Signaling Regulates Ciliary Localization of Mouse Odorant Receptors. *Proc. Natl. Acad. Sci. USA* **2017**, *114*, E9386–E9394. [CrossRef] [PubMed]

International Journal of
Molecular Sciences

Case Report

Extraenteric Malignant Gastrointestinal Neuroectodermal Tumor of the Neck: A Diagnostic Challenge

Manuel Tousidonis [1,2,*], Maria J. Troulis [3], Carolina Agra [1,4], Francisco Alijo [1,4], Ana Alvarez-Gonzalez [1,5], Carlos Navarro-Cuellar [1,2], Saad Khayat [1], Gonzalo Ruiz-de-Leon [1], Ana Maria Lopez-Lopez [1,2], Jose Ignacio Salmeron [1,2] and Santiago Ochandiano [1,2]

[1] Department of Oral and Maxillofacial Surgery, Gregorio Marañón University Hospital, Calle Doctor Esquerdo 46, 28006 Madrid, Spain; ruizdeleong@gmail.com (G.R.-d.-L.)

[2] Gregorio Marañón Research Institute, 28009 Madrid, Spain

[3] Department of Oral and Maxillofacial Surgery, Massachusetts General Hospital, Harvard School of Dental Medicine, 188 Longwood Avenue, Boston, MA 02115, USA; mtroulis@mgb.org

[4] Department of Pathology, Gregorio Marañón University Hospital, Calle Doctor Esquerdo 46, 28006 Madrid, Spain

[5] Department of Radiation Oncology, Gregorio Marañón University Hospital, Calle Doctor Esquerdo 46, 28006 Madrid, Spain

* Correspondence: manuel@tousidonisrial.com

Abstract: Malignant gastrointestinal neuroectodermal tumor (MGNET) and clear cell sarcoma (CCS) of soft tissue represent related, extremely rare, malignant mesenchymal neoplasms. Both entities are genetically characterized by the same molecular alterations, *EWSR1::CREB1* fusions. Malignant gastrointestinal neuroectodermal tumor has significant morphological overlap with CCS, although it tends to lack overt features of melanocytic differentiation. Recently, rare MGNET cases were reported in extragastrointestinal sites. The diagnosis represents a major challenge and significantly impacts therapeutic planning. In this study, we reported the clinicopathologic features of a molecularly confirmed MGNET of the neck and provided a review of the pertinent literature.

Keywords: malignant gastrointestinal neuroectodermal tumor; clear cell sarcoma; soft-tissue neoplasm; neck cancer; *EWSR1::CREB*

1. Introduction

Malignant gastrointestinal neuroectodermal tumor (MGNET) [1], previously referred to as clear cell sarcoma (CCS)-like tumor of the gastrointestinal tract [2–4], is an exceedingly rare and aggressive mesenchymal neoplasm. Malignant gastrointestinal neuroectodermal tumor (MGNET) and clear cell sarcoma (CCS) are rare soft tissue neoplasms that, despite sharing molecular features such as EWSR1 gene rearrangements (*EWSR1-CREB1* in MGNET and *EWSR1-ATF1* in CCS), represent distinct pathological entities. Initially described in the gastrointestinal (GI) tract, particularly in the small intestine and stomach, this tumor has recently been identified in extraintestinal locations, including the head and neck. Occurrences in the head/neck (H/N) region are exceptionally rare, with less than 100 cases cited in the literature worldwide. The first case of MGNET in the H/N was reported by Alpers and Beckstead [1] as a "malignant neuroendocrine tumor of the jejunum with osteoclast-like giant cells". The etiopathogenesis of MGNET remains poorly understood due to its rarity [5], though several key mechanisms have been identified. MGNET is believed to arise from neuroectodermal cells, which show early neural differentiation [6]. Central to its development are genetic fusions [7], particularly involving

EWSR1::CREB1, which drive tumor proliferation. The exact environmental or external factors contributing to MGNET are still unknown, although its aggressive behavior and occurrence in both gastrointestinal and extraintestinal sites, including the head/neck, may suggest multiple pathways of tumorigenesis. According to the World Health Organization (WHO) classification of soft tissue tumors [8], MGNET and CCS are categorized within the group of neoplasms with uncertain differentiation. Despite shared molecular features, including *EWSR1::CREB1* fusions and co-expression of S100 protein and SOX10, MGNET and CCS exhibit distinct morphological and immunohistochemical characteristics [9–12]. Notably, CCS expresses melanocytic markers such as HMB-45, Melan-A, tyrosinase, and MiTF, which are absent in MGNET [13–16]. This immunophenotypic divergence, combined with distinct anatomical localization and lack of melanocytic differentiation in MGNET, supports the classification of these tumors as separate entities despite occasional diagnostic challenges. These differences have led to the preference for the nomenclature of "MGNET" over "clear cell sarcoma-like tumor" [17].

The identification of MGNETs in the head/neck region presents a novel and complex challenge in oncopathology, given the unique embryological origin and intricate anatomy of this area (Table 1). These tumors are characterized by their neuroectodermal differentiation, distinctive histopathological features, and aggressive clinical behavior, underscoring the importance of prompt diagnosis and intervention.

Table 1. Clinicopathologic and molecular genetic features of published neck E-MGNET cases.

Case	Age/Sex	Location/Size	IHC Results	Molecular Results	Therapy	Local Recurrence	Metastasis	Outcome
1 [18]	9/M	Right neck/ 5.5 cm	Positive: S100 protein, SOX10, synaptophysin, CD56. Negative: HMB45, Melan A, MART-1, TTF1, EMA, pancytokeratin, cytokeratin AE1/AE3, CAM5.2, CK5/6, p63, CD31, SMA, desmin, synaptophysin, chromogranin, GFAP, CD68	EWSR1 gene locus was attempted on the specimen but was technically unsatisfactory	Chemotherapy	No	No	ANED at 12 mo
2 [19]	14/M	Right neck/ 3.7 cm	Positive: S100 protein, SOX10, synaptophysin, CD56; Negative: Keratins, HMB-45, Melan-A, MiTF, chromogranin A, CD117, DOG1, ALK	EWSR1 exon 8::ATF1 exon 4	Resection with + margin and radiotherapy	No: persistent local disease, undergoing radiotherapy	No	AWD at 11 mo
3 [19]	30/M	Right neck/NA	Positive: S100 protein, SOX10, synaptophysin, CD56; Negative: Keratins, HMB-45, Melan-A, MiTF, chromogranin A, CD117	EWSR1 exon 8::ATF1 exon 4	No radical resection and radiotherapy	Yes: 57 mo	No	ANED at 70 mo
4 [19]	48/F	Right neck/ 5.5 cm	Positive: S100 protein, SOX10; Negative: Keratins, synaptophysin, MiTF, HMB-45, Melan-A, chromogranin A, CD117, ALK	EWSR1 exon 7::ATF1 exon 5	No radical resection	No	No	ANED at 10 mo
5 [our case]	58/F	Left neck/ 4 cm	Positive: S100, SOX10, CD99 (weak/focal), Fli-1, and synaptophysin (focal). Negative: Melan-A, HMB-45, CKIT, AE1/AE3	EWSR1 gene translocation	Surgery (radical resection) and radiotherapy	Yes: 48 mo	Yes: 36 mo	Died: 54 mo

Abbreviations: ANED: alive with no evidence of disease; AWD: alive with disease; IHC: immunohistochemistry; E-MGNET: extraenteric malignant gastrointestinal neuroectodermal tumor; mo: months.

Histologically, both CCS and MGNET consist of epithelioid and spindle cells with prominent nucleoli, organized into large nodules separated by thick fibrous septa. Contrary to the name, the clear cell component is minimal, with most cells displaying eosinophilic cytoplasm. Multinucleated giant cells are observed in approximately half of CCS cases, and melanin pigment can occasionally be detected in soft tissue CCS. In contrast, MGNET is distinguished by a pseudopapillary and/or pseudoalveolar growth pattern. In this report, a rare case of extraenteric MGNET (E-MGNET) located at the neck is presented, and the clinical, morphological, and molecular features are discussed.

2. Case Report

A 58-year-old woman, with no relevant medical history, was evaluated at another hospital for a rapidly growing painful left cervical mass, biopsied, and diagnosed with "clear cell sarcoma" in October 2019. No cutaneous lesions were identified, and the patient had no history of malignant melanoma. Subsequently, she was referred for evaluation and a second opinion at the Sarcoma Reference Unit of the Gregorio Marañon University Hospital. A new biopsy was conducted, and histological, immunohistochemical, cyto-genetic, and next-generation sequencing studies were performed. The results were that a tumor originating from the left laterocervical region was examined, consisting of two paraffin blocks with nine histological slides (H&E and IHC). Microscopically, the neoplasm infiltrated adjacent soft tissues and lymph nodes, displaying perineural and vascular invasion. The tumor exhibited a solid, pseudoalveolar growth pattern with oval to fusiform cells, eosinophilic cytoplasm, vesicular chromatin, and prominent nucleoli. Mitotic count was assessed in five non-overlapping high-power fields (HPFs) corresponding to a total area of 5 mm^2 (each HPF $\approx$ 0.2 mm^2). An average of 1.4 mitoses/mm^2 (SD $\pm$ 0.5) was identified, with focal tumor necrosis observed. This assessment was performed according to current recommendations for reproducibility in mitotic index evaluation [20,21]. Immunohistochemically, the tumor was positive for S100, SOX10, CD99 (weak/focal), Fli-1, and synaptophysin (focal), but negative for melanocytic and epithelial markers (e.g., Melan-A, HMB-45, CKIT, and AE1/AE3) (Figure 1). The Ki-67 proliferation index was assessed by manual counting in three selected hot spot areas of 1 mm^2 each, with an average labeling index of 15% (SD $\pm$ 3%). Quantification was performed in accordance with current recommendations for reproducible assessment of proliferation indices in diagnostic pathology [20,22]. PAS staining was negative. Fluorescence in situ hybridization (FISH) analysis for *EWSR1* was performed using break-apart probes (*EWSR1* Break Apart Probe, TITAN FISH PROBES, OACP IE LTD, Phoenix House, Monahan Road, T12H1XY, Cork, Ireland) on paraffin-embedded sections (Figure 2A). Fluorescence in situ hybridization (FISH) analysis using a break-apart probe targeting *EWSR1* revealed red–green split signals in 93% of tumor cell nuclei, indicating the presence of an *EWSR1* gene rearrangement (Figure 2B). The *EWSR1* rearrangement was confirmed at the Centro Nacional de Investigaciones Oncologicas (CNIO). However, it was not possible to confirm the identity of the fusion partner. Specifically, RT-PCR analysis using *CREB1*-specific primers did not yield amplification products, suggesting that *CREB1* was not involved as the fusion partner gene in this case. FISH analysis confirmed the presence of an EWSR1 gene translocation, supporting a diagnosis of a malignant gastrointestinal neuroectodermal tumor (MGNET) with neuroectodermal differentiation and no melanocytic differentiation. Imaging showed a single left laterocervical mass that invaded the left common carotid artery with no vascular plane of separation (Figure 3A–D). PET-CT confirmed the absence of distant metastasis, and the decision was made to perform surgical excision. One week before the oncological surgery, a left common carotid occlusion test was performed. The test confirmed that the circle of Willis and the collateral pathways were able to compensate for the loss of blood flow through the affected carotid artery before surgery and that, therefore, ligation of the left common carotid artery infiltrated by the tumor could be performed. Treatment with surgery and intraoperative radiotherapy was planned. Surgery with radical left dissection was performed, including resection of the common carotid artery, internal jugular vein, sternocleidomastoid muscle, spinal nerve, and vagus nerve. After surgical resection, intraoperative radiotherapy was performed on the surgical bed with 10 Gy with 6 MeV electrons (Figure 4A).

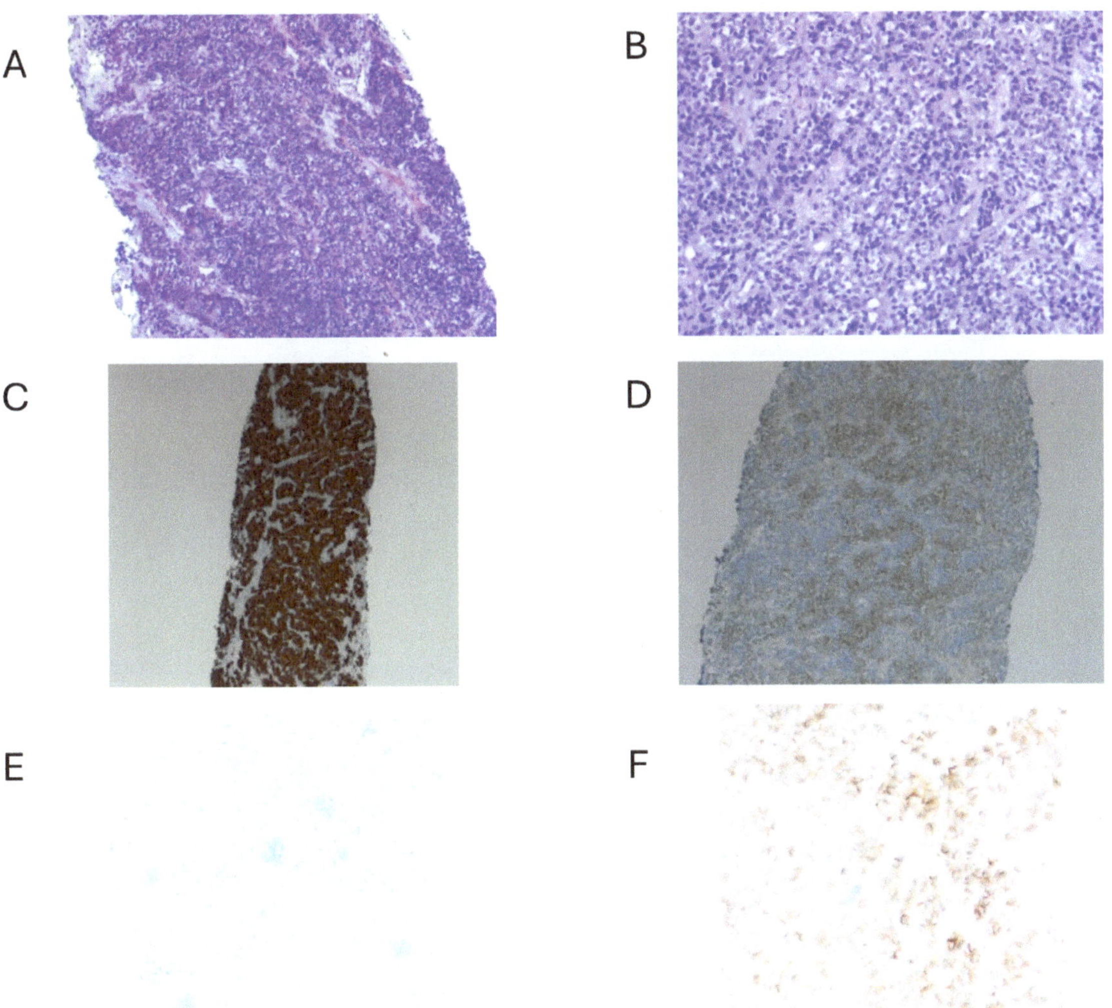

Figure 1. (**A**) Low-power view showing a multinodular, lobulated growth pattern [10×]. (**B**) The tumor is composed of nests and sheets of moderately pleomorphic epithelioid cells with eosinophilic to clear cytoplasm, set in a fibrotic and lymphoplasmacytic stroma [20×]. (**C,D**) Immunohisto-chemically, the tumor cells show diffuse nuclear positivity for SOX10 [10×] (**C**) and cytoplasmic expression of S100 protein [10×] (**D**). (**E**) The neoplastic cells were negative for melanocytic markers, including MiTF, HMB45, and Melan-A [20×]. (**F**) Diffuse membranous expression of CD56 supports neuroectodermal differentiation, consistent with the diagnosis of MGNET [20×].

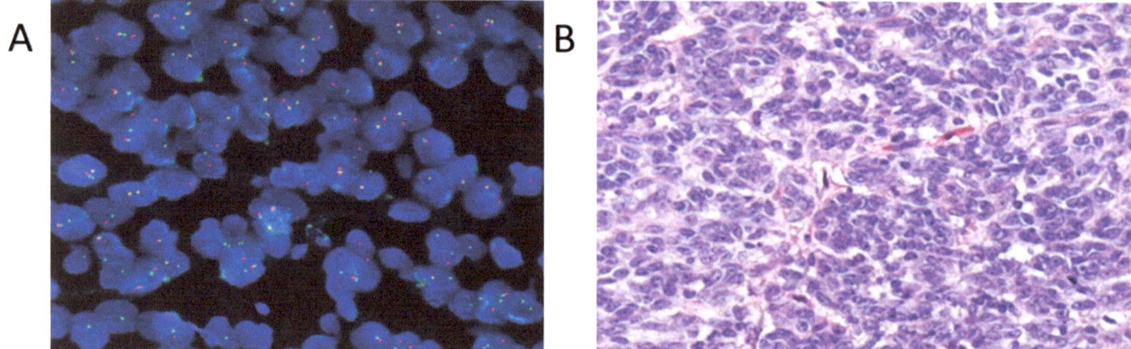

Figure 2. (**A**) At high power, neoplastic cells show vesicular chromatin with small nucleoli (H&E, 400×). (**B**) FISH probe targeting the EWSR1 rearrangement, which revealed red–green split signals in 94% of the tumor cell nuclei, indicating EWSR1 gene rearrangement [20×].

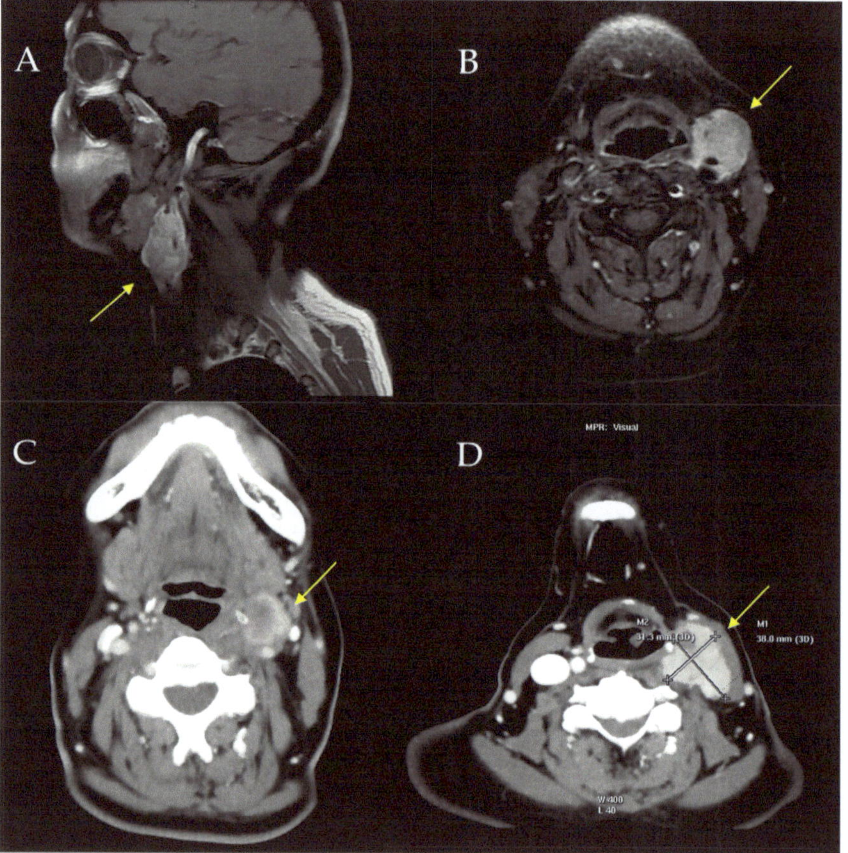

Figure 3. Extraenteric malignant gastrointestinal neuroectodermal tumor involving the soft tissues of the neck in a 58-year-old woman: (**A**) Sagittal MRI image at diagnosis. (**B**) Axial MRI image of the tumor at diagnosis. (**C**) Axial CT image in which vascular invasion by the tumor is observed at diagnosis. (**D**) Tumor measurements in the preoperative CT scan.

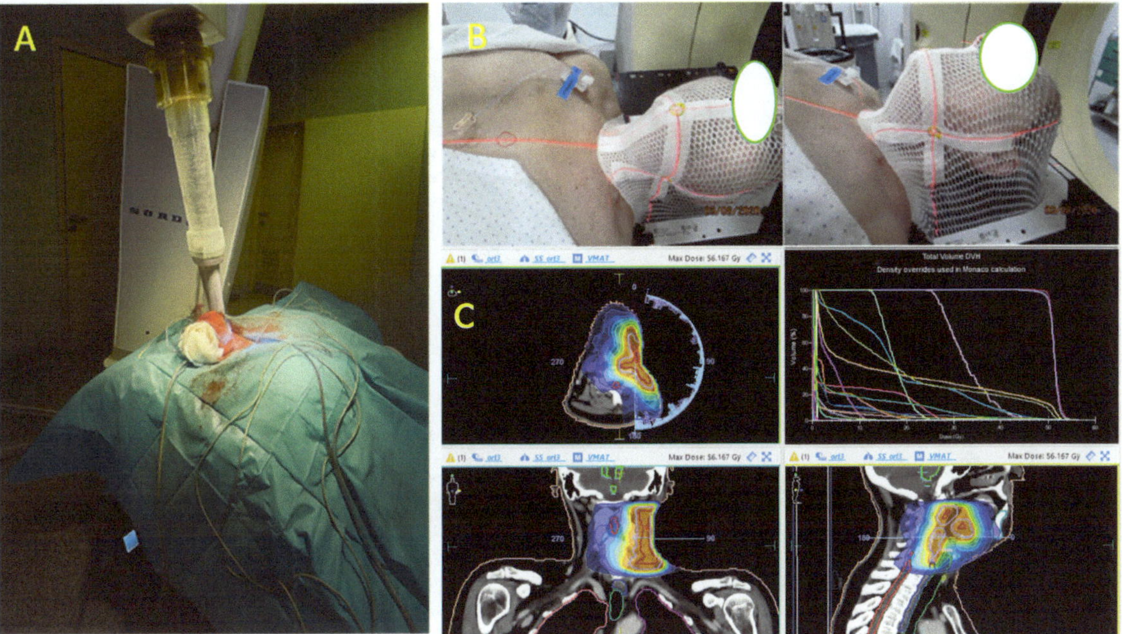

Figure 4. Mixed radiotherapy protocol combining intraoperative radiotherapy (IORT), administered directly to the tumor bed during surgery, followed by postoperative external beam radiotherapy (EBRT) to optimize local disease control. (**A**) Patient setup under general anesthesia for point-of-care photon radiotherapy. (**B**) Placement and fitting of a customized thermoplastic mask on the patient's face, ensuring stable fixation to the anatomical landmarks. Facial features have been masked for privacy. (**C**) Radiotherapy planning and dose distribution over axial, sagittal, and coronal views, including dose–volume histogram analysis.

The surgical specimen, "left radical neck dissection", showed a malignant gastrointestinal neuroectodermal tumor (4 × 2.5 cm), with involvement of the vagus nerve, adventitial layer of the internal carotid artery, and sternocleidomastoid muscle. The surgical resection margin is clear (at >1 mm). Twenty-two lymph nodes measuring between 0.6 and 0.1 cm were isolated, showing nonspecific reactive lymphadenitis, negative for tumor (0/22), as was the submandibular gland. No fusions were detected in *ALK*, *ROS1*, *RET*, *MET*, *N-TRK1*, *N-TRK2*, or *N-TRK3* genes in the Ydilla Genefusion Test. The tumor was composed of epithelioid eosinophilic neoplastic cells with focal cytoplasmic clearing, organized in solid nests, often exhibiting a pseudoalveolar growth pattern (Figure 3A,B) and diffusely expressing S100. The diagnosis MGNET was confirmed. The patient received full-dose postoperative radiotherapy (50 Gy in 25 fractions) with the VMAT technique (Figure 4B,C). During the follow-up period, the patient presented distant metastases in the right gluteal and left paravertebral regions (Figure 5B,C) 3 years after the initial surgery. The patient started systemic chemotherapy with six different lines of treatment due to poor response. The patient died of massive left supraclavicular recurrence with lung involvement (Figure 5A) 54 months after her initial surgery.

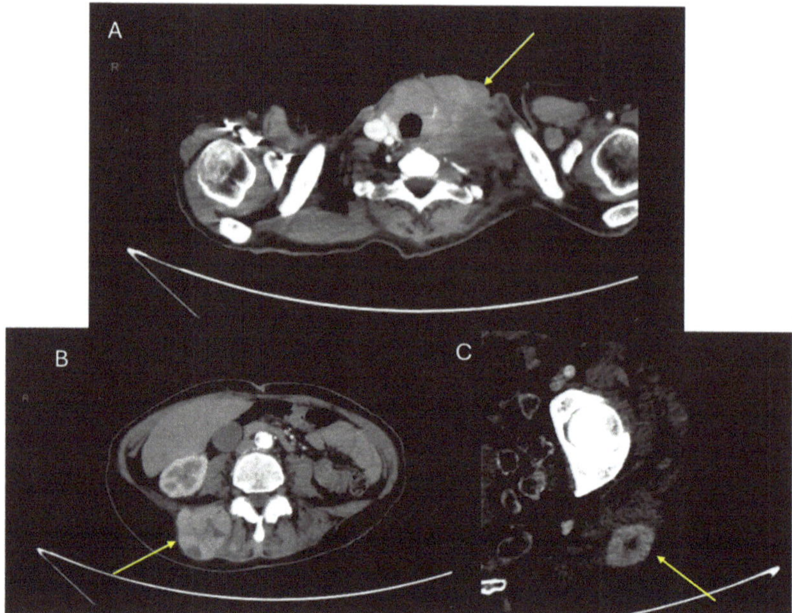

Figure 5. CT images of recurrences during the follow-up period: (**A**) Extensive supraclavicular recurrence 4 years after radical surgery. Absence of the vascular bundle of the neck, resected in radical cervical dissection surgery. (**B**) Contralateral paravertebral metastasis 3 years after diagnosis. (**C**) Ipsilateral gluteal metastasis 3 years after initial diagnosis.

3. Discussion

Malignant gastro neuroectodermal tumor of the head/neck region is an exceptionally rare neoplasm, with less than 100 cases cited worldwide. Like all rare tumors, diagnosis is significantly challenging due to overlap in histopathological features with other tumors of neuroectodermal origin. The presentation of MGNETs in the head/neck region is often subtle, given that these tumors can grow insidiously without producing specific symptoms until they reach an advanced stage. Patients may present with a mass or swelling, pain, dysphagia, or other region-specific symptoms depending on the exact anatomical location of the tumor. As with all rare tumors, MGNET in this anatomical region further complicates the clinical diagnosis, as it is more commonly associated with the gastrointestinal tract. When histopathology reveals neuroectodermal differentiation but lacks melanocytic markers, MGNET should be included in the differential diagnosis.

The diagnosis of MGNET relies heavily on a combination of histological evaluation and molecular testing. MGNETs display a high degree of cellular atypia and frequent mitotic activity, consistent with their aggressive nature. Necrosis is also common, which contributes to the rapid clinical progression of these tumors. Histologically, MGNET demonstrates a pseudoalveolar or pseudopapillary growth pattern with oval to spindle-shaped cells, eosinophilic cytoplasm, and vesicular nuclei. However, these features overlap with other tumors, such as clear cell sarcoma (CCS), synovial sarcoma, and melanoma, making a definitive diagnosis based on morphology alone difficult. This often delays definitive diagnosis and management, contributing to the poor prognosis associated with these tumors. Radiologically, MGNET may appear as non-specific soft tissue masses, and advanced imaging techniques such as magnetic resonance imaging (MRI) or positron emission tomography (PET) scans are often required to assess the extent of the disease. Immunohistochemistry (IHC) plays a critical role in differentiating MGNET from other entities. MGNET typically shows positivity for S100 and SOX10 while being negative for melanocytic markers like HMB-45, Melan-A, and MiTF, which are usually present in melanoma and CCS. Molecular studies, specifically fluorescence in situ hybridization (FISH), are essential to confirm the presence of EWSR1::CREB1, which are characteristic of MGNET and further distinguish it from other neuroectodermal tumors.

Given its aggressive clinical behavior, early and accurate diagnosis is critical for appropriate treatment planning. Surgical resection remains the cornerstone of treatment for MGNET, aiming for complete excision with negative margins. However, given the anatomical complexity of the head/neck region and the potential for local invasion into critical structures, achieving clear surgical margins can be challenging. Adjuvant therapies, including radiation and chemotherapy, may be considered in cases where complete resection is not possible or in patients with high-risk features such as vascular invasion, perineural spread, or lymph node involvement. Unfortunately, there is no standardized nomenclature or standardized treatment protocols due to the scarcity of cases, and treatment decisions are often based on institutional experience and extrapolation from similar sarcomas. In the future, the fusion itself may name, classify, categorize, and specify treatment.

The prognosis of MGNET in the head/neck is generally poor due to its aggressive nature and high potential for local recurrence and distant metastasis. Factors influencing prognosis include tumor size, mitotic rate, and the ability to achieve complete resection. Long-term survival is uncommon, and patients require close monitoring for recurrence and metastasis. Further research and case studies are needed to improve diagnostic techniques and treatment outcomes for this extremely rare entity.

The management of MGNETs in the head/neck is challenging due to their rarity and aggressive behavior. Surgical resection remains the mainstay of treatment, with the goal of achieving complete excision with clear margins. However, due to the complex anatomy of the head and neck region, radical resection is often difficult without significant functional morbidity. Adjuvant radiotherapy is commonly employed post-surgery to reduce the risk of local recurrence, although its efficacy remains uncertain due to the limited number of cases reported. In the management of MGNET, achieving negative surgical margins remains challenging due to the tumor's aggressive biological behavior and frequent involvement of adjacent critical anatomical structures. Neither radiation nor chemotherapy has been adequately studied in these extremely rare tumor cases. As intraoperative radiotherapy (IORT) emerges, offering direct delivery of high-dose radiation precisely to the tumor bed, it optimizes local control, particularly in scenarios of microscopic residual disease or in recurrent cases, and may potentially improve long-term oncological outcomes. The role of chemotherapy in MGNET is less defined, largely because these tumors are typically resistant to conventional sarcoma regimens. Nonetheless, some reports suggest that targeted therapies against EWSR1 fusion proteins or other molecular targets may hold promise in the future [19]. Emerging therapies such as immune checkpoint inhibitors or tyrosine kinase inhibitors are currently under investigation in clinical trials, but no definitive treatment guidelines have yet been established [23–25].

Furthermore, with this and other rare tumors, precision therapy may be the primary treatment [26]. The prognosis for patients with MGNETs is generally poor due to the tumor's aggressive behavior and high likelihood of metastasis, particularly to the lungs and regional lymph nodes. Recurrence rates are high, and despite aggressive surgical management, the long-term survival rates remain low. Studies suggest that the 5-year survival rate for patients with MGNETs is less than 30%, emphasizing the need for early detection and multimodal treatment. All information must be shared and disseminated broadly for all Refractory Atypical Resistant Extremely (RARE) tumors. This way similarities and differences may be determined [27–29]. Furthermore, for RARE tumors, precision targeted medicine is most important to affect the outcome.

4. Conclusions

Extraenteric malignant gastrointestinal neuroectodermal tumor is a rare but highly aggressive malignancy that can present in the head/neck region, posing significant diagnostic and therapeutic challenges. The rarity of this tumor, combined with its overlapping histological features with other neuroectodermal tumors, complicates its diagnosis. Advances in molecular pathology, particularly the identification of EWSR1 translocations, have been crucial in distinguishing MGNET from other malignancies. However, the prognosis remains poor, and further research is needed to better understand the pathogenesis and to develop more effective therapeutic strategies. With the ongoing development of targeted therapies and molecular diagnostics, there is hope that the outcomes for patients with this devastating disease may improve in the future.

Author Contributions: Conceptualization, M.T. and M.J.T.; methodology, M.T. and C.N.-C.; software, M.T. and A.M.L.-L.; validation, M.T., M.J.T. and S.O.; formal analysis, M.T. and S.O.; investigation, M.T. and M.J.T.; resources, M.T., A.A.-G., C.A. and F.A.; data curation, S.K. and G.R.-d.-L.; writing—original draft preparation, M.T.; writing—review and editing, M.J.T. and S.O.; visualization, M.T.; supervision, M.J.T. and S.O.; project administration, A.M.L.-L. and M.T.; funding acquisition, M.T., J.I.S. and S.O. All authors have read and agreed to the published version of the manuscript.

Funding: This research was funded by Instituto de Salud Carlos III (ISCIII) through the project PI22/00601 Craniofacial Customized Image-Guided Surgery and co-funded by the European Union and Bertarelli Foundation, MGH/MGB, Boston, MA.

Institutional Review Board Statement: This study was conducted in accordance with the Declaration of Helsinki and approved by the Institutional Review Board of Comité de Ética de la Investigación con Medicamentos Hospital General Universitario Gregorio Marañón (protocol code CMF2, approval date: 19/10/2023).

Informed Consent Statement: Informed consent was obtained from all subjects involved in this study.

Data Availability Statement: The data presented in this study are available on request from the corresponding author.

Conflicts of Interest: The authors declare no conflicts of interest. The funders had no role in the design of this study; in the collection, analyses, or interpretation of data; in the writing of the manuscript; or in the decision to publish the results.

Abbreviations

The following abbreviations are used in this manuscript:

MGNET	malignant gastrointestinal neuroectodermal tumor
CCS	clear cell sarcoma
GI	gastrointestinal
WHO	World Health Organization
IHC	immunohistochemistry
PAS	periodic acid–Schiff (staining)
FISH	fluorescence in situ hybridization
CNIO	Centro Nacional de Investigaciones Oncológicas (National Cancer Research Center, Spain)
PET-CT	positron emission tomography–computed tomography
HPF	high-power field (used in microscopy)
ANED	alive with no evidence of disease
AWD	alive with disease
E-MGNET	extraenteric malignant gastrointestinal neuroectodermal tumor

References

1. Alpers, C.E.; Beckstead, J.H. Malignant neuroendocrine tumor of the jejunum with osteoclast-like giant cells. Enzyme histochemistry distinguishes tumor cells from giant cells. *Am. J. Surg. Pathol.* **1985**, *9*, 57–64. [CrossRef] [PubMed]
2. Stockman, D.L.; Miettinen, M.; Suster, S.; Spagnolo, D.; Dominguez-Malagon, H.; Hornick, J.L.; Adsay, V.; Chou, P.M.; Amanuel, B.; Vantuinen, P.; et al. Malignant gastrointestinal neuroectodermal tumor: Clinicopathologic, immunohistochemical, ultrastructural, and molecular analysis of 16 cases with a reappraisal of clear cell sarcoma-like tumors of the gastrointestinal tract. *Am. J. Surg. Pathol.* **2012**, *36*, 857–868. [CrossRef]
3. Washimi, K.; Takagi, M.; Hisaoka, M.; Kawachi, K.; Takeyama, M.; Hiruma, T.; Narimatsu, H.; Yokose, T. Clear cell sarcoma-like tumor of the gastrointestinal tract: A clinicopathological review. *Pathol. Int.* **2017**, *67*, 534–536. [CrossRef] [PubMed]
4. Zambrano, E.; Reyes-Mugica, M.; Franchi, A.; Rosai, J. Clear cell sarcoma-like tumor of the gastrointestinal tract: An immunohistochemical, ultrastructural, and molecular study of a distinctive GI neoplasm. *Mod. Pathol.* **2003**, *16*, 502–511.
5. Antonescu, C.R.; Nafa, K.; Segal, N.H. Genomic hallmarks of GNET and its clinical implications. *Clin. Cancer Res.* **2016**, *22*, 4620–4628.
6. Kuo, C.T.; Kao, Y.C.; Huang, H.Y.; Hsiao, C.H.; Lee, J.C. Malignant gastrointestinal neuroectodermal tumor in head and neck: Two challenging cases with diverse morphology and different considerations for differential diagnosis. *Virchows Arch.* **2022**, *481*, 131–136. [CrossRef] [PubMed]
7. French, C.A. EWS-ATF1 in malignant clear cell tumors. *Oncogene* **2010**, *29*, 5043–5051.
8. Fletcher, C.D.; Unni, K.K.; Mertens, F. *World Health Organization Classification of Tumours. Pathology and Genetics of Tumours of Soft Tissue and Bone*; IARC Press: Lyon, France, 2002.
9. Breton, S.; Dubois, M.; Geay, J.F.; Gillebert, Q.; Tordjman, M.; Guinebretière, J.M.; Denoux, Y. Clear cell sarcoma or gastrointestinal neuroectodermal tumor (GNET) of the tongue? Case report and review of the literature of an extremely rare tumor localization. *Ann. Pathol.* **2019**, *39*, 167–171. [CrossRef]
10. Sbaraglia, M.; Zanatta, L.; Toffolatti, L.; Spallanzani, A.; Bertolini, F.; Mattioli, F.; Lami, F.; Presutti, L.; Tos, A.P.D. Clear cell sarcoma-like/malignant gastrointestinal neuroectodermal tumor of the tongue: A clinicopathologic and molecular case report. *Virchows Arch.* **2021**, *478*, 1203–1207. [CrossRef]
11. Allanson, B.M.; Weber, M.A.; Jackett, L.A.; Chan, C.; Lau, L.; Ziegler, D.S.; Warby, M.; Mayoh, C.; Cowley, M.J.; Tucker, K.M.; et al. Oral malignant gastrointestinal neuroectodermal tumour with junctional component mimicking mucosal melanoma. *Pathology* **2018**, *50*, 648–653. [CrossRef]
12. Yang, Y.; Chen, Y.; Chen, S.; Han, A. Malignant gastrointestinal neuroectodermal tumour in soft tissue. *Pathology* **2021**, *53*, 276–278. [CrossRef] [PubMed]
13. Kraft, S.; Antonescu, C.R.; Rosenberg, A.E.; Deschler, D.G.; Nielsen, G.P. Primary clear cell sarcoma of the tongue. *Arch. Pathol. Lab. Med.* **2013**, *137*, 1680–1683. [CrossRef] [PubMed]
14. Baus, A.; Culie, D.; Duong, L.T.; Ben Lakhdar, A.; Schaff, J.B.; Janot, F.; Kolb, F. Primary clear cell sarcoma of the tongue and surgical reconstruction: About a rare case report. *Ann. Chir. Plast. Esthet.* **2019**, *64*, 98–105. [CrossRef] [PubMed]
15. Donzel, M.; Zidane-Marinnes, M.; Paindavoine, S.; Breheret, R.; de la Fouchardière, A. Clear cell sarcoma of the soft palate mimicking unclassified melanoma. *Pathology* **2019**, *51*, 331–334. [CrossRef]
16. Argani, P.; Perez-Atayde, A.R.; Fletcher, C.D. Clear cell sarcoma of soft parts and gastrointestinal clear cell sarcoma-like tumor: Two distinct clinicopathologic entities. *Am. J. Surg. Pathol.* **2000**, *24*, 699–707.
17. Chi, P.; D'Angelo, S.P.; Ryan, C.W. Gastrointestinal neuroectodermal tumor: A rare tumor with unique molecular characteristics. *Cancer* **2018**, *124*, 3460–3468.

18. Çomunoglu, N.; Dervisoglu, S.; Elçin, B.; Tekant, G.; Apak, H. Malignant extragastrointestinal neuroectodermal tumor located at right cervical region. *Open J. Pathol.* **2015**, *5*, 125e128. [CrossRef]

19. Ulici, V.; Hornick, J.L.; Davis, J.L.; Mehrotra, S.; Meis, J.M.; Halling, K.C.; Fletcher, C.D.M.; Kao, E.; Folpe, A.L. Extraenteric Malignant Gastrointestinal Neuroectodermal Tumor-A Clinicopathologic and Molecular Genetic Study of 11 Cases. *Mod. Pathol.* **2023**, *36*, 100160. [CrossRef]

20. WHO Classification of Tumours Editorial Board. *Soft Tissue and Bone Tumours*, 5th ed.; WHO Classification of Tumours, Volume 3; International Agency for Research on Cancer: Lyon, France, 2020.

21. van Diest, P.J.; van der Wall, E.; Baak, J.P. Prognostic value of proliferation in invasive breast cancer: A review. *J. Clin. Pathol.* **2004**, *57*, 675–681. [CrossRef]

22. Dowsett, M.; Nielsen, T.O.; A'Hern, R.; Bartlett, J.; Coombes, R.C.; Cuzick, J.; Ellis, M.; Henry, N.L.; Hugh, J.C.; Lively, T.; et al. Assessment of Ki67 in breast cancer: Recommendations from the International Ki67 in Breast Cancer Working Group. *J. Natl. Cancer Inst.* **2011**, *103*, 1656–1664. [CrossRef]

23. Kalia, V.; Levine, P.; Ludwig, K. Malignant gastrointestinal neuroectodermal tumor: Treatment outcomes and review of the literature. *J. Surg. Oncol.* **2019**, *120*, 492–498.

24. Schöffski, P.; Wozniak, A.; Stacchiotti, S.; Rutkowski, P.; Blay, J.Y.; Lindner, L.H.; Strauss, S.J.; Anthoney, A.; Duffaud, F.; Richter, S.; et al. Activity and safety of crizotinib in patients with advanced clear-cell sarcoma with MET alterations: European Organization for Research and Treatment of Cancer phase II trial 90101 'CREATE'. *Ann. Oncol.* **2017**, *28*, 3000–3008. [CrossRef] [PubMed]

25. Subbiah, V.; Holmes, O.; Gowen, K.; Spritz, D.; Amini, B.; Wang, W.L.; Schrock, A.B.; Meric-Bernstam, F.; Zinner, R.; Piha-Paul, S.; et al. Activity of c-Met/ALK inhibitor crizotinib and multi-kinase VEGF inhibitor pazopanib in metastatic gastrointestinal neuroectodermal tumor harboring EWSR1-CREB1 fusion. *Oncology* **2016**, *91*, 348–353. [CrossRef] [PubMed]

26. Cornillie, J.; van Cann, T.; Wozniak, A.; Hompes, D.; Schöffski, P. Biology and management of clear cell sarcoma: State of the art and future perspectives. *Expert Rev. Anticancer. Ther.* **2016**, *16*, 839–845. [CrossRef] [PubMed]

27. El-Salam, M.A.A.; Troulis, M.J.; Pan, C.-X.; Rao, R.A. Unlocking the potential of organoids in cancer treatment and translational research: An application of cytologic techniques. *Cancer Cytopathol.* **2024**, *132*, 96–102. [CrossRef]

28. Rivera, C.M.; Faquin, W.C.; Thierauf, J.; Afrogheh, A.H.; Jaquinet, A.; Iafrate, A.J.; Rivera, M.N.; Troulis, M.J. Detection of EWSR1 fusions in CCOC by targeted RNA-seq. *Oral Surg. Oral Med. Oral Pathol. Oral Radiol.* **2022**, *134*, 240–244. [CrossRef]

29. Xhori, O.; Deol, N.; Rivera, C.M.; Zavras, J.; Weil, S.; Zafari, H.; Thierauf, J.H.; Faquin, W.C.; Choy, E.; Rivera, M.N.; et al. A comparison of Clear Cell Sarcoma to Jaw and Salivary Tumors bearing EWS Fusions. *Head. Neck Pathol.* **2024**, *18*, 25. [CrossRef]

MDPI AG

Grosspeteranlage 5

4052 Basel

Switzerland

Tel.: +41 61 683 77 34

International Journal of Molecular Sciences Editorial Office

E-mail: ijms@mdpi.com

www.mdpi.com/journal/ijms

www.ingramcontent.com/pod-product-compliance
Lightning Source LLC
LaVergne TN
LVHW071938160726
843515LV00011B/2640